Soft Computing Techniques and Applications in Mechanical Engineering

Mangey Ram
Graphic Era University, India

J. Paulo Davim
University of Aveiro, Portugal

A volume in the Advances in Mechatronics and Mechanical Engineering (AMME) Book Series

IGI Global
DISSEMINATOR OF KNOWLEDGE

Published in the United States of America by
IGI Global
Engineering Science Reference (an imprint of IGI Global)
701 E. Chocolate Avenue
Hershey PA, USA 17033
Tel: 717-533-8845
Fax: 717-533-8661
E-mail: cust@igi-global.com
Web site: http://www.igi-global.com

Library of Congress Cataloging-in-Publication Data

Names: Ram, Mangey, editor. | Davim, J. Paulo, editor.
Title: Soft computing techniques and applications in mechanical engineering / Mangey Ram and J. Paulo Davim, editors.
Description: Hershey, PA : Engineering Science Reference, [2018] | Includes bibliographical references.
Identifiers: LCCN 2017013547| ISBN 9781522530350 (hardcover) | ISBN 9781522530367 (ebook)
Subjects: LCSH: Mechanical engineering--Data processing. | Soft computing.
Classification: LCC TJ153 .S638 2018 | DDC 620.00285/63--dc23 LC record available at https://lccn.loc.gov/2017013547

This book is published in the IGI Global book series Advances in Mechatronics and Mechanical Engineering (AMME) (ISSN: 2328-8205; eISSN: 2328-823X)

British Cataloguing in Publication Data
A Cataloguing in Publication record for this book is available from the British Library.

Advances in Mechatronics and Mechanical Engineering (AMME) Book Series

J. Paulo Davim
University of Aveiro, Portugal

ISSN:2328-8205
EISSN:2328-823X

Mission

With its aid in the creation of smartphones, cars, medical imaging devices, and manufacturing tools, the mechatronics engineering field is in high demand. Mechatronics aims to combine the principles of mechanical, computer, and electrical engineering together to bridge the gap of communication between the different disciplines.

The **Advances in Mechatronics and Mechanical Engineering (AMME) Book Series** provides innovative research and practical developments in the field of mechatronics and mechanical engineering. This series covers a wide variety of application areas in electrical engineering, mechanical engineering, computer and software engineering; essential for academics, practitioners, researchers, and industry leaders.

Coverage

- Bioengineering Materials
- Mechanisms and machines
- Design and Manufacture
- Control Methodologies
- Micro and nanomechanics
- Manufacturing Methodologies
- Computer-Based Manufacturing
- Tribology and surface engineering
- Control Systems Modelling and Analysis
- Intelligent Sensing

IGI Global is currently accepting manuscripts for publication within this series. To submit a proposal for a volume in this series, please contact our Acquisition Editors at Acquisitions@igi-global.com or visit: http://www.igi-global.com/publish/.

The Advances in Mechatronics and Mechanical Engineering (AMME) Book Series (ISSN 2328-8205) is published by IGI Global, 701 E. Chocolate Avenue, Hershey, PA 17033-1240, USA, www.igi-global.com. This series is composed of titles available for purchase individually; each title is edited to be contextually exclusive from any other title within the series. For pricing and ordering information please visit http://www.igi-global.com/book-series/advances-mechatronics-mechanical-engineering/73808. Postmaster: Send all address changes to above address.

Titles in this Series

For a list of additional titles in this series, please visit: www.igi-global.com/book-series

Numerical and Analytical Solutions for Solving Nonlinear Equations in Heat Transfer
Davood Domiri Ganji (Babol Noshirvani University of Technology, Iran) and Roghayeh Abbasi Talarposhti (University of Mazandaran, Iran)
Engineering Science Reference • copyright 2018 • 275pp • H/C (ISBN: 9781522527138) • US $195.00 (our price)

Nanofluid Technologies and Thermal Convection Techniques
Ramesh Chand (Government Arya Degree College Nurpur (HP), India)
Engineering Science Reference • copyright 2017 • 229pp • H/C (ISBN: 9781683180067) • US $185.00 (our price)

Mathematical Concepts and Applications in Mechanical Engineering and Mechatronics
Mangey Ram (Graphic Era University, India) and J. Paulo Davim (University of Aveiro, Portugal)
Engineering Science Reference • copyright 2017 • 495pp • H/C (ISBN: 9781522516392) • US $215.00 (our price)

Handbook of Research on Advances and Applications in Refrigeration Systems and Technologies
Pedro Dinis Gaspar (University of Beira Interior, Portugal) and Pedro Dinho da Silva (University of Beira Interior, Portugal)
Engineering Science Reference • copyright 2015 • 924pp • H/C (ISBN: 9781466683983) • US $545.00 (our price)

Engineering Creative Design in Robotics and Mechatronics
Maki K. Habib (The American University in Cairo, Egypt) and J. Paulo Davim (University of Aveiro, Portugal)
Engineering Science Reference • copyright 2013 • 366pp • H/C (ISBN: 9781466642256) • US $195.00 (our price)

Computational Methods for Optimizing Manufacturing Technology Models and Techniques
J. Paulo Davim (University of Aveiro, Portugal)
Engineering Science Reference • copyright 2012 • 395pp • H/C (ISBN: 9781466601284) • US $195.00 (our price)

Multi-Agent Systems for Traffic and Transportation Engineering
Ana Bazzan (Instituto de Informática, Universidade Federal do Rio Grande do Sul, Brazil) and Franziska Klügl (Örebro University, Sweden)
Information Science Reference • copyright 2009 • 446pp • H/C (ISBN: 9781605662268) • US $195.00 (our price)

IGI Global
DISSEMINATOR OF KNOWLEDGE

701 East Chocolate Avenue, Hershey, PA 17033, USA
Tel: 717-533-8845 x100 • Fax: 717-533-8661
E-Mail: cust@igi-global.com • www.igi-global.com

Editorial Advisory Board

List of Reviewers

Table of Contents

Detailed Table of Contents

of these characteristics enables the student to appreciate the effect of the different types of turbulence models on turbulent mixing and related phenomena. Finally, recommendations of turbulence model for rod bundle are finalized. The inclusion of related references provides a starting point for the interested reader / researchers /industrialists.

Summability methods are a useful tool in dealing with the problems in the soft computing like in filtering of the signals and for stabilizing the systems. Signals can be in the form of various types of series (Infinite Series, Fourier series, etc.) and hence, summability theory is applicable in finding the error of approximation and degree of approximation of such signals. In this chapter, the authors gave an introductory discussion on summability theory and approximation of the signals. Further, they explained about the stability of the frequency response of the system. Also, they used the Fourier approximation in the soft computing models (multilayer perceptrons; radial basis function (RBF) or regularization networks, and fuzzy logic models) and found the output data of requirement.

In today's scenarios, the utilization of simulation and optimization in the field of designing is achieving wider recognition in the various zones of commerce as the computational competences of computers upsurge day by day. The result is that the uses for numerical optimization have increased tremendously. Design process in engineering is a distinct practice of solving the problems where a group of recurrently indistinct objectives has to be well-adjusted deprived of violating any given circumstances. Consequently, it seems quite ordinary to consider a design process as an optimization process. The design process could be articulated as to allocate values to the system parameters to confirm that the state variables and the characteristics are as suitable as possible through an inclusive range of operating and environmental variables. This is a complex multi-objective optimization problem (MOOP). This chapter discusses the use of MOO algorithms in mechanical engineering.

Preface

In recent years, soft computing techniques have an extraordinary growth in various engineering systems. The principal components of soft computing techniques are probabilistic reasoning, fuzzy set theory, neural nets, neuro fuzzy system, adaptive neuro fuzzy inference system (ANFIS), neuro-computing, genetic algorithms, hybrid fuzzy logic genetic algorithms (HFLGA) belief networks, chaotic systems, evolutionary computing, probabilistic computing, Computational intelligence (CI) as well as learning theory. The book *Soft Computing Techniques and Applications in Mechanical Engineering* covers a comprehensive range of soft computing techniques applied in various fields of mechanical engineering problems.

ORGANIZATION OF THE BOOK

The book is organized into 14 chapters. A short description of each of the chapters follows:

Chapter 1 presents the soft computing application to water pollution risk assessment and water quality analysis using data of satellite observation of water bodies, soil cover and air pollutant transfer.

Chapter 2 provides a multi-objective fixed charged solid transportation model with criterion, e.g. transportation penalty, amounts, demands, carriages and budget constraints as type-2 triangular fuzzy variables with condition on fewer components.

Chapter 3 examines how to optimize the machining parameters of EDM using Artificial Neural Network (ANN) and Particle Swarm Optimization (PSO) techniques for maximization of Material Removal Rate (MRR) and minimizing the surface Roughness (SR).

Chapter 4 analyses the Wire-Electric Discharge Machining (WEDM) to machine AISI-D3 material in order to measure the performance of multi-objective responses like high material removal rate and low roughness. The contradictory objective is accomplished by the control parameters like Pulse on Time (Ton), Pulse off Time (Toff), Wire Feed (W/Feed) and Wire Tension (W/Ten) employing brass wire. Artificial Neural Network is employed to train and validate the experimental result. Primarily the individual responses are optimized by employing Firefly algorithm (FA).

Chapter 5 optimizes the work-piece on PAC machining process using Genetic Algorithm and Fuzzy Logic Algorithm for maximization of Material Removal Rate (MRR) and minimizing the surface Roughness (SR). The Gas Pressure (A), Arc Current (B) and Torch Height (C) were considered with equally spaced three levels within the operating range for each of the process parameters as the input parameters.

Chapter 6 describes with the comparison of the most used back propagation training algorithms, neural networks, mainly Levenberg-Marquardt, conjugate gradient and Resilient back propagation using radial overcoat prediction as illustrations, comparisons are made based on the effectiveness and efficiency of three training algorithms on the networks.

The Chapter 7 recommends that Genetic Algorithms are highly useful to increase the High-Performance Areas for Engineering Design. To use Genetic Algorithms to a large number of design areas and delivered a comprehensive conversation on the use, scope and its applications in mechanical engineering.

Chapter 8 identifies the ten challenges in the SFSCM (and sustainable food supply chain management) on the basis of literature review and expert opinion. The approach is an integration of fuzzy with DEMATEL (decision-making trial and evaluation laboratory) which can be used for dividing the challenges into cause and effect group.

The Chapter 9 addresses a hybrid genetic scatter search algorithm for solving multi-objective FJSP (flexible job-shop problem). Makespan and flow time are the objective functions considered in this chapter. The computational results provided for the effectiveness of the proposed algorithm for solving the flexible job-shop scheduling problem.

In Chapter 10, the electrical discharge machining (EDM), a thermo-mechanical machining process, is used in producing a complicated intrinsic cavity in difficult-to- machine materials with excellent surface finish. Copper and tungsten powders in weight percentage of 30 and 70 respectively are used for the preparation of the tool electrode by varying the PM process parameters like compaction pressure and sintering temperature.

Chapter 11 presents the novel approaches to predictions of the number of failures that will be observed during a future inspection of a sample of units, based only on the results of the previous in-service inspections of the same sample. The methodology described here is based on the use of order statistics from an underlying model, which have such properties as bivariate dependence and conditional predictability.

Chapter 12 presents an overview of various types of turbulence model and their effect on thermal-hydraulic characteristics of nuclear fuel bundle, both past and present using Computational Fluid Dynamic (CFD) approach. It includes the mathematical definition related to fuel bundle in nuclear energy.

Chapter 13 discusses the benefits of summability methods in dealing with the problems in the soft computing and explained about the stability of the frequency response of the system. Also, the Fourier approximation in the soft computing models (multilayer perceptrons; radial basis function (RBF) or regularization networks, and fuzzy logic models) and found the output data of requirement is used.

Chapter 14 discusses the use of MOO (multi-objective optimization) algorithms in mechanical engineering.

This research book can be used as a support book for the final undergraduate engineering course (for example, mechanical, mechatronics, industrial, computer science, information technology, etc.) or as a subject on soft computing applications at the postgraduate level. Also, this book can serve as a valuable reference for academics, mechanical, mechatronics, computer science, information technology and industrial engineers, as well as, researchers in related subjects with computing applications in various engineering fields.

Acknowledgment

The editors acknowledge IGI Global and the editorial team for their adequate and professional support during the preparation of this book. In addition, we would like to acknowledge all the chapter authors, the reviewers and the editorial advisory board members for their availability for work on this editorial project.

Mangey Ram
Graphic Era University, India

J. Paulo Davim
University of Aveiro, Portugal

Chapter 1
Decision Making Under Deep Uncertainty With Fuzzy Algorithm in Framework of Multi-Model Approach:
Water Pollution Risk Assessment Using Satellite Data

Yuriy V. Kostyuchenko
Scientific Centre for Aerospace Research of the Earth, National Academy of Sciences of Ukraine, Ukraine

Yulia Stoyka
Institute of Hydrobiology, National Academy of Sciences of Ukraine, Ukraine

Iurii Negoda
Institute of Geological Sciences, National Academy of Science of Ukraine, Ukraine

Ivan Kopachevsky
Scientific Centre for Aerospace Research of the Earth, National Academy of Sciences of Ukraine, Ukraine

ABSTRACT

Task of soft computing for decision support in field of risk management is analyzed in this chapter. Multi-model approach is described. Interrelations between models, remote sensing data and forecasting are described. Method of water quality assessment using satellite observation is described. Method is based on analysis of spectral reflectance of aquifers. Correlations between reflectance and pollutions are quantified. Fuzzy logic based approach for decision support in field of water quality degradation risk is discussed. Decision on water quality is making based on fuzzy algorithm using limited set of uncertain

DOI: 10.4018/978-1-5225-3035-0.ch001

parameters. It is shown that this algorithm allows estimate water quality degradation rate and pollution risks. Using proposed approach, maps of surface water pollution risk from point and diffuse sources are calculated. Conclusions concerned soft computing in risk management are proposed and discussed. It was demonstrated, that basing on spatially distributed measurement data, proposed approach allows to calculate risk parameters with resolution close to observations.

INTRODUCTION

Modern development of applied mathematics and computational methods allows to formulate and solve new problems in mechanical and civil engineering, first of all tasks in the field of control and decision theory including tasks, aimed to risk management and security control. New types of data and new monitoring instrument can be harnessed, more wide areas of human activity and life may be analyzed. Since innovative technologies should serve to increasing of life quality, applied computing approaches in engineering directed to risk assessment toward disaster threats is important task.

Water pollution is one of the biggest environmental problems, as well as linked with it soil and air pollution. Nitrate is among the most common and widespread pollutants in surface water and groundwater. Diffuse pollution through soils and air from agricultural activities and livestock are the main sources of increased nitrate concentrations in groundwater and surface water bodies (European Commission, 2005). Nitrogen is a vital nutrient to enhance plant growth, which has motivated intensive use of nitrogen-based fertilizers to boost up the crop production. But increased fertilizer use also has social and environmental costs. The fertilizers deteriorate the water quality inducing economic and ecological problems. In the last century automation of agriculture and the introduction of high yield crops has raised the use of fertilizers, increasing nitrate concentration in groundwater.

In some cases, when through the natural and artificial circumstances the nitrogen application exceeds demand and the denitrification capacity of the soil, nitrogen can leach to the water, usually as nitrate.

The nitrogen average fertilizer use in Europe is 70 kg/ha (EEA (European Environment Agency, 2003). In some regions with intensive irrigation the water bodies reach nitrate concentrations between 50-100 mg/l (Martínez, Albiac, 2006). The monitoring studies in Ukraine indicate that about 50% of groundwater suffers nitrates concentration over 50 mg/l, and 70% over 25 mg/l. Nitrogen from agricultural sources accounts for from 50 to 80% of the nitrates entering Europe's water (European Commission, 2005).

Therefore, the assessment of water pollution and water quality is an urgent task of socio-ecological security. In this task, from viewpoint of applied mathematics, the situation of uncontrolled uncertainty is frequent.

It is quite easy to control one selected parameter with required accuracy. But, if we need control large set of spatially distributed parameters on the long-time intervals, the uncertainties – both aleatorical and epistemical - are increasing critically. This is a typical situation for water quality analysis, where wide range of parameters should be controlled simultaneously. In this situation, the decision making and risk analysis are complicated problems (Polasky et al., 2011).

There are few methods of uncertainty control, including uncertainties, generating by complex multi-physics systems (Ermoliev et al., 2012). But also, may be formulated task about using the soft computing method for assessment of risk with limited set of parameters. In such cases we can reduce a number of controlled variables without neglecting the systems processes, drivers and feedbacks.

This approach should include formulation of correct methodology, method of variables selection, indicators selection method, and method of risk assessment with limited set of parameters.

This chapter is dedicated to soft computing application to water pollution risk assessment and water quality analysis using data of satellite observation of water bodies, soil cover and air pollutant transfer.

METHODOLOGY: MULTI-MODEL APPROACH

How to estimate risks using the limited set of parameters derived from different sources with different drivers, scales and nature?

First of all, we need consider the problem of the model application to selection of the optimal set of remote sensing indicators in risk assessment tasks (Kostyuchenko et al., 2015, Kostyuchenko, 2018).

At the initial stage, the process of forecast generation should be based on the set of initial assumptions, captured in a vector *x* $\left(x = \left(x^1, x^2, \ldots, x^s\right)\right)$. These could be a priori assumptions, observed or measured values.

The further step is the modeling: model recalculation these values into a group of core hydrological, bio-physical, and climatological series (with $F(x, \varepsilon, y) = 0$, $(x, \varepsilon) \rightarrow y$), collected in a vector *y*. next, basing on the information from the pair *(x, y)*, we calculate values for a list of parameters, grouped into what referred as the vector of satellite indicators based models: $z = (z^1, z^2, \ldots, z^s)$ (with $z^s = g^s(x, y)$, $(x, y) \rightarrow z^s$).

As the result of the integrated modeling we gets parameters summarized in the triplet *(x,y,z)*. This combined vector is a starting point for the modeling of socio-economic, socio-ecological and risk parameters (Figure 1).

Usually, we consider a group of *S* satellite data based models, labeled $s \in \{1, 2, \ldots, s\}$. Each of these equations is such that the endogenous variables z^s can be obtained as an explicit mapping of the core variables $z^s = g^s(x, y)$.

Figure 1. Multi-model approach to socio-ecological risks assessment

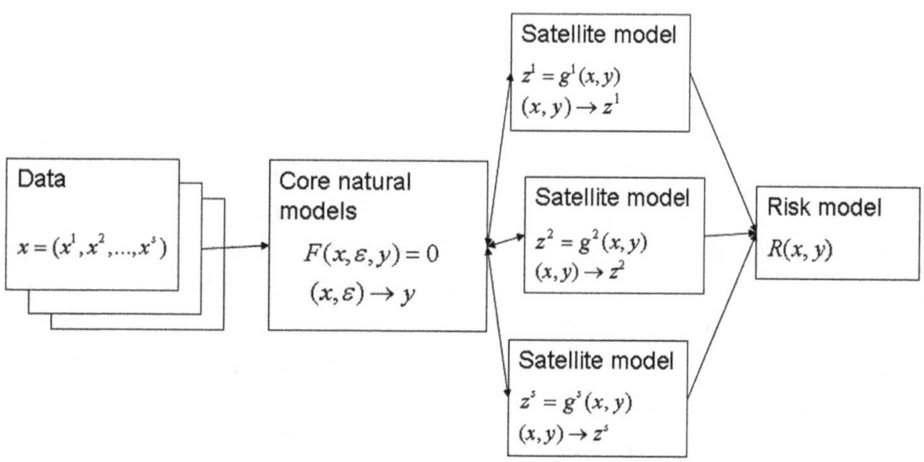

Therefore the satellite model might be conceptually presented by time-series of *(x,y,z)*, which will determine the behavior of z_t^s, with a residual term μ_t^s:

$$z_t^s = f\left(x_t, x_{t-1}, \dots, x_{t-L}, y_t, y_{t-1}, \dots, y_{t-L}, z_{t-T}^s, \mu_t^s\right) \quad (1)$$

As z_t^s is the only variable on the left-hand side of the equation, the relationship is unidirectional: from *(x,y)* to *z*. This simple time-series satellite data based model formally also allows for no interactions with other satellite variables nor any feedback between z_t^s and the core assumptions in *(x,y)* (Campbell and Wynne, 2011), so we can use both methodology utilizing as separate as well interlinked indicators.

Traditional time-series models such as autoregressive moving average models are good examples of satellite data based models. This representation may include, for example, autoregressive lags and/or moving average components. A number of standard methodologies (for example, Box-Jenkins method, or nonlinear autoregressive exogenous model) are followed to find the most usable model of the data-generating process for a given risk metric Z_t:

$$Z_t = c + \sum_{l=0}^{N} \beta_l X_{t-l} + \sum_{l=0}^{P} \rho_l Y_{t-l} + \sum_{l=1}^{L} \partial_l Z_{t-l} + \sum_{k=0}^{K} \theta_k \varepsilon_{t-k} \quad (2)$$

where Z_t is a satellite variable, X_t is a row vector of initial exogenous core variables, Y_t is a row vector of the layer of core parameters series, and ε_t is the value of the stochastic error term. The parameters c, β, ρ, ∂ are unknown and should be estimated.

However, including autoregressive terms in the model often results in a muted impact of core drivers on a target variable. Thus, it is a common and recommended practice is to exclude autoregressive terms from the supplementary variable equations (Engle and Russell, 1998). Therefore, depending of risk metric Z_t and of type of supplementary variable may be applied different form of equation (2). For example, for analysis of climate related risk an approach based on copulas utilization may be used (Kostyuchenko et al., 2013).

A key aspect of satellite model development is variable selection to identify which core drivers best explain the dynamic behavior of the studied socio-ecological risk variable (Kostyuchenko et al., 2013). In accordance with modern principles of Earth sciences, our approach toward variable selection is based on a combination of ecology, climatology, hydrogeology, hydrology, and geostatistics as consideration of the statistical properties of the estimated model (Kostyuchenko, 2015).

Models built using pure data-mining techniques or principles such as machine learning, neural network, etc., though they may fit the existing data well, are more likely to fail in a changing external environment because they lack theoretical underpinnings. The best analytical and prediction models employ a combination of statistical rigor with physical principles (Kostyuchenko et al., 2013). Hence, our models combine geo-ecological models with statistical optimization (Figure 2). Models built this way have an additional benefit of ease of interpretation.

The satellite model development process consists of selecting optimal exogenous drivers X_t, Y_t in equation (2) from a set of potential drivers. Once the final model is selected and estimated, the conditional dynamic forecasts of Z_t are generated given the sets of final parameter estimates and the forecasts

Figure 2. Data utilization in framework of multi-modeling approach for socio-ecological risks assessment

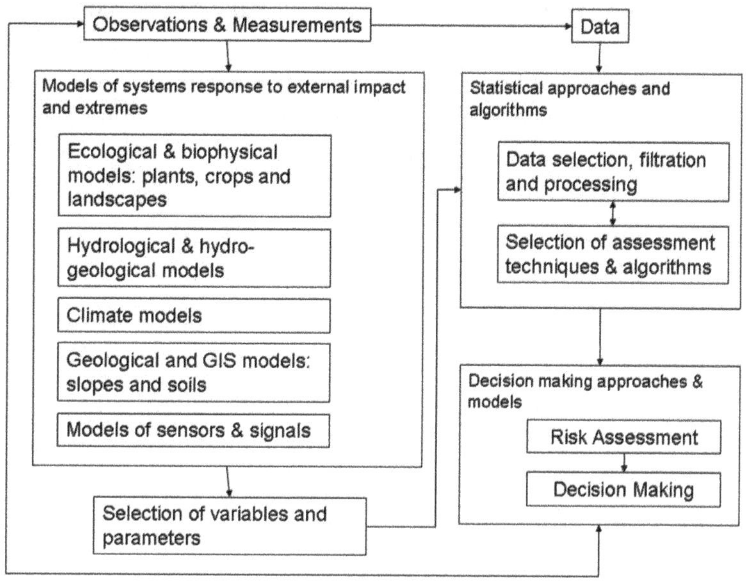

of the core variables from the first stage (Figure 2). The final step is to validate the final model in and out of sample.

The procedure of selecting optimal drivers is the following. Potential drivers are identified based on relevant theory and ensuring with calibration measurements or priory assumptions. This ensures that we obtain the most robust and predictive model available from the tested variables. To avoid model over-fitting, uncorrelated core drivers are preferably selected. The selected drivers should be significant at a conventional level and have the analyzed parameters of distribution. To obtain a required distribution a regularization procedure should be applied (Kostyuchenko, 2015). The final models selected by the search procedure are reviewed for consistency with initial assumptions.

So, the problem is the selection of variables for each model type $(x,\varepsilon) \to y$, the search of the relevant type of formal relationship $(x,y) \to z^s$ between the physical and observable variable parameter, and development of the total distribution for each type of risk investigated (2). There also should be separately considered the problem of regularization of initial distributions of variables (Kostyuchenko et al 2015).

After we obtain distributions of parameters that determine the state of the system, we need to estimate the distribution of risk and make the management decisions.

MODELS OF RISK ASSESSMENT AND DECISION SUPPORT

Basing on the obtained sets of indicators, the methodology of risk assessment, based on the optimal decisions can be proposed (Ermoliev et al., 2012).

In the framework of task of risk assessment and risk management, as opposed to the classic example of the resource management, reducing or non-increasing of losses should be used as quantitative characteristics. Variables that affect the characteristics of the management system (or decision-making system) can be or we can influence (controlled) and independent from our influence (unmanageable). Thus, the controlled variables are the parameters of decision-making under the influence of information (input data) on the behavior of unmanageable variables. Analysis of the effectiveness of full process of collecting, processing, interpretation of information about the system studied, decision making, analysis of system's response to decisions may be considered as part of the "information – response" formalization.

Such formalization can be made as follows (Schlaifer and Raiffa, 1961): define as $I(x,y,z)$ (presumably stochastic) information obtained from direct (field) measurements, observations and model forecasts; $H_I(i|\theta)$ - is probability distribution function and where θ – is a state of the studied natural object or system. In general, the state of the system cannot be determined with certainty, and thus we should define the appropriate probability distribution $p(\theta)$ and distribution $H_I(i|\theta)$, which describing a priori incompleteness of information available about the studied system.

Making and implementation of management decisions will formalize as a response to incoming information as decision function $d(I)$. Classical approach assuming that in the case of certain specific strategy of decision making in conditions of constant state of natural systems, θ, or with a defined change this state, the losses are defined as $l(d(I), \theta)$. For decision function d expected losses or risks associated with the development of dangerous processes, connected with the management decisions based on the information received, can be described as:

$$R(I,d) = R\left[H_I(\bullet),d),d\right] = \iint l\left(d(i),\theta\right)dH_I(i|\theta)p(\theta)d\theta \tag{3}$$

This risk is minimized by optimal decision function d^*, entitled as Bayes decision function, and it is determined by information I:

$$R(I(z^s),d^*) = \min_{d(i)} \iint l(d(i),\theta(x,y))dH_I(i|\theta)p(\theta)d\theta\,. \tag{4}$$

So the risk is a simple functional of decision function. It is important that the minimization of losses requires an intention to completeness of information on the studied system, i.e. definition of θ states, for each of which can be defined a solution $\{a\}$ (which build up the set of possible decisions or administrative actions A).

Let's consider the realization of the set of data (information obtained from direct measurements (field), observations and model forecasts) i^*, which optimizes the decision function d^* and minimizes appropriate risk, therefore nominally makes information (I) formally completed (I^*).

From a formal viewpoint, $I^* \equiv I$ (information is nominally full), if there exists a function $\varphi(i)$, that when $H_I(i|\theta) \neq 0$, $\theta = \varphi(i)$. In other words, completeness formally means that there is a single state of studied natural object or system that meets all of the set of data - a separate realization of information I. Here it means that we have to develop a set of models $(x,\varepsilon) \rightarrow y$ that have operated the set of parameters $(x,y) \rightarrow z^s$, which can be controlled by certain technological tools within the framework of a sustainable methodology of measurement. In our case, these requirements correspond to the data and methods of satellite observation of the Earth's surface z_t^s.

In the case described an optimal decision function may be defined as follows:

$$d^*(i^*) = b\,,\ l\big(b, \varphi(i^*)\big) = \min_a l(a, \varphi(i^*)) \tag{5}$$

Then, under optimal decision function and nominally complete information about the studied system, the risk will be defined as:

$$R(I^*, d^*) = \int \min_{a \in A} l(a, \theta) p(\theta) d\theta \tag{6}$$

For each case should be described models aimed to analysis of behavior the distribution of $H_I(i|\theta)$ та *p(θ)*, and so to the determination of the realization of i^* of set *I*. As an optimal decision function in this approach may be used stochastic (Kopachevsky et al., 2016), Bayesian (Kostyuchenko et al., 2012) or fuzzy operators, (Kostyuchenko et al., 2016) depending on the task, data availability and properties of their distributions.

Considered complex of analytical models is aimed to the calculation of a unique set of parameters that should be obtained from determined observation systems, using defined tools of processing and interpretation of data. Equation (6) allows to estimate the distributions of risk of disasters and also to develop a basement for a system of risk management decision-making.

As it was demonstrated in (Kostyuchenko, 2017), modeling of geo-systems should be an integral part both of remote sensing interpretation methods, as well as of the risk assessment systems based on remote sensing data utilization. It requires of increased level of our knowledge in the field of Earth sciences, as well by increased requirements in the area of decision-making. New challenges define new methodological requirements.

Firstly, the methodology proposed allows to expand the problem definition of using the satellite observations in tasks of socio-ecological security. In addition to traditional statistical analysis directed to surface change detection, it is possible to analyze and predict state of the studied systems, basing on the models of geo-systems.

This certainly expands the scope and sphere of application of approach, and could positively affect the reliability of the results obtained through the using of different sources of data.

Second, the proposed methodology includes feedbacks between management decisions and the systems state. Thus, it is postulated that the state of the system depends on the observer: risks depend on the decision made and management impacts (past, current and planned) to the system.

This could positive affecting to the effectiveness of management decisions and to the quality of risk assessment (Ermoliev and Winterfeldt, 2012).

CORE NATURAL MODELS: BASINS AND TRANSPORT

Basing on the described general methodology of modeling, the problem-oriented multi-model approach can be proposed for investigated problem (Figure 3).

Core models should include models of water bodes and fluxes oriented to analysis of pollution transport and modeling of hydro-chemical balance in water ecosystems. Besides, models of pollutions, including air transport and crop nitrate pollution production should be considered. Set of variables describing the water quality is the result of this stage of analysis. Satellite data models should include models of signal

Figure 3. Problem-oriented multi-model approach to water quality risk assessment

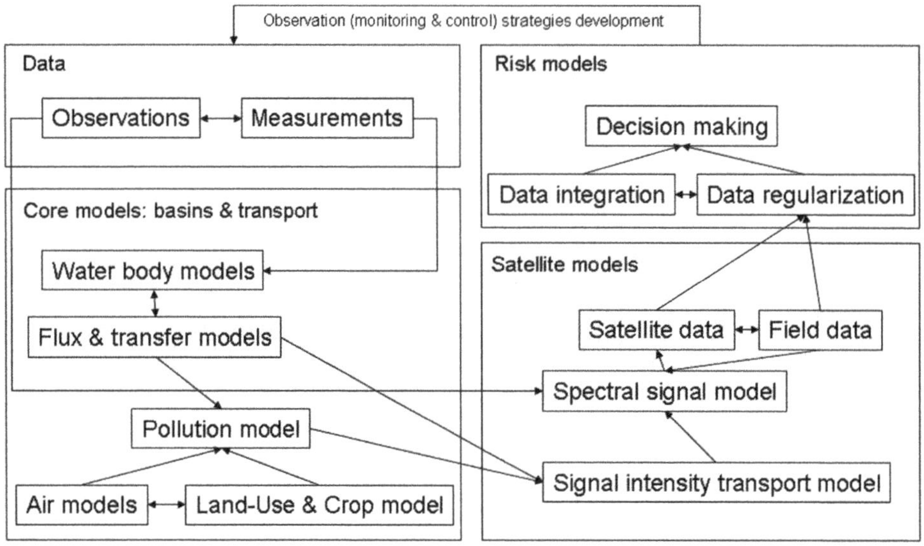

forming, and analysis of links between satellite and ground data. Set of core models based selected indicators is the result of this stage of modeling. Risk models should be directed to construction of algorithm of risk assessment using the set of selected indicators.

Water Body Model: General Model of Spatially Distributed Water Ecosystem

Let define $c(x, y, z, t)$ as concentration of spatially distributed matter. The general mathematical representation of dynamics of matter j will be described by following equation:

$$\begin{gathered}\frac{\partial c_j}{\partial t} = F_j(c_1, c_2, ..., c_n, x, y, z, t) + \\ +D\left(\frac{\partial^2 c_j}{\partial x^2} + \frac{\partial^2 c_j}{\partial y^2} + \frac{\partial^2 c_j}{\partial z^2}\right) - \frac{\partial}{\partial x}(V_x c_j) - \frac{\partial}{\partial y}(V_y c_j) - \frac{\partial}{\partial z}(V_z c_j)\end{gathered} \tag{7}$$

where V_x, V_y, V_z - are the components of velocity vector.

System $\frac{\partial c_j}{\partial t} = F_j(c_1, c_2, ..., c_n, t)$ with $j = 1, 2, ..., n$ is a point model with distributed parameters; function $F_j(c_1, c_2, ..., c_n, x, y, z, t)$ may be linear or non-linear with variable coefficients, depending of time t.

General Model of Fluxes and Transport of Pollution in Ecosystem

Let analyze volume element of studied environment Δx, Δy, Δz (Figure 4)

Figure 4. Transfer of pollutions in water environment

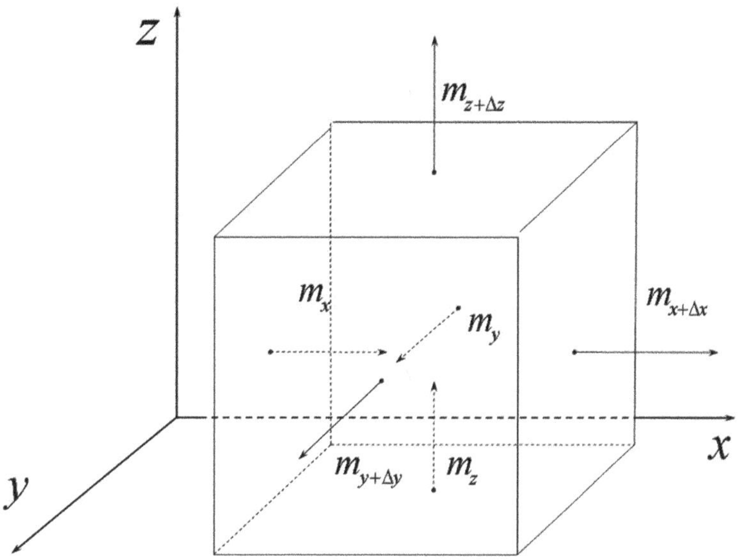

Mass of transported pollution matters through unit square during unit time interval along (*Ox*), (*Oy*) and (*Oz*) are m_x, m_y and m_z correspondingly. Therefore, we can write the following equation:

$$m_x = -D\frac{\partial c}{\partial x},\ m_y = -D\frac{\partial c}{\partial y},\ m_z = -D\frac{\partial c}{\partial z} \tag{8}$$

where m_x, m_y, m_z – flux density, *D* – diffusion coefficient.

If our environment is changing with velocity $V(V_x, V_y, V_z)$, we can write:

$$m_x = cV_x - D\frac{\partial c}{\partial x},\ m_y = cV_y - D\frac{\partial c}{\partial y},\ m_z = cV_z - D\frac{\partial c}{\partial z} \tag{9}$$

where V_x, V_y, V_z are the components of velocity vector.

In the unit volume $\Delta W = \Delta x \Delta y \Delta z$ during time interval Δt will be deposited mass:

$$\Delta M = -\frac{\partial m_x}{\partial x}\Delta W \Delta t - \frac{\partial m_y}{\partial y}\Delta W \Delta t - \frac{\partial m_z}{\partial z}\Delta W \Delta t \tag{10}$$

The distribution of matters in static environment will be described as:

$$\frac{\partial c}{\partial t} = D\left(\frac{\partial^2 c}{\partial x^2} + \frac{\partial^2 c}{\partial y^2} + \frac{\partial^2 c}{\partial z^2}\right) \tag{11}$$

and in the varied environment as the following:

$$\frac{\partial c}{\partial t} = D\left(\frac{\partial^2 c}{\partial x^2} + \frac{\partial^2 c}{\partial y^2} + \frac{\partial^2 c}{\partial z^2}\right) - \frac{\partial}{\partial x}(V_x c) - \frac{\partial}{\partial y}(V_y c) - \frac{\partial}{\partial z}(V_z c) \tag{12}$$

If we equate $\frac{\partial c}{\partial t} = 0$, we obtain stationary distribution of matters in static:

$$\frac{\partial^2 c}{\partial x^2} + \frac{\partial^2 c}{\partial y^2} + \frac{\partial^2 c}{\partial z^2} = 0 \tag{13}$$

and varied environment:

$$D\left(\frac{\partial^2 c}{\partial x^2} + \frac{\partial^2 c}{\partial y^2} + \frac{\partial^2 c}{\partial z^2}\right) - \frac{\partial}{\partial x}(V_x c) - \frac{\partial}{\partial y}(V_y c) - \frac{\partial}{\partial z}(V_z c) = 0. \tag{14}$$

If in the environment presents source of matters G, or chemical or biological transformations $F(c, x, y, z, t)$, these equations will be presented as follow:

$$\frac{\partial c}{\partial t} = F(c, x, y, z, t) + D\left(\frac{\partial^2 c}{\partial x^2} + \frac{\partial^2 c}{\partial y^2} + \frac{\partial^2 c}{\partial z^2}\right) \tag{15}$$

$$\frac{\partial c}{\partial t} = F(c, x, y, z, t) + D\Delta c - V_x \frac{\partial c}{\partial x} - V_y \frac{\partial c}{\partial y} - V_z \frac{\partial c}{\partial z} \tag{16}$$

Model of Water and Hydro-Chemical Balance of Water Ecosystems

In the varied water environment, many ways of matter transportation are exist (Baird and Wilby, 1999). Let define as D_x, D_y i D_z the components of turbulent diffusion coefficient. In this case the corresponding components of pollutant flow with concentration $c(x, y, z, t)$ will be defined as:

$$V_x c - D_x \frac{dc}{dx},\ V_y c - D_y \frac{dc}{dy} \text{ та } V_z c - D_z \frac{dc}{dz}. \tag{17}$$

Equation for transport of non-conservative matters (pollutants) in the water flow:

$$\frac{\partial c}{\partial t}=\frac{\partial}{\partial x}\left(D_x\frac{\partial c}{\partial x}-V_x c\right)+\frac{\partial}{\partial y}\left(D_y\frac{\partial c}{\partial y}-V_y c\right)+\frac{\partial}{\partial z}\left(D_z\frac{\partial c}{\partial z}-V_z c\right)-\alpha * c+f\left(x,y,z,t\right), \tag{18}$$

where $\alpha*$ – coefficient of sedimentation velocity, $f(x, y, z, t)$ – internal sources and run-off function.

Boundary conditions may be defined as:

$$c(x, y, z, t_0) = c(x, y, z), \tag{19}$$

$$c\left(x,y,z,t\right)\Big|_{x=x_1}=c_1\left(y,z,t\right)\frac{\partial c}{\partial x}\Big|_{x=x_2}=0,\ \frac{\partial c}{\partial n}=0 \tag{20}$$

where $x=x_1$ – usually upper site of water body, $x=x_2$ – lower site, n – external normal to surface.

In case of presence of influx of pollutants with precipitation:

$$\left[V_z c-D_z\frac{\partial c}{\partial z}\right]_{z=0}=c_{pr}q_{pr} \tag{21}$$

where q_{pr} - precipitation to unit square of water body, c_{pr} - concentration of pollutant in precipitation.

In case of presence of pollutants in bottom sediments, in particular with ground water filtration:

$$c\left(x,y,h_b,t\right)=c_f;\ \left[V_z c-D_z\frac{\partial c}{\partial z}\right]_{z=h_b}=c_f v_f \tag{22}$$

where c_f – pollutant concentration in ground waters; v_f – vertical component of velocity vector of ground water filtration; $\left(h_b\right)$ – depth of water body.

To solve this problem let's divide our environment to n cells. For every cell $i = 1, 2, 3, \ldots, n$ of studied environment lets define: q^i, A^{ij} – flux of water from cell (i) to cell (i+1), and resulted concentration of matter j in cell i; q^{i-1}, $c^{i-1,j}$ – flux of water from cell (i-1) to cell (i), and resulted concentration of matter j in cell (i-1); q^i_k, c^{ij}_k – flux of water from tributary (k) to cell (i), and resulted concentration of matter j in cell (i), ($k=1,2,3,\ldots,n_i;4$); q^i_{f}, A^{ij}_{f} – infiltrations of ground waters with pollutant (j), and concentrations of matter (j); q^i_{pr}, c^{ij}_{pr} – precipitation to cell (i) with pollutant (j), and concentration of (j); q^i_{wd} – water demand; q^i_{ev} – evaporation from cell (i-1); m^i_{bs} (A^{ij}, c^{ij}_{bs}) – mass of matter (j) in cell (i) transported from bottom sediments, and its concentration c^{ij}_{bs}; α^{ij}, λ^j – coefficients of sedimentation, chemical and biological decomposition; $W^i(t)$, W^i_0 – volume of cell (i) at the moment t and initial time $t = t_0$; F^{ij} (A^{ij}, c^{ij}_{hb}, t) – function of mass exchange of matter (j) in cell (i) with hydro-biota; t – time.

Mass of matter (pollutant) (j) we can estimate using differential equation:

$$\frac{dM^{ij}}{dt} = q^{i-1}c^{i-1,j} + \sum_{k=1}^{n_i} q_k^i c_k^{ij} + q_f^i A_f^{ij} + q_{pr}^i A_{pr}^{ij} + m_{bs}^{ij}(A^{ij}, c_{bs}^{ij}, t) - (q^i + q_{wd}^i)A^{ij} - \\ -(\alpha^{ij} + \lambda^j)W^i c^{ij} - F^{ij}(c^{ij}, c_{hb}^{ij}, t), \tag{23}$$

Mass of matter (*j*) also can be expressed through water volume and concentration:

$$M^{ij}(t) = c^{ij}(t)W^i(t) = c^{ij}(t)\left(W_0^i + \left(q^{i-1} + \sum_{k=1}^{n_i} q_k^i + q_f^i + q_{pr}^i - q^i - q_{wd}^i - q_{ev}^i\right)t\right) \tag{24}$$

So, we can calculate:

$$\frac{dc^{ij}}{dt} = \frac{1}{W^i}\left(q^{i-1}c^{i-1,j} + \sum_{k=1}^{n_1} q_k^i c_k^{ij} + q_f^i A_f^{ij} + q_{pr}^i c_{pr}^{ij} + m_{bs}^{ij}(c^{ij}, c_{bs}^{ij}, t) - \left(q^i + q_{wd}^i + \frac{dW^i}{dt}\right)c^{ij}\right) - . \\ -(\alpha^{ij} + \lambda^j)c^{ij} - f^{ij}(c^{ij}, c_{hb}^{ij}, t), \tag{25}$$

Also the equation of water balance of cell (*i*) should be added:

$$\frac{dW^i}{dt} = q^{i-1} + q_f^i + \sum_{k=1}^{n_i} q_k^i + q_{pr}^i - q^i - q_{wd}^i - q_{ev}^i \tag{26}$$

If q^{i-1}, q_f^i, q_k^i, q_{pr}^i, q^i, q_{wd}^i, q_{ev}^i are continuous, we can integrate:

$$W^i = W_0^i + \left(q^{i-1} + q_f^i + \sum_{k=1}^{n_i} q_k^i + q_{pr}^i - q^i - q_{wd}^i - q_{ev}^i\right)t \tag{27}$$

and if $c^{i-1,j}$, A_f^i, c_k^i, c_{pr}^i are also continuous and differentiated, and $m_{bs}^{ij} = f^{ij} = \alpha^{ij} = \lambda^{ij} = 0$, we can integrate:

$$c^{ij}(t) = c^{ij*} + (c_0^{ij} - c^{ij*})\left(\frac{W_0^i}{W^i(t)}\right)^{\frac{q_{in}^i}{q_{in}^i - q_{out}^i}}$$

where

$$q_{in}^i = q^{i-1} + \sum_{k=1}^{n_i} q_k^i + q_f + q_{pr}$$

and

$$q_{out}^i = q^i + q_{wd}^i + q_{ev}^i$$

A^{ij*} - is the equilibrium concentrations:

$$c^{ij*} = \frac{q^{i-1}c^{i-1,j} + q_f^i c_f^{ij} + \sum_{k=1}^{n_i} q_k^i c_k^{ij} + q_{pr}^i c_{pr}^{ij} + m_{bs}^{ij}}{q^{i-1} + q_f^i + \sum_{k=1}^{n_i} q_k^i + q_{pr}^i - q_{ev}^i} \tag{28}$$

If water level in studied body is changing not significantly, we can propose:

$$c^{ij}(t) = c_\beta^* + (c_0^{ij} - c_\beta^*)\exp\left(-\frac{1+\beta^{ij}\tau_0^i}{\tau_0^i}t\right) \tag{29}$$

where

$$c_\beta^* = \frac{q^{i-1}c^{i-1,j} + q_f^i c_f^{ij} + \sum_{k=1}^{n_i} q_k^i c_k^{ij} + q_{pr}^i c_{pr}^{ij} + m_{bs}^{ij}}{(1+\beta^{ij}\tau_0^i)\left(q^{i-1} + q_f^i + \sum_{k=1}^{n_i} q_k^i + q_{pr}^i - q_{ev}^i\right)} \tag{30}$$

$$\tau_0^i = \frac{W_0^i}{q^{i-1} + q_f^i + \sum_{k=1}^{n_i} q_k^i + q_{pr}^i - q_{ev}^i}, \ \beta^{ij} = \alpha^{ij} + \lambda^{ij}, \tag{31}$$

$$\alpha^{ij} \neq 0, \ \lambda^{j} \neq 0, \text{ and } f^{ij}(c^{ij}, c_{hb}^{ij}, t) = 0. \tag{32}$$

This model has high efficiency in analysis of hydro-chemical and water balance of water bodies (Koch and Grünewald, 2009).

Air Pollution Dispersion Models

Risk assessment and security management tasks require to analyze a transport of pollutants in the atmosphere. There are many ways to analyze distribution of atmosphere pollutions. Key mathematical models of atmospheric air contamination and pollution dispersion are described in (Moussiopoulos et al., 1997).

Distribution of pollutants in the atmosphere can be analyzed with models of particles dispersion and models of air contamination (Zannetti, 1970).

There are few basic dispersion models:

- Eulerian dispersion model (numerical solution of atmospheric diffusion equation);
- Gaussian dispersion model (concentration of pollutant describing as a Gaussian distribution);
- Lagrangian dispersion model (analysis of processes in dynamic air environment or imitation of dispersion with conditional particles).

The models of air contamination may be divided to semi-empirical (based on empiric parameterization), stochastic, and imitational (voxel) models. To forecast a pollutant distribution the equation of molecular and convective diffusion can be applied. This equation describes matter transport in different environments with certain boundaries constrains.

In the simplest case, the one-dimensional empiric model may be applied:

$$v(t) = \frac{dx}{d(t - t_0)} \tag{33}$$

where v is the velocity of pollution, t – time from initial moment t_0; x – is a coordinate.

More accurate calculations can be executed using three-dimensional parabolic equation (National Research Council, 1992):

$$Q(T,t) = \frac{\partial C(x,y,z^*,t)}{\partial t} + v(x,y,z^*,t) gradC(x,y,z^*,t) - divDgradC(x,y,z^*,t) \tag{34}$$

where C is concentration, D – coefficient of turbulent transport, t – time from initial moment t_0, Q – intensity of contamination source, v is the velocity of pollution; z^* – vertical coordinate reduced to terrain z_0 taking into account effective depth of vegetation, T – temperature.

Pollution Model: Crop and Nitrate Contamination

Management of nitrogen contaminations requires a model of contamination, based on understanding of nitrogen cycle in environment (Figure 5) and, first of all, on the crop yield model, since agriculture is a main source of diffuse nitrate pollutions (Zheng and Bennett, 2002).

The crop yield, required to estimate the nitrate pollution, can be calculated according to the following polynomial equation:

$$Y_{s,y} = a + b \cdot W_{s,y} + c \cdot W_{s,y}^2 + d \cdot N_{s,y} + e \cdot N_{s,y}^2 + f \cdot W_{s,y} N_{s,y}, \tag{35}$$

where $Y_{s,y}$ is the crop yield located at site source s for a year y (kg/ha), $W_{s,y}$ is the water applied to the crop located at source s (m3 /ha), and $N_{s,y}$ is the fertilizer applied to the crop located at source s (kg/ha) within the year y. The coefficients $a,\ldots, f$ are calibrated (for example by a least-square fitting technique) with the in-field measurement data.

Figure 5. General model of nitrogen cycle

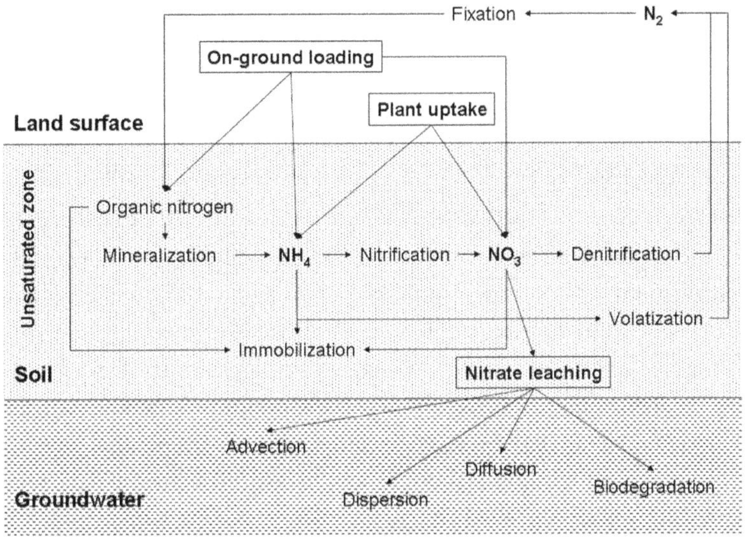

The amount of leaching and hence the amount of nitrates in groundwater have been found to be a function of the timing of fertilizer application, vegetative cover, soil porosity, fertilizer application method, irrigation rate (Canter, 1996). Once the nitrogen is applied to the crop it suffers some transformation. After the plant uptake and the transformation, some of that nitrogen applied is converted into nitrate that can leach to the aquifer. The amount of nitrogen leached was introduced into the management model as quadratic functions as follows:

$$L_{s,y} = g + h \cdot W_{s,y} + i \cdot W_{s,y}^2 + j \cdot N_{s,y} + k \cdot N_{s,y}^2 + l \cdot W_{s,y} N_{s,y} \tag{36}$$

where $L_{s,y}$ is the nitrogen leached (kg/ha), and the coefficients $g,..., l$ are calibrated (for example by a least-square fitting technique) with the in-field measurement data. The nitrate leached in (kg/ha) is diluted by the irrigation water recharge, therefore the nitrate concentration $cr_{s,x,y}$ entering the aquifer is:

$$cr_{s,x,y} = \frac{L_{s,x,y}}{r_{s,x,y}} \tag{37}$$

where $r_{s,x,y}$ is the water that recharges the aquifer (m^3 /ha) at planning period *y*, and $L_{s,x,y}$ is the nitrogen leached from each crop area *s* (kg/ha) at planning period *y*. The subindex *y* in the formulation refers to the year within the planning horizon or the number of successive years in which the fertilizer is applied. The vector of *n* elements $cr_{s,x,y}$ correspond to concentration recharge (the product of the concentrations of the source waters times the volumetric water fluxes), during the management period *y* and the disposal site location *s*.

And the nitrate concentration $\rho_{t,x,y}$ at the moment *t* in site *s* can be calculated as:

$$\rho_{t,x,y} = \alpha_0 + \alpha_1 \rho_0 + \alpha_2 \sum_{s,x,y} (A_{s,t} cr_{s,x,y} + \vartheta_s) + \varepsilon \quad (38)$$

where $\alpha_0, \alpha_1, \alpha_2$ are the regression coefficients, ρ_0 is the initial concentration, A_s is the area cultivated for crop located at source *s*; ϑ_s is a coefficient of infiltration, and ε is an uncertainty factor.

From viewpoint of socio-ecological security control, this algorithm should be a part of hydro-economic approach, of an optimization model, which should be developed to define efficient and secure fertilizer allocation in agriculture. It should be assessed when, where and by how much fertilizer reductions have to be applied to meet the ambient standards (groundwater quality) in specific control sites in the aquifer.

In framework of this hydro-economic model the non-point pollution abatement problem should be stated as the maximization of welfare from crop production subject to constraints that control the environmental impacts of the decisions in the study region. Welfare was measured as the private net revenue, calculated through crop production functions and data on crops, nitrogen and water prices. The hydro-economic model integrates the environmental impact of fertilization by simulation of soil nitrogen dynamics and fate and transport of nitrate in groundwater with the economic impact (agricultural income losses) of water and fertilization restrictions, assessed through agronomic functions representing crop yields and crop prices. The decision variables of the problem are the sustainable quantities of nitrogen per hectare applied in the different crop areas (diffuse pollution sources) to meet the environmental constraints. The management model for groundwater pollution control may be formulated as:

$$\max\{G\} = \sum_{s,y} \frac{A_s (p_s Y_{s,y} - p_n N_{s,y} - p_w W_{s,y} - C_s + S_s)}{(1+r)^y} \quad (39)$$

with constrain:

$$(\sum_{s,x,y} cr_{s,x,y} \frac{\rho_{x,y} t}{c_{s,y}}) \le q_{ct}, \ \forall c, t, y \quad (40)$$

where *G* is the objective function to be maximized and represents the present value of the net benefit from agricultural production defined as crop revenues minus fertilizer and water variable costs, A_s is the area cultivated for crop located at source *s*; p_s is the crop price (per kg); $Y_{s,y}$ is the production yield of crop located at source *s* at planning year *y* (kg/ha), that depends on the nitrogen fertilizer and irrigation water applied; p_n is the nitrogen price (per kg); $N_{s,y}$ is the fertilizer applied to crop located at source *s* at year *y* (kg/ha), p_w is the price of water, and $W_{s,y}$ is the water applied to crop located at source *s* at each planning year *y* (m^3); C_s is the aggregation of the remaining per hectare costs for crop located at source *s*; S_s are the subsidies for the crop located at source *s*; *r* is the annual discount rate, $\rho_{x,y}$ is the nitrate concentration for each crop area *s*, *t* is the time (number of years within the planning horizon *y*), $c_{s,y}$ is the number of control sites for each crop area *s* during the period *y*; q_{ct} is a vector of water quality standard imposed at the control sites; $cr_{s,x,y}$ is above described three component vector, which corresponds to the nitrate concentration recharge reaching groundwater from a crop located at source *s*.

SATELLITE DATA MODELS

Model of Reflectance Variation Conditioned by Pollutant Transport in Dynamic Environment

From viewpoint of satellite data utilization, the propagation of pollutants in water and air environment may be modeled through assessment of variations of intensity distribution on satellite images. Therefore, to analyze pollutants distribution we can study spatial-temporal distributions of spectral reflectance using satellite time-series.

Basing on the equation of transfer and non-stationary Navier-Stokes equation may be constructed approach to iterative minimax assessment of velocity of variations of optical flux (Herlin et al., 2012). Using this approach it is possible to solve a problem of observation/measurements data discreteness, and to control the uncertainties. Model of signal intensity connected with pollutant distribution in changing environment may be proposed basing on this approach.

Intensity *I* varying in the velocity field $\mathbf{v} = \left(u(x,y,t), v(x,y,t)\right)^T$, which can be described by Navier-Stokes equations:

$$\begin{aligned} &\partial_t u + u\partial_x u + v\partial_y u + \partial_x p = \nu\Delta u + e_u^m, \\ &\partial_t v + u\partial_x v + v\partial_y v + \partial_y p = \nu\Delta v + e_v^m, \\ &\partial_x u + \partial_y v = 0, \quad (x,y,t) \in \Omega_T, \\ &\mathbf{v}(x,y,t) = 0, \quad (x,y) \in \partial\Omega, \\ &\mathbf{v}(x,y,t_0) = \mathbf{v}_0(x,y) + \mathbf{e}^b(x,y), \quad (x,y) \in \Omega, \end{aligned} \tag{41}$$

where $\mathbf{v}_0$ – initial velocity field, e_u^m, e_v^m, $\mathbf{e}^b$ - members to description of epistemic uncertainty (model).

Distribution of observed values $I(x_k, y_l, t_s)$ of intensity function $I(x,y,t)$ in the points $\{(x_k, y_l)\}_{k,l=1}^{N_x,N_y}$ of the image site Ω can be expressed as:

$$Y_s^{kl} = \int_{\Omega_T} g_s^{kl}(x,y,t) I(x,y,t)\,dx\,dy\,dt + \eta_s^{kl} \tag{42}$$

$$k = \overline{1,N_x}, \; l = \overline{1,N_y}, \; s = \overline{1,S}$$

where g_s^{kl} is image receiving procedure function, η_s^{kl} – aleatiroc uncertainty (observation error).

If according to assumption every element of image corresponds to velocity field $\mathbf{v} := (u(x,y,t), v(x,y,t))^T$, and its intensity is constant along full trajectory ($I(x,y,t) \approx \text{const}$), intensity function will comply with constraints (Horn and Schunck, 1981):

$$\frac{d}{dt} I(x.y.t) = \partial_t I + u(x.y.t)\partial_x I + v(x.y.t)\partial_y I = e^o(x.y.t) \tag{43}$$

Here we assume that exist all partial derivatives of I, and there is $e^o \in \mathbb{L}_2(t_0, T, \mathbb{L}_2(\Omega))$, which describe an uncertainty.

Minimax assessment of velocity field $\mathbf{v}$ is building using the time-series of discrete images Y_s^{kl}, basing on assumption that uncertain parameters e^o, e_u^m, e_v^m, $\mathbf{e}^b$ belong to the limited convex set $L_2(\Omega_T)$, and η_s^{kl} is an independent scalar random values with zero mean and covariations R_s^{kl}.

For specified velocity field $\mathbf{v}^*$ can be constructed assessment of intensity $\hat{I}^*$, which determined by this field and constraints. Assessment $\hat{I}^*$ should correspond to observation data Y_s^{kl}, with aleatoric errors η_s^{kl} and epistemic uncertainty of $\mathbf{v}^*$.

Assessment of gradient ∇I^* calculating using constructed assessment of intensity function $\hat{I}^*$. Using substitution of $\mathbf{v}^*$ and ∇I instead ∇I^*, a system of linear equations for I and $\mathbf{v}$ can be obtained for further construction of $\mathbf{v}^{**}$ – optimal assessment of velocity field $\mathbf{v}$ (for which corresponding I^{**} has optimal correlation with data Y_s^{kl}).

Iterative method consists of two stages. First, assessment of velocity field $\hat{\mathbf{v}}^i = (\hat{u}^i, \hat{v}^i)^T$ should be substituted into the equation of optical flux constrains:

$$\partial_t I = -\hat{u}^i(x,y,t)\partial_x I - \hat{v}^i(x,y,t)\partial_y I + e^o(x,y,t) \tag{44}$$

Defining this equation as an equation of state, the minimax assessment $\hat{I}^i$ of the intensity function I and assessment ∇I^* of the image gradient ∇I can be calculated.

Second, taking into account ∇I^* and $\hat{\mathbf{v}}^i$, the next equation of state can be constructed:

$$\begin{aligned}
&\partial_t I = -u(x,y,t)\partial_x I^i - \nu(x,y,t)\partial_y I^i + e^o(x,y,t) \\
&\partial_t u + \hat{u}^i\partial_x u + \hat{\nu}^i\partial_y u + \partial_x p = \nu\Delta u + e_u^m, \\
&\partial_t \nu + \hat{u}^i\partial_x \nu + \hat{\nu}^i\partial_y \nu + \partial_y p = \nu\Delta\nu + e_\nu^m, \\
&\partial_x u + \partial_y \nu = 0, (x,y,t) \in \Omega_T, \\
&\nu(x,y,t) = 0, (x,y) \in \partial\Omega
\end{aligned} \tag{45}$$

This system of linear parabolic equations may be used for calculation of minimax assessment of $\hat{\mathbf{v}}^{i+1}$. Also for uncertainty assessment it is important to obtain a linear assessment of intensity function I.

Minimax assessment $\hat{I}$ of the function I is the solution of the system of equations:

$$\begin{aligned}
&\frac{d\hat{I}}{dt} = L_\varepsilon(t,\mathrm{v})\hat{I} + BQ^{-1}B^*\hat{p},\\
&\hat{I}(x,y,t_0) = \overline{I}_0 + B_0Q_0B_0^*\hat{p}(x,y,t_0), \quad N_\Gamma \hat{I} = 0,\\
&-\frac{d\hat{p}}{dt} = L_\varepsilon^*(t,\mathrm{v})\hat{p} + \sum_{s=1}^{N}\sum_{k,l=1}^{M} g_s^{kl}(R_s^{kl})^{-1}(Y_s^{kl} - \int_{\Omega_T} g_s^{kl}\hat{I}dxdydt),\\
&\hat{p}(x,y,T) = 0, \quad N_\Gamma \hat{p} = 0.
\end{aligned} \tag{46}$$

where $L_\varepsilon(t,\mathbf{v}) := -u(t,x,y)\partial_x - v(t,x,y)\partial_y + \varepsilon^2\Delta$, $\varepsilon > 0$. $L_\varepsilon(t,\mathbf{v})$ is a linear differential operator; $\overline{I}_0 \in L^2(\Omega)$ is an initial state (initial image of time-series); B_0 and B – linear limited operators in $L^2(\Omega)$, which are the additional constrains to uncertain parameters; e^o, e^b - realizations of independent random processes:

$$\begin{aligned}
&m_0(x,y) := Ee^b(x,y), \quad m_0 = L^2(\Omega),\\
&m(x,y,t) := Ee^o(x,y,t), \quad m \in L^2(\Omega_T).
\end{aligned} \tag{47}$$

$$\begin{aligned}
&Q_0(x,y,x',y') := Ee^b(x,y)e^b(x',y') \in L^2(\Omega \times \Omega),\\
&Q(x,y,t,x',y',t') := Ee^o(x,y,t)e^o(x',y',t') \in L^2(\Omega_T \times \Omega_T).
\end{aligned} \tag{48}$$

Solutions of this system of equations are the following:

$$\hat{I} = \overline{I} + \sum_{s=1}^{N}\sum_{k,l=1}^{M}(R_s^{kl})^{-1}(Y_s^{kl} - \beta_s^{kl})I_s^{kl}, \tag{49}$$

$$\hat{p} = \sum_{s=1}^{N}\sum_{k,l=1}^{M}(R_s^{kl})^{-1}(Y_s^{kl} - \beta_s^{kl})p_s^{kl}, \tag{50}$$

where β_s^{kl} ($k = \overline{1,N_x}$, $l = \overline{1,N_y}$, $s = \overline{1,S}$) is a solution of:

$$\begin{aligned}
&\beta_s^{kl} + \sum_{s'=1}^{N}\sum_{k',l'=1}^{M}[\int_{\Omega_T} g_s^{kl}(R_{s'}^{k'l'})^{-1}I_{s'}^{k'l'}dxdydt]\beta_{s'}^{k'l'}\\
&= \int_{\Omega_T} g_s^{kl}\overline{I}dxdydt + \sum_{s'=1}^{N}\sum_{k',l'=1}^{M}[\int_{\Omega_T} g_s^{kl}(R_{s'}^{k'l'})^{-1}I_{s'}^{k'l'}dxdydt]Y_{s'}^{k'l'}
\end{aligned} \tag{51}$$

As a result, we obtain distribution $\hat{I}$, which is a minimax assessment of intensity function, and minimax assessment of image gradient ∇I^i can be calculated on this base.

In the field of security management and control the distribution of extremums is important. In this class of tasks, the linear minimax assessment $\hat{I}$ of function *I*, which allows to determine extremes of distribution of spectral reflectance is an optimal solution of presented system of equations.

Therefore, transfer of pollutants can be assessed by the sampling of intensity distribution in separate spectral ranges.

Model of Spectral Reflectance Toward Water Quality

As an experience of satellite observations shows (Zilioli and Brivio, 1997; McFeeters, 1996; Kostyuchenko et al., 2012), spectral reflectance indices can be used for as indicators of ecological state of water eco-systems. In particular, Normalized Difference Vegetation Index *NDVI* can serves as indicator of water quality of inland and offshore marine water ($NDVI_W$). Distributions of *NDVI* have good correlations with concentrations of chlorophyll-α, mineral suspension, with water clarity and other quality characteristics.

General model of signal formation of water surface can be presented as superposition of luminances (Figure 6) (Kogan et al., 2004; Gordon et al., 1988):

$$L_t(\lambda) = \sum_{\lambda}\left(L_R^{(\lambda)} + L_A^{(\lambda)} + T_\omega^{(\lambda)} L_G^{(\lambda)} + T_\nu^{(\lambda)} L_C^{(\lambda)} + T_\nu^{(\lambda)} L_W^{(\lambda)}\right) \tag{52}$$

where $L_t(\lambda)$ is the luminance on the sensor, $L_W^{(\lambda)}$ is the signal on water surface, $L_R^{(\lambda)}$ is the luminance corresponding to atmospheric Rayleigh scattering, $L_A^{(\lambda)}$ is the luminance corresponding to atmospheric aerosol scattering, $L_G^{(\lambda)}$ and $L_C^{(\lambda)}$ are luminance connected with surface glaring and foaming, $T_\omega^{(\lambda)}$ is the spectral directional transmittance, $T_\nu^{(\lambda)}$ is the spectral diffuse transmittance. Luminance *L* is connected with intensity *I* through the projection of source area by the simple relation $L = dI/dA\cos\theta$, where *A* is the square of observed source area, and *θ* is the angle between line of observation and the surface normal.

In framework of this model the $NDVI_W$ index can be presented as:

$$\mathrm{NDVI_W} = \frac{L_W^{\lambda_1} - L_{WW}^{\lambda_2} + L_A^{\lambda_1} - L_A^{\lambda_2}}{L_W^{\lambda_1} + L_W^{\lambda_2} + L_A^{\lambda_1} + L_A^{\lambda_2}}, \tag{53}$$

where $L_{W,A}$ is a luminance of radiance from water surface with the wavelength λ and luminance of molecular and aerosol scattering, λ_1 corresponds wavelength range [0,84-0,88] mkm, λ_2 - to range [0,63-0,68] mkm.

Taking into account a molecular scattering:

$$\mathrm{NDVI}_{\mathrm{W}} = \frac{L_A^{\lambda_1} - L_A^{\lambda_2}}{L_A^{\lambda_1} + L_A^{\lambda_2}} = \frac{L_{AZ}^{\lambda_1} + L_M^{\lambda_1} - L_{AZ}^{\lambda_2} - L_M^{\lambda_2}}{L_{AZ}^{\lambda_1} + L_M^{\lambda_1} + L_{AZ}^{\lambda_2} + L_M^{\lambda_2}} \tag{54}$$

Figure 6. Signal model of water surface ($L(\lambda)$ is a luminance, C_S - concentration of water suspensions)

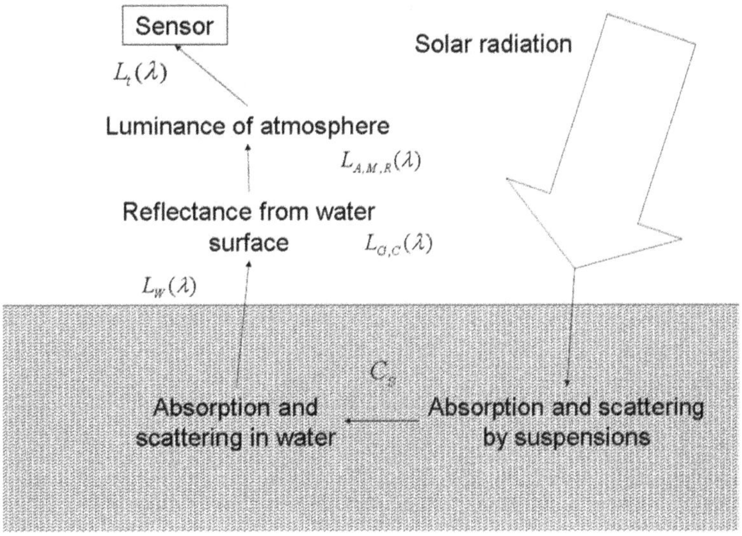

here L_{AZ} – luminance of aerosol scattering, L_M - luminance of molecular scattering, according to: $L_M(\lambda) = \tau_R(\lambda) F_0'(\lambda) R(\alpha_S, \alpha_V, \phi_S, \phi_V) \big/ 4\pi$, where $\tau_R(\lambda)$ is optic thickness of molecular scattering layer; τ_{0Z} is optic thickness of ozone layer; $F_0{}^* = F_0 \exp[-\tau_{0Z}(1/\cos\alpha_S + 1/\cos\alpha_v)]$ – irradiance on the water level; F_0 – irradiance on the level of upper atmosphere; $\alpha_S, \alpha_v, \phi_S, \phi_v$ are the zenith and azimuth angles of Sun and sensor.

Calculated values of $L_M^{\lambda_{1,2}}$, measured with zenith and azimuth angles of sensor and Sun in diapason [0^0; 50^0], are in interval of $L_M^{\lambda_1} \in [0,4;0,48]$, $L_M^{\lambda_2} \in [1,81;2,15]$ (Wt/m^{2}mkm·sr), which is less of 10% of average atmospheric scattering ($L^{\lambda_{1,2}} \in [2,6;25,1]$). So, we can assume, that in our model $L_M \to 0$ and is omissible.

Component of aerosol scattering can be calculated as: $L_{AZ}(\lambda) = F_0'(\lambda) \cdot C_{AZ} \cdot \lambda^{-\alpha}$. So, taking into account previous equation and assuming $L_M \to 0$, we can propose:

$$NDVI_W = \frac{F_0'^{\lambda_1} \cdot C_{AZ} \cdot (\lambda_1)^{-\alpha} - F_0'^{\lambda_2} \cdot C_{AZ} \cdot (\lambda_2)^{-\alpha}}{F_0'^{\lambda_1} \cdot C_{AZ} \cdot (\lambda_1)^{-\alpha} + F_0'^{\lambda_2} \cdot C_{AZ} \cdot (\lambda_2)^{-\alpha}} \quad (55)$$

Therefore

$$\alpha = -1,56 - 3,5 \ln \frac{1 + NDVI_W}{1 - NDVI_W}$$

Value of $NDVI_W$ for clear water varied in interval [-0,41; -0,6]; and Angstrom exponent varied in [1,8; 3,28]. Variation of α is probably connected with diversity of particles: higher values of α correspond to bigger particles (Kaufman and Tanre, 1992).

In the inland and marine water, there are two types of particles, which can scatter a light and to form a distribution of intensity and spectral pattern of reflected radiance. These are mineral and organic suspensions. Interrelation of energy reflected by two particles with different refractive indices n_1 i n_2:

$$\frac{E_1}{E_2} \propto \left(\frac{n_1^2 - 1}{n_1^2 + 2}\right)^2 \Bigg/ \left(\frac{n_2^2 - 1}{n_2^2 + 2}\right)^2 \tag{56}$$

Mean value for mineral suspension is $n_1 = 1{,}15$, and for organic – $n_2 = 1{,}02$. So, scattered energy on mineral particles is in 50-70 times more than on organic particles.

Backscattering in the water with varied size particles can be described as (Mueller et al 2002): $b_b(\lambda) = 0{,}5b_W(\lambda) + B_S b_{PS}(\lambda)P_S + B_1 b_{Pl}(\lambda)P_1$, where indices *w, s, l,* correspond clear water, small and large particles; probabilities of scattering by small and large particles are: B_s=0,039 and B_l=0,00064; P_s and P_l are the concentrations of mineral and organic particles (g/m^3); scattering coefficients for clear water $b_w(\lambda) = 5{,}826 \cdot 10^{-3}(400/\lambda)^{4{,}322}$, small particles $b_{ps}(\lambda) = 1{,}1513(400/\lambda)^{1{,}7}$, and large particles $b_{pl}(\lambda) = 0{,}3411(400/\lambda)^{0{,}3}$; λ is the wavelength, nm.

Usual way to obtain a relation between particles concentration and distribution of spectral indices is analysis of empiric data in framework of described model assumptions (Figures 7 and 8).

For calibration of obtained regressive relationships were used data of ground spectrometric measurements on test-sites with FieldSpec spectoradiometer (Lyalko et al., 2015). In period 2010-2015 on 6 test-sites have been collected 263 spectral signatures of water objects, which were used for calibration

Figure 7. Luminance of water surface with NDVI for different wavelengths

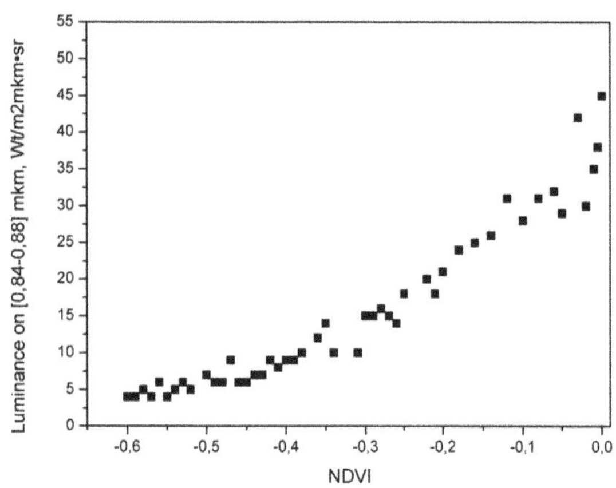

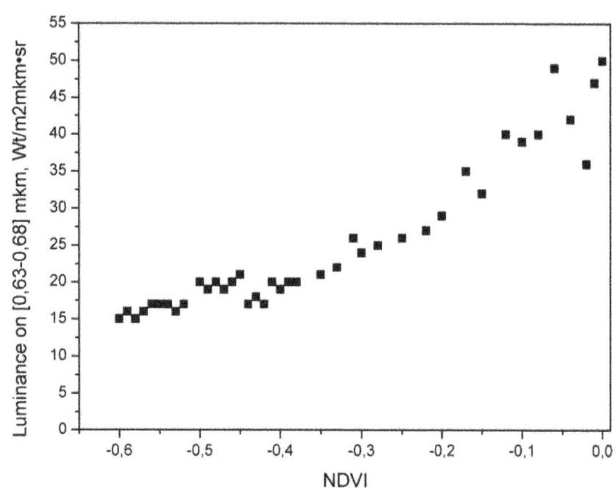

Figure 8. Total concentration of suspensions in water environment toward spectral index $NDVI_W$ *(empirical relation according to satellite data calibrated by ground data verified by lad measurements)*

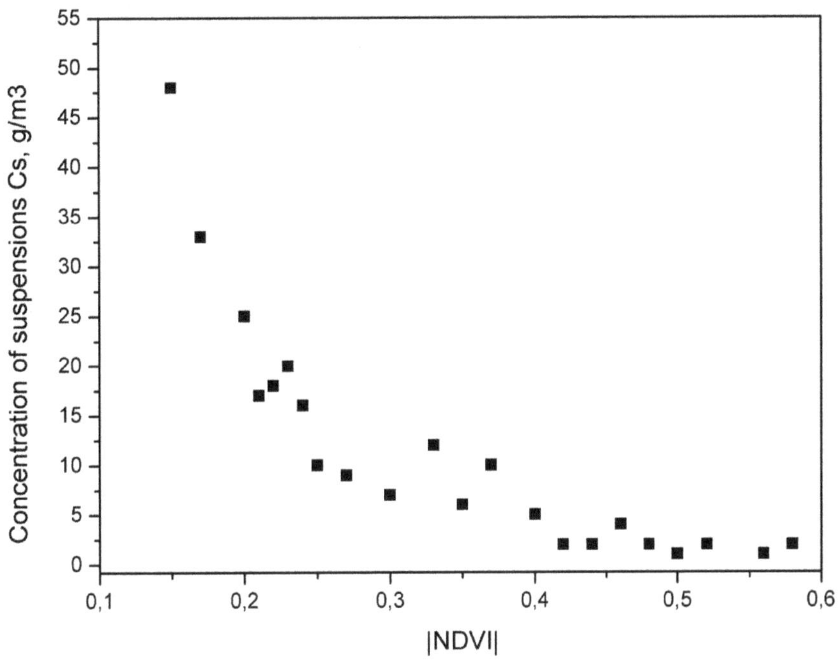

of models and satellite data. For verification of regressions were used lab measurements provided by Institutes of Hydrobiology, Hydrometeorology and Colloid Chemistry of National Academy of Sciences of Ukraine (Romanenko et al., 2015, Osadchy et al., 2008) on selected test-sites in 2009-2014.

Distributions $L_A^{\lambda_1}$ and $L_A^{\lambda_2}$ can be analyzed as a minimum of function $L_t(\lambda)$, and so determined as $L_A^{\lambda_1} \approx 4{,}5$, $L_A^{\lambda_2} \approx 16{,}3$ (Wt/m²mkm·sr). Basing on empirical relations $L_t = f(NDVI)$ and $NDVI = -0{,}605(P_s + P_l)^{-0{,}326}$, we can propose:

$$P_S + 0{,}01P_1 = 11{,}43(P_S + P_1)^{-0{,}652} - 21{,}46(P_S + P_1)^{-0{,}326} + 10{,}06 \quad (57)$$

Corresponding regressions for studied water bodies can be presented as follow:

$$L_t^{\lambda_1} = 113{,}4(NDVI)^2 + 133{,}5NDVI + 43{,}67; \quad R^2 = 0{,}91 \quad (58)$$

$$L_t^{\lambda_2} = 96{,}85(NDVI)^2 + 109{,}9NDVI + 47{,}5; \quad R^2 = 0{,}86 \quad (59)$$

Measurements demonstrate that empirical relation of chlorophyll-α for concentrations lass 3 mg/l may be presented as relation of particles in form: $P_s = 0,059P_l^{1,17}$.

Satellite observations, calibrated on data of in-field spectrometry, verified by lab measurements, allow determine an empirical relation for sum suspension concentration:

$$C_S = 0,21 \cdot \left|NDVI_W\right|^{-3,05} \tag{60}$$

Composition of suspensions is highly variable and depends of season and human impact. In the majority of studied cases about 80% of suspensions were composed from organic matters (plankton, detritus, other organic matters), and less than 20% is mineral matters. Large organic particles are statistically dominated both in freshwaters and marine environment.

So, we obtain a possibility to determine distributions of concentrations of separate components of suspensions by spectral reflectance. This is the methodical basement for assessment of ecological state of water environment and water quality by satellite derived spectral indices.

RISK MODEL

This section directed to development of formal model of integration of heterogeneous datasets for assessment of water quality in terms of risks, i.e. as a probability of water quality degradation with presence of pollutants.

According to regulatory documents, water quality estimating by classes (from I to IV) and categories (from I to VII) (Table 1). For each class and category water quality determined by non-identical sets of indices (see Table 2), which grouped to three clusters: salt composition (3 parameters), ecological and sanitarian criteria (20 parameters), and specific toxic contamination (up to 15 parameters). These indices interconnected and linked with quality classes more methodically, than genetically (Liu et al., 2013).

In real situations in large territories on long time intervals we can control (measure, observe or calculate) only limited number of parameters, which non-directly connected with mentioned indices. For example, satellite derived water spectral indices may correspond to water quality.

It means that in the most common case, with limited sets of indirect data, the problem of complete and accurate assessment of water quality is methodically very difficult. However, we can estimate the probability of changes of the water quality class (or category) according to the changes of observed indicators that correlate with the quality indices (Landis and Thomas, 2009).

Therefore, the problem is to create a formal algorithm to obtain dimensionless interval estimates using the sets of ranging criteria. (Jason and Shepard, 2006).

Assessment of risk connected with water pollution, using a limited set of data with specified criteria and classification schemes, can be formulated as a complex fuzzy problem. The appropriate approach in the simplest case can be built basing on the fuzzy sets theory (Li et al., 2007).

Risk assessment algorithm may be divided to few stages, and it can be presented in relatively simple form.

Set of indices for risk assessment will be defined as:

Table 1. Water quality classes and categories: freshwater ecological classification scheme

Class of water quality	I	II		III		IV	V
Water quality categories	1	2	3	4	5	6	7
Name of classes and categories by state of water bodies	Excellent	Good		Satisfactory		Bad	Very bad
	Excellent	Very good	Good	Satisfactory	Mean	Bad	Very bad
Name of classes and categories by purity / contamination rate	Very clear	Clear		Polluted		Polluted	Very Polluted
	Very clear	Clear	Quite clear	Mild contaminated	Moderate polluted	Polluted	Very Polluted
Trophic types (preferential)	Oligotrophic	Mesotrophic		Eutrophic		Polytrophic	Hypertrophic
	Oligotrophic - Oligomesotrophic	Mesotrophic	Mesoeutrophic	Eutrophic	Eupolytrophic	Polytrophic	Oligotrophic - Oligomesotrophic
Saprophytes	Oligo- Saprobic		β-meso- Saprobic		α- meso- Saprobic		Poly- Saprobic
	β- Oligo- Saprobic	α- Oligo- Saprobic	β'- meso- Saprobic	β''- meso- Saprobic	α'- meso- Saprobic	α''- meso- Saprobic	Poly- Saprobic

Table 2. Water quality criteria and indicators

Indexes	Salt composition index	Ecological and sanitarian criteria index					Specific toxic contaminants content index
Groups and types of parameters	Pollutants content	Hydro-physical	Hydro-chemical	Hydro-biological	Bacteriological	Bio-indicators	Pollutants content
Parameters, units	Ions sum, mg/dm^3	Clarity, m	pH	Phytoplankton Biomass, mg/dm^3	Bacterial plankton, $bln.cl/cm^3$	P-B Index	Hg, mg/dm^3
	Chlorides, mg/dm^3	Suspended matters, mg/dm^3	Nitrogen of Ammonium, mgN/dm^3	Self-purification /Contamination Index	Saprophyc Bacteria, bln.cl/cm3	G-W Index	Ka, mg/dm^3
	Sulfates, mg/dm^3		Nitrogen of Nitrites, mgN/dm^3			Saprophytes	Cu, mg/dm^3
			Nitrogen of Nitrates, mgN/dm^3			Trophic types (preferential)	Zn, mg/dm^3
			Phosphorus of Phosphates, mgP/dm^3				Pb, mg/dm^3
			Oxygen dissolved, mgO_2/dm^3				Cr (general), mg/dm^3
			Oxygen saturation, %				Ni, mg/dm^3
			Permanganate oxidation, mgO/dm^3				Ar, mg/dm^3
			Bio-chromatic oxidation, mgO/dm^3				Fe (general), mg/dm^3
			Biochemical Oxygen Demand BOD_5, mgO_2/dm^3				Mn, mg/dm^3
							Fluorides, mg/dm^3
							Cyanides, mg/dm^3
							Oils, mg/dm^3
							Phenols, mg/dm^3
							Synthetic surfactants, mg/dm^3

Types of indicators:	Directly observable parameters	Calculated and indirectly observable parameters	Measurable parameters

$$M = (x_1, x_2, ..., x_n) = \{x_i\}, i = 1, 2, ..., n \tag{61}$$

where x_i is number of selected assessed parameters, x_i - parameter from set of $x_{\min}$ parameters of risk/ pollutions (in main part of real cases this is limited number of known pollutants, for example, measurable or observable parameters: clarity, suspension concentrations, phytoplankton biomass, trophicity, surfactants, i.e. $n = 6$).

Following the water quality criteria introduced in European Water Directive and national regulation, set of risk assessment criteria D should correspond to number of introduced classes (or categories):

$$D = (d_1, d_2, ..., d_m) = \{d_i\}, j = 1, 2, ..., m \tag{62}$$

where m is the number of water quality classes or categories d_j, which correspond to assessed pollutions i^{th}, usually $m = 5$ (number of classes).

Let divide water contamination risks into finite number of intervals: small risk, acceptable risk, unacceptable risk, high risk and catastrophic risk.

The membership matrix Z will connect indices of risk M and water quality criteria D as:

$$Z = \begin{bmatrix} z_{11} \cdots W \cdots z_{1m} \\ A \cdots C \cdots T \\ z_{n1} \cdots E \cdots Z_{nm} \end{bmatrix} \tag{63}$$

where z_{ij} is risk assessment degree of separate parameter i^{th} on the water quality criteria j^{th}.

To determine the weight of indexes V_i the coefficient of variation may be used (Wang et al., 2011):

$$V_i = \frac{\sqrt{\frac{1}{n}\sum_{i=1}^{n}(x_i - \frac{1}{n}\sum_{i=1}^{n} x_i)}}{\frac{1}{n}\sum_{i=1}^{n} x_i} \tag{64}$$

where $0 \le V_i \le 1$.

In this case, the risk evaluation matrix ***F*** may be presented as:

$$F = V \bullet Z = (f_1 \dots f_2 \cdots, \dots f_m) \tag{65}$$

And the risk assessment quantitative indices RI can be calculated according to relatively simple algorithm (Geng et al., 2011):

$$RI = \frac{\sum_{j=1}^{n} f_i \times j}{\sum_{j=1}^{n} f_i} \tag{66}$$

Therefore, the relatively simple algorithm for calculating of interval assessments of water contamination risk in accordance with water quality criteria and using sets of observed and measured data may be constructed. Proposed method requires further calibration on wide range of basins using ground spectrometry.

Using proposed algorithm, a distribution of water quality degradation risk in terms of most probable classes of quality resulted by the impact of uncertain pollutions may be calculated (Figure 9).

There are two questions, which are important from viewpoint of water quality impact to socio-ecological security: inter-annual and intra-annual dynamics of risk parameters. Because no average values determine the impacts, but variations and diversity.

Basing on point spatially distributed data, proposed approach allows to calculate the risk parameters with temporal resolution close to temporal resolution of observations. At the same time, taking into account a possibility of land-cover classification, risk parameters may be calculated for different types of water bodies during the selected period (Figure 10).

On the criteria of a multidimensional distribution-free multivariate Kolmogorov–Smirnov test (Justel et al., 1997), calculated distributions of risk parameters corresponds to distributions of point measurements on network of hydrological and hydro-chemical sites (Nabyvanets et al., 2007, Osadchy et al., 2008) with level α better 5%.

The comparing test was provided according to (Marsaglia et al., 2003). If parameter $\rho(x) \le \alpha$ (where α is the level of statistical significance) we will assume that these distributions ($F_1(x), F_2(x), \ldots, F_n(x)$) are the same relative to the average regional distribution, presented by the reference sample $F(x)$. $F_n(x)$ is the empirical distribution functions of measured parameter ξ on sample $X = (X_1, \ldots, X_n)$. Target parameter can be found as: $\rho(x) = \sup_x \mid F_n(x) - F(x) \mid$, and calculated as $\rho(x) = \frac{\sqrt{2\pi}}{x} \sum_{k=1}^{\infty} e^{-(2k-1)^2 \pi^2 / (8x^2)}$.

Connected with water quality degradation the socio-ecological risks may be assessed separately. Let define a probability of water quality degradation in point *(x,y)* as P^{deg}. This probability may be calculated in every point of surface using set of calculated values using a Bayes rule:

Figure 9. Risk of water quality degradation

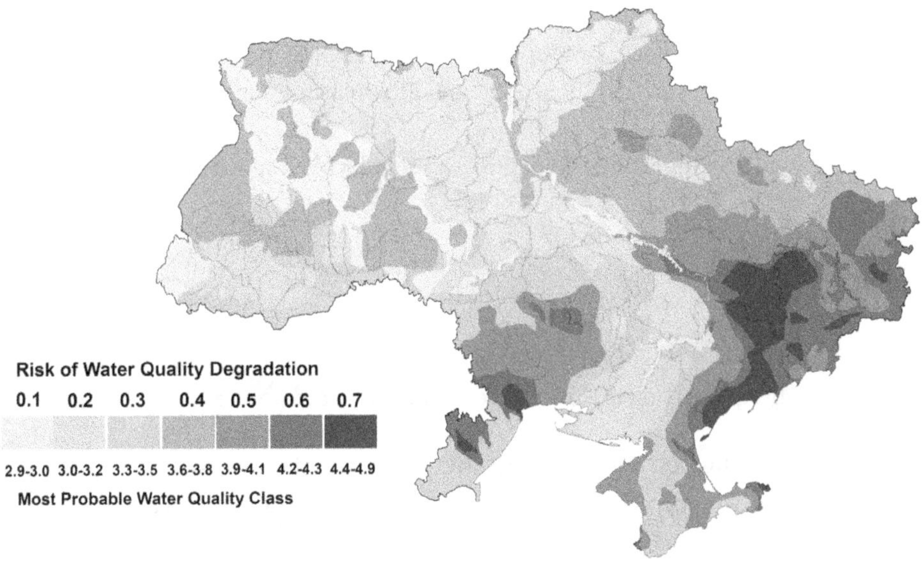

Figure 10. Annual dynamics of water contamination risk

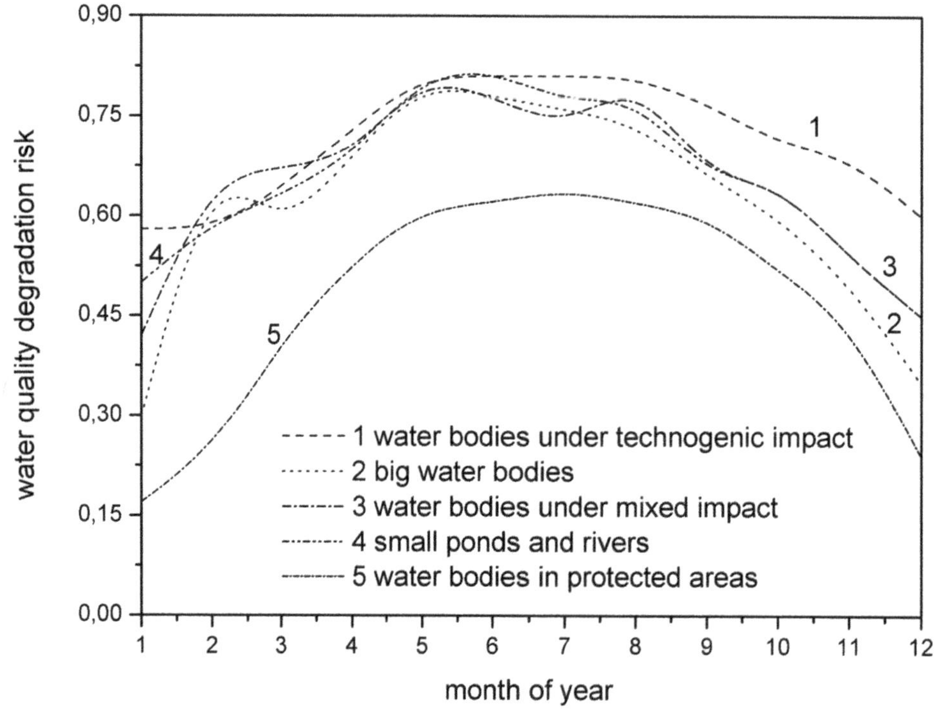

$$P_{x,y}^{\deg}(\Omega_{x,y} \mid Q_{\deg}) = \frac{P_s(x,y) \cdot \prod_N P_N(\Omega_{x,y} \mid Q_{\deg})}{\int_{x,y} P_N(\Omega_{x,y} \mid Q) dP_s(x,y)} = \frac{P_S(x,y) \cdot P_N(\Omega_{x,y} \mid Q_{\deg})}{P_N(\Omega_{x,y} \mid Q_{\deg}) P_S(x,y) + P_N(\Omega_{x,y} \mid Q_0) P_0(x,y)} \quad (67)$$

where $\Omega_{x,y}$ is the distribution of detected water pollution reduced on observation period, Q_{deg} is the sites with degraded quality of water, verified by ground and lab measurements, Q_0 is the sites with no water quality degradation.

Distribution of $\Omega_{x,y}$ calculating using the data *h(x,y)* regularization procedure:

$$\Omega_{x,y} = \sum_{m=1}^{n} w_{x,y}\left(\tilde{h}_m\left(x,y\right)\right) h_m(x,y) \quad (68)$$

where $w_{x,y}(\tilde{h}_m)$ is weighting coefficient, calculated according: $\min\{\sum_{m=1}^{n} \sum_{f_m \in F} w_{x,y}(\tilde{h}_m)(1 - \frac{h_m}{\tilde{h}_m})^2\}$, where *m* is the observation points number; *n* is the observation series number; h_m is the distribution of measurements; *F* general set of measurements; $\tilde{h}_m$ is the mean distribution of measured parameters.

Probability $P_S(x,y)$ is calculating semi-empirically. Relation between $P_S(x,y)$ and $P_0(x,y)$ determining as $\lim_{x,y,\tau}(P_S(x,y)_\tau + P_0(x,y)_\tau) = 1$ (on the long observation periods $P_S(x,y) = 1 - P_0(x,y)$). Probability $P_S(x,y)$ may be defined using Gauss weighing function:

$$P_S(x,y) = P_{\min} + \left(P_{\max} - P_{\min}\right) \cdot e^{d_s^2/2\sigma_p^2} \tag{69}$$

where $P_S(x,y)$ is the probability of pollution during the observation period calculated by proposed approach; P_{max} is the maximal possible probability of pollution calculated by core models and regional statistics (Bernardo and Smith, 2001) (P_{max} for used sensors might be assessed on the level 0,75-0,78); P_{min} is the minimal probability of pollution calculated using regional statistics (P_{min} might be defined about 0,02-0,04); $d_s(x,y)$ is the distance from pollution source to assessed point; σ_p is the empirical parameter, depending of sensor and local features (σ_p may be assessed on the level about 0,6 km).

To estimate impact to socio-ecological security, according to (Straub, 2009), a robustness coefficient β may be proposed, through probability density function, as a Gaussian measure of possible damage:

$$\beta = -\Phi^{-1}\left(P_{x,y}^{sub}\right) \cdot k = \left(\frac{0 - \tilde{P}^{sub}}{\sigma_{x,y}\left(P_{x,y}^{sub}\right)}\right) \cdot k \tag{70}$$

where k is the population structural coefficient, empirically describing a vulnerability of different population groups toward the water quality changes. Average values of the coefficient β for study region lie in the interval [0,5; 3,5], which is correspond to pollutions of mean intensity (average water pollution risk parameters 0,35 - 0,45). This parameter has a sense of integrated influence of water quality change on population and may be used as impact factor to the population distribution and structure. With the increasing of the calculated risk level, the value of robustness coefficient β is decreasing.

It is a method of determination of regional water quality degradation risks and socio-ecological impact using the data of satellite observations and ground calibration. This method allows obtain regularized spatial-temporal distributions of risks parameters with smoothed reliability.

Such kind of spatial and temporal distributions of risk parameters is more adequate base for decision making in field of socio-ecological security than a set of point measurement.

CONCLUSION

Idea of this research is to establish a hypothesis of applicability and utility of soft computing in security management: if there is an a priori information about initial state of the system, and researcher is able to model key processes in this system, which forming the key interlinked parameters of socio-ecological security, it is possible to select and control a limited set of indicators, and basing of which to calculate a most probable evolution of system. So, it is possible to reduce a number of controlled variables, obtain a possibility to harness a multi-source data, simplify a problem of uncertainty control, receive a spatially and temporally distributed assessments with a scale close to the observation resolution.

In situation of limited access to data the soft computing techniques becomes especially useful: not so many parameters are directly measurable with required spatial, temporal, spectral and radiometric resolutions, at the same time many parameters are observable, and almost everything we can simulate. So, soft computing techniques are especially important in the field of socio-ecological security.

Complex security management requires an integrated approach to socio-ecological risk assessment, including modeling of water, land and air environment. Whole set of existing data, methods and instruments should be used for solution of this complicated task. Proposed methodology aimed to expanding the set of accessible for analysis data among the volume of available data.

It was demonstrated, that basing on point spatially distributed measurement data, proposed approach allows to calculate the risk parameters with temporal resolution close to temporal resolution of observations. Besides, using existing land-cover classification, risk parameters may be calculated for different types of water bodies during the selected intra-annual and interannual periods. This model of spatial and temporal distributions of risk parameters is more useful and adequate base for decision making in field of socio-ecological security than a heterogenious set of point measurement. Therefore, proposed approach is a useful applied instrument.

The proposed approach and the results obtained can be applied in aerospace industry for development of sensors and systems for Earth's aerospace monitoring, as well as for the development of automatic design systems of integrated control systems in civil engineering.

At the same time the main problem, which should be decided at the next stage of the study, is the analysis of uncertainties, connected with land-cover classification, as well as the development of problem-oriented land-cover classification approach, adapted for the algorithm proposed. Besides, further research should be directed to analysis of water quality on local scale.

ACKNOWLEDGMENT

The authors are grateful to anonymous referees for constructive suggestions that resulted in important improvements to the chapter, to colleagues from the International Institute for Applied Systems Analysis (IIASA), from the American Statistical Association (ASA), American Meteorological Society (AMS), and from the International Association for Promoting Geoethics (IAPG) for their critical and constructive comments and suggestions. Particular thanks authors express to Science and Technology Center in Ukraine (STCU) for partial support of this study in the framework of research project #6165 "Information and Technological Support for Greenhouse Effect Impact Assessment on Regional Climate using Remote Sensing", as well as to National Academy of Sciences of Ukraine for partial support of this study in the framework of joint Ukraine-IIASA research program "Integrated modeling of food, energy and water security management for sustainable social, economic and environmental developments".

REFERENCES

Baird, A. J., & Wilby, R. L. (1999). *Eco-hydrology: plants and water in terrestrial and aquatic environments*. Psychology Press.

Bernardo, J. M., & Smith, A. F. (2001). *Bayesian theory*. Academic Press.

Campbell, J. B., & Wynne, R. H. (2011). *Introduction to remote sensing*. Guilford Press.

Canter, L. W. (1996). *Nitrates in groundwater*. CRC Press.

Elachi, C., & Van Zyl, J. J. (2006). *Introduction to the physics and techniques of remote sensing* (Vol. 28). John Wiley & Sons. doi:10.1002/0471783390

Engle, R. F., & Russell, J. R. (1998). Autoregressive conditional duration: A new model for irregularly spaced transaction data. *Econometrica, 66*(5), 1127–1162. doi:10.2307/2999632

Ermoliev, Y., Makowski, M., & Marti, K. (2012). Robust management of heterogeneous systems under uncertainties. In *Managing Safety of Heterogeneous Systems* (pp. 1–16). Springer Berlin Heidelberg. doi:10.1007/978-3-642-22884-1_1

Ermoliev, Y., Makowski, M., & Marti, K. (Eds.). (2012). *Managing safety of heterogeneous systems: decisions under uncertainties and risks* (Vol. 658). Springer Science & Business Media. doi:10.1007/978-3-642-22884-1

Ermoliev, Y., & von Winterfeldt, D. (2012). Systemic risk and security management. In *Managing safety of heterogeneous systems* (pp. 19–49). Springer Berlin Heidelberg. doi:10.1007/978-3-642-22884-1_2

European Commission. (2005). *Common implementation strategy for the water framework directive* (2000-60-EC). Groundwater summary report. Technical report on groundwater body characterization, monitoring and risk assessment issues as discussed at the WG C workshop in 2003-2004.

European Environment Agency. (2003). *Europe's Water: An Indicator-based Assessment*. Topic Rep. 1. Copenhagen, Denmark: Author.

Geng, Y., Zhang, J., Zhou, Q., Xu, C., & Zhao, Y. Q. (2011). *Fuzzy synthetic evaluation of Weihe water quality*. Academic Press.

Gordon, H. R., Brown, J. W., & Evans, R. H. (1988). Exact Rayleigh scattering calculations for use with the Nimbus-7 coastal zone color scanner. *Applied Optics, 27*(5), 862–871. doi:10.1364/AO.27.000862 PMID:20523701

Herlin, I., Nakonechnyi, O., & Zhuk, S. (2012, September). Minimax optical flow estimation from a sequence of 2D images. *XX International Conference on Problems of Decision Making under Uncertainties (PDMU)*.

Horn, B. K., & Schunck, B. G. (1981). Determining optical flow. *Artificial Intelligence, 17*(1-3), 185–203. doi:10.1016/0004-3702(81)90024-2

Jason, R. E., & Shepard, C. N. (2006). *Water Suppliers*. Academic Press.

Justel, A., Peña, D., & Zamar, R. (1997). A multivariate Kolmogorov–Smirnov test of goodness of fit. *Statistics & Probability Letters, 35*(3), 251–259. doi:10.1016/S0167-7152(97)00020-5

Kaufman, Y. J., & Tanre, D. (1992). Atmospherically resistant vegetation index (ARVI) for EOS-MODIS. *IEEE Transactions on Geoscience and Remote Sensing, 30*(2), 261–270. doi:10.1109/36.134076

Koch, H., & Grünewald, U. (2009). A comparison of modelling systems for the development and revision of water resources management plans. *Water Resources Management*, *23*(7), 1403–1422. doi:10.1007/s11269-008-9333-x

Kogan, F., Stark, R., Gitelson, A., Jargalsaikhan, L., Dugrajav, C., & Tsooj, S. (2004). Derivation of pasture biomass in Mongolia from AVHRR-based vegetation health indices. *International Journal of Remote Sensing*, *25*(14), 2889–2896. doi:10.1080/01431160410001697619

Kopachevsky, I., Kostyuchenko, Y. V., & Stoyka, O. (2016). Land Use Drivers of Population Dynamics in Tasks of Security Management and Risk Assessment. *International Journal of Mathematical, Engineering and Management Sciences*, *1*, 18–24.

Kostyuchenko, Y. V. (2015). Geostatistics and remote sensing for extremes forecasting and disaster risk multiscale analysis. In *Numerical Methods for Reliability and Safety Assessment* (pp. 439–458). Springer International Publishing. doi:10.1007/978-3-319-07167-1_16

Kostyuchenko, Y. V. (2018). On the Methodology of Satellite Data Utilization in Multi-Modeling Approach for Socio-Ecological Risks Assessment Tasks: A Problem Formulation. *International Journal of Mathematical, Engineering and Management Sciences*, *3*(1), 1–8.

Kostyuchenko, Y. V., Bilous, Y., Kopachevsky, I., & Solovyov, D. (2013). Coherent risk measures assessment based on the coupled analysis of multivariate distributions of multisource observation data. *Proceedings of 11-th International Probabilistic Workshop*, 183-192.

Kostyuchenko, Y. V., Kopachevsky, I., Yuschenko, M., Solovyov, D., Marton, L., & Levynsky, S. (2012). Spectral reflectance indices as indirect indicators of ecological threats. In Sustainable Civil Infrastructures–Hazards, Risk, Uncertainty. Research Publishing.

Kostyuchenko, Y. V., Movchan, D., Artemenko, I., & Kopachevsky, I. (2016). Stochastic Approach to Uncertainty Control in Multiphysics Systems: Modeling of Carbon Balance and Analysis of GHG Emissions Using Satellite Tools. In Mathematical Concepts and Applications in Mechanical Engineering and Mechatronics. IGI Global. DOI: doi:10.4018/978-1-5225-1639-2.ch017

Kostyuchenko, Y. V., Movchan, D., Kopachevsky, I., & Bilous, Y. (2015, November). Robust algorithm of multi-source data analysis for evaluation of social vulnerability in risk assessment tasks. In *SAI Intelligent Systems Conference (IntelliSys)* (pp. 944-949). IEEE. doi:10.1109/IntelliSys.2015.7361256

Kostyuchenko, Y. V., Yuschenko, M., Movchan, D., Zagorodny, A. G., & Yermoliev, Y. M. (2013). Regional risk analysis based on multisource data statistics of natural disasters. In Integrated modeling of food, energy and water security management for sustainable social, economic and environmental developments, 229-238.

Landis, W. G., & Thomas, J. F. (2009). Regional risk assessment as a part of the long-term receiving water study. *Integrated Environmental Assessment and Management*, *5*(2), 234–247. doi:10.1897/IEAM_2008-062.1 PMID:19127980

Li, J., Huang, G. H., Zeng, G., Maqsood, I., & Huang, Y. (2007). An integrated fuzzy-stochastic modeling approach for risk assessment of groundwater contamination. *Journal of Environmental Management*, *82*(2), 173–188. doi:10.1016/j.jenvman.2005.12.018 PMID:16574309

Lillesand, T., Kiefer, R. W., & Chipman, J. (2014). *Remote sensing and image interpretation*. John Wiley & Sons.

Liu, Y. G., Wang, N. L., Wang, L. G., Zhao, Y. Q., & BoWu, X. (2013). Application of GIS in regional ecological risk assessment of water resources. *Environmental Engineering & Management Journal, 12*(7).

Lyalko, V. I., Kostyuchenko, Yu. V., Elistratova, L., & Artemenko, I. (2015). *The greenhouse effect and climate change in Ukraine: assessment and effect*. Kiev: Naukova Dumka. (in Ukrainian)

Marsaglia, G., Tsang, W. W., & Wang, J. (2003). Evaluating Kolmogorovs Distribution. *Journal of Statistical Software*, *8*(18), 1–4. doi:10.18637/jss.v008.i18

Martínez, Y., & Albiac, J. (2006). Nitrate pollution control under soil heterogeneity. *Land Use Policy*, *23*(4), 521–532. doi:10.1016/j.landusepol.2005.05.002

McFeeters, S. K. (1996). The use of the Normalized Difference Water Index (NDWI) in the delineation of open water features. *International Journal of Remote Sensing*, *17*(7), 1425–1432. doi:10.1080/01431169608948714

Moussiopoulos, N., Berge, E., Bohler, T., Leeuw, F., Gronskei, K. E., Mylona, S., & Tombrou, M. (1997). Ambient Air Quality, Pollutant Dispersion and Transport Models, European Topic Centre on Air Quality, Topic Report 19. European Environment Agency, EU Publications.

Mueller, J., Mueller, J. L., Pietras, C., Hooker, S. B., Clark, D. K., Frouin, A. M. R., & Fargion, G. S. (2002). Ocean Optics Protocols For Satellite Ocean Color Sensor Validation, Revision 3, volumes 1 and 2. *NASA tech. memo, 210004*.

Nabyvanets, B. I., Osadchy, V. I., Osadcha, N. M., & Nabyvanets, Yu. B. (2007). *Analytical chemistry of surface water*. Kiev: Naukova Dumka. (in Ukrainian)

National Research Council. (1992). *Nuclear Power: Technical and institutional options for the future*. National Academies Press.

Osadchy, V. I., Nabyvanets, B. I., Osadcha, N. M., & Nabyvanets, Yu. B. (2008). *Hydrochemical guide. Surface water of Ukraine. Hydrochemical calculations. Methods of analysis*. Kiev: Naukova Dumka. (in Ukrainian)

Polasky, S., Carpenter, S. R., Folke, C., & Keeler, B. (2011). Decision-making under great uncertainty: Environmental management in an era of global change. *Trends in Ecology & Evolution*, *26*(8), 398–404. doi:10.1016/j.tree.2011.04.007 PMID:21616553

Qiu, H. L., Sanchez-Azofeifa, A., & Gamon, J. A. (2007). Ecological applications of remote sensing at multiple scales. In *Functional Plant Ecology* (2nd ed.). CRC Press. doi:10.1201/9781420007626.ch22

Romanenko, O. V., Arsan, O. M., Kipnis, L. S., & Sytnyk, Yu. M. (2015). *Ecological problems of the Kiev reservoirs and adjoining territories*. Kiev: Naukova Dumka. (in Ukrainian)

Schlaifer, R., & Raiffa, H. (1961). *Applied statistical decision theory*. Academic Press.

Straub, D. (2009). Stochastic modeling of deterioration processes through dynamic Bayesian networks. *Journal of Engineering Mechanics*, *135*(10), 1089–1099. doi:10.1061/(ASCE)EM.1943-7889.0000024

Suter, G. W. II. (2016). *Ecological risk assessment.* CRC Press.

Wang, T. R., Sun, G. N., & Liu, S. Y. (2011). Relationship between Spatiotemporal Variation of Water Pollution and Runoff Volume of Mainstream Section of the Weihe River in Shaanxi Province. *Arid Zone Research, 28*(4), 599–615.

Zannetti, P. (1970). Numerical simulation modeling of air pollution: an overview. *WIT Transactions on Ecology and the Environment, 1.*

Zheng, C., & Bennett, G. D. (2002). *Applied contaminant transport modeling* (Vol. 2). New York: Wiley-Interscience.

Zilioli, E., & Brivio, P. A. (1997). The satellite derived optical information for the comparative assessment of lacustrine water quality. *The Science of the Total Environment, 196*(3), 229–245. doi:10.1016/S0048-9697(96)05411-3 PMID:9260309

Chapter 2
Fixed Charged Solid Transportation Problem With Budget Constraints in Type-2 Fuzzy Variables:
Multi-Objective Solid Transportation Problem

Dhiman Dutta
National Institute of Technology Silchar, India

Mausumi Sen
National Institute of Technology Silchar, India

ABSTRACT

A multi-objective fixed charged solid transportation model with criterion e.g. transportation penalty, amounts, demands, carriages and budget constraints as type-2 triangular fuzzy variables with condition on few components and carriages is proposed here. With the critical value based reductions of corresponding type-2 fuzzy variables, a nearest interval approximation model and a chance constrained programming model applying generalized credibility measure for the constraints is proposed for this particular problem. The credibility measure is also applied to the objective functions of the chance constrained programming model. The model is then transformed into the corresponding crisp deterministic form by these two methods. A numerical example is provided to explain the model with hypothetical data and is then worked out by applying a gradient based optimization - Generalized Reduced Gradient technique (applying LINGO 16). The corresponding objective function values are compared numerically by two approaches after transforming it to crisp form by these two methods.

DOI: 10.4018/978-1-5225-3035-0.ch002

1. INTRODUCTION

In today's highly competitive market, the pressure on organizations to find better ways to create and deliver value to customers becomes stronger. How and when to send the products to the customers in the quantities in a cost effective manner becomes more challenging. Transportation models provide a powerful framework to meet this challenge. They ensure the efficient movement and timely availability of raw materials and finished goods. The transportation problem (TP) is a special class of linear programming problem in which the objective is to transport a commodity from various plants called origins to different places called destinations at a minimum cost. There are two types of constraints in classical TP namely origin and destination constraints introduced by Balinski (1961). In practical applications, there exists another constraint such as carriage mode constraint besides origin and destination constraints. That is why, the classical transportation problem takes the form solid transportation problem (STP), an exclusive form of linear programming problem, where we deal with condition of sources, destinations and carriages. There may be several other types of constraints besides source, destination and conveyance. It may be budget constraints arising out of limited budget. A few important criterions in the STP are always treated as uncertain variables to fit the realistic positions due to complex situation during the transportation movement. It is impossible to form a transportation plan for the future months because there is no definite data for the amount of quantity necessary at every origin, the requirement at every destination and the carriage quantity and these values all are uncertain. And as such it is much better to explore this issue by applying fuzzy or stochastic optimization models. It is difficult to predict the exact transportation cost for a certain time period. Soft computing is the method to handle uncertainty, partial truth, and approximation to attain practicability, robustness and low solution cost. The principal component of Soft Computing includes Neural Network (NN), Fuzzy Logic (FL), and Genetic Algorithms (GA). Fuzzy set theory is the one of the popular approaches to deal with this uncertainty. Transportation model is sometimes associated with additional costs along with shipping cost that may arise due to toll charges, road taxes etc. In this case it is called fixed charged transportation problem (FCTP).

Multi-item STP is a problem of shipping multiple items from multiple sources to multiple destinations through some conveyance. A situation may arise while transporting some items from origin to some destination where all types of items cannot be transported through all types of conveyances due to nature of items (e.g. liquid, solid, breakable etc.). Multi-item fixed charge solid transportation problem (FCSTP) with restriction on conveyances is a problem of transporting goods to some destinations through a particular conveyance with additional fixed charge for that particular route. The models that are used to find optimal solutions of multiple objective functions of shipping multiple components from multiple sources to multiple destinations over a few carriages are called as multi-item multi-objective solid transportation problem. The authors wish to solve multi-objective multi-item fixed charged solid transportation problem with budget constraints in type-2 fuzzy variables in this chapter.

This chapter has 6 Sections: background is given in Section 2, Section 2.2 contains some basic preliminaries relating to the notions of reductions of type-2 fuzzy variables. Also, CV-based reduction methods for type-2 fuzzy variables are discussed in this Section. Section 3 describes the main focus of the chapter. The authors have formulated a multi-item multi objective fixed charged solid transportation problem with budget constraints and conditions on a few brands and carriages in the sense that a few specific brands are restricted to be shipped over a few particular carriages in Section 3.1. The problem follows some transportation guidelines e.g. unit transportation penalty, fixed costs, amounts, requirements, budget amount and carriage quantities as type-2 triangular fuzzy variables. The model is investigated by developing a chance constrained programming model and nearest interval approximation applying the CV based reduction method in Section 4.3. Finally the model is solved numerically in Section 4.3.2 applying fuzzy programming technique and LINGO 16 solver.

2. BACKGROUND

2.1 Literature Review

The notion of type-2 fuzzy set was defined by Zadeh (1975) as a generalization of type-1 fuzzy set (1965). The membership function of a type-1 fuzzy set is a real number in [0,1], while the membership function of a type-2 fuzzy set is a type-1 fuzzy set. Type-2 fuzzy sets and fuzzy inference system is being applied in many areas of science and technology such as: computing with words (2007), human resource management (2012), forecasting of time-series (1999), pattern recognition (2009), fuzzy logic controller (2006), industrial application (2011), simulation (2011), and transportation problem (2014,2015,2016,2017). The motivation behind this chapter is to study solid transportation model with type-2 fuzzy parameters. In recent years, the STP obtained sufficient interest and several problems and algorithms under both crisp environment and uncertain environment have been studied. For example, Jiménez and Verdegay (1998) investigated a balanced fuzzy STP taking into account fuzzy supplies, fuzzy demands and fuzzy conveyance capacities. Liu (2006) presented a non-balanced STP by considering supply capacities, demands, and conveyance capacities are all convex fuzzy numbers. Yang and Liu (2007) studied the FCSTP by considering the unit costs, fixed charges, quantities, requirements and carriage capacities are fuzzy variables. Molla et al. (2013) presented a FCSTP under a fuzzy environment and used three meta-heuristics to find out the solution of the problem. Yang and Feng (2007) investigated a bi-criteria STP, where the parameters are random variables, and considered a hybrid algorithm to produce an approximate optimal solution. Gen et al. (1995) studied a bicriteria FCSTP and used genetic algorithm for solving it. Baidya et al. (2013) solved a multi-item STP with safety measure. Bit et al. (1993) investigated the fuzzy programming model for a multi-objective STP. Ojha et al. (2009) presented entropy based STP by considering general fuzzy costs and time. Nagarjan and Jeyaraman (2010) studied the chance constrained programming model for a multi-objective STP with the parameters as stochastic intervals. Kundu et al. (2014) investigated multi-objective solid transportation problems with budget constraint in uncertain environment. Pandian and Anuradha (2010) found out a new method to solve the STP. Li et al. (1997) discussed a neural network method to express bi-criteria STP, and Li et al. (1997) also studied multi-objective STP with fuzzy numbers and used improved genetic algorithm to solve it. Kundu et al. (2014) has investigated two models namely (i) model-I: fixed charge transportation problem (FCTP) with type-2 fuzzy cost parameters and (ii) model-II: FCTP with costs, supplies and demands as type-2 fuzzy variables. Maiti et al. (2015) has solved one model with multi-item solid transportation problem with restriction on conveyances with the type-2 fuzzy variables. Kundu et. al. (2012) investigated the multi-objective STP under various uncertain environments. Amrit et. al. (2016) investigated multi stage STP under budget with Gaussian type-2 fuzzy parameters. Anushree et al. (2017) discussed STP under type 2 trapezoidal fuzzy environment. Kundu et al. (2015), Liu et al. (1998), Yang et al. (2007) developed chance-constrained programming with type-1 fuzzy parameters using credibility measure. Zhang et al. (1999) introduced deterministic forms of the uncertain constraints applying the notion of possibility degree of interval number defining definite degree by which one interval is greater or lesser than another.

Zimmermann (1978) established that fuzzy linear programming technique regularly provides useful solutions and an optimal compromise solution for multiple objective problems. Dhiman et al. (2017) investigated a multi-item multi-objective fixed charged solid shipment model with type-2 fuzzy variables.

2.2 Preliminaries

Let us demonstrate some preliminary definitions and theorems in this section to solve our proposed model.

2.2.1. Definitions

Definition 2.2.1.1 (Chen et al. (2013): A pair $(\eta^{'}, q^{'}, Pos)$ is termed as a possibility space, where $\eta^{'}$ is non-empty set of points, $q^{'}$ is power set of $\eta^{'}$ and $Pos : \eta^{'} \to [0,1]$, called possibility measure defined as:

1. $Pos(\phi) = 0$ and $Pos(\eta^{'}) = 1$.
2. For any $\{A_k \mid k \in K\} \subset \eta^{'}$, $Pos(\bigcup A_k) = \sup_k Pos(A_k)$.

Definition 2.2.1.2 (Nahmias (1978)): The possibility measure of a fuzzy event $\{\tilde{\tau}^{'} \in C^{'}\}$, $C^{'} \subset R$ is explained as $Pos\{\tilde{\tau}^{'} \in C^{'}\} = \sup_{y^{'} \in C^{'}} \mu_{\tilde{\tau}^{'}}(y^{'})$, where $\mu_{\tilde{\tau}^{'}}(y^{'})$ is defined as a possibility distribution of $\tilde{\tau}^{'}$.

Definition 2.2.1.3 (Liu et al. (2002)): The necessity (Nec) and credibility measure (Cr) of a normalized fuzzy variable ($\sup_{y^{'} \in \mathbf{R}} \mu_{\tilde{\tau}^{'}}(y^{'}) = 1$) is defined as follows

$$Nec\{\tilde{\tau}^{'} \in B^{'}\} = 1 - Pos\{\tilde{\tau}^{'} \in B^{'c}\} = 1 - \sup_{y^{'} \in B^{'c}} \mu_{\tilde{\tau}^{'}}(y^{'})$$

$$Cr\{\tilde{\tau}^{'} \in B^{'}\} = (Pos\{\tilde{\tau}^{'} \in B^{'}\} + Nec\{\tilde{\tau}^{'} \in B^{'}\}) / 2$$

Definition 2.2.1.4 (Liu et al. (2010)): For a possibility space $(\eta^{'}, q^{'}, Pos)$, a regular fuzzy variable (RFV) $\tilde{\tau}^{'}$ is denoted by $\eta^{'} \to [0,1]$ in the notion that for every $s^{'} \in [0,1]$, one has $\{\delta^{'} \in \eta^{'} \mid \mu_{\tilde{\tau}^{'}}(\delta^{'}) \le s^{'}\} \in q^{'}$.

Definition 2.2.1.5 (Liu et al. (2010)): If $(\eta^{'}, q^{'}, Pos)$ is a fuzzy possibility space then a type-2 fuzzy variable (T2 FV in short) $\tilde{\tau}^{'}$ is expressed as $\eta^{'} \to \mathbb{R}$, such that for any $s^{'} \in \mathbb{R}$, the set $\{\delta^{'} \in \eta^{'} \mid \mu_{\tilde{\tau}^{'}}(\delta^{'}) \le s^{'}\} \in q^{'}$.

Definition 2.2.1.6 (Zadeh (1975)): A type-2 fuzzy set $\tilde{D}$ explained on the universe of discourse Z is described by a membership function $\tilde{\mu}_{\tilde{D}} : Z \to F([0,1])$ and is expressed by the following set notation: $\tilde{D} = \{\left(z, \tilde{\mu}_{\tilde{D}}(z)\right) : z \in Z\}$.

Example 2.2.1.7 (Liu et al. (2010)): A type-2 triangular fuzzy variable $\tilde{\tau}^{'}$ is expressed by $(r_1^{'}, r_2^{'}, r_3^{'};$ $\theta_l^{'}, \theta_r^{'})$ where $r_1^{'}, r_2^{'}, r_3^{'} \in \mathbf{R}$ are real numbers and $\theta_l^{'}, \theta_r^{'}$ are two criterions defining the degree of ambiguity that $\tilde{\tau}^{'}$ takes a value x and the secondary possibility distribution function $\tilde{\mu}_{\tilde{\tau}^{'}}(x^{'})$ of $\tilde{\tau}^{'}$ is denoted as

$$\tilde{\mu}_{\tilde{\tau}'}(x') = \begin{cases} (\frac{x'-r_1'}{r_2'-r_1'} - \theta_l' \frac{x'-r_1'}{r_2'-r_1'}, \frac{x'-r_1'}{r_2'-r_1'}, \frac{x'-r_1'}{r_2'-r_1'} + \theta_r' \frac{x'-r_1'}{r_2'-r_1'}), \text{if} x' \in [r_1', \frac{r_1'+r_2'}{2}]; \\ (\frac{x'-r_1'}{r_2'-r_1'} - \theta_l' \frac{r_2'-x'}{r_2'-r_1'}, \frac{x'-r_1'}{r_2'-r_1'}, \frac{x'-r_1'}{r_2'-r_1'} + \theta_r' \frac{r_2'-x'}{r_2'-r_1'}), \text{if} x' \in (\frac{r_1'+r_2'}{2}, r_2']; \\ (\frac{r_3'-x'}{r_3'-r_2'} - \theta_l' \frac{x'-r_2'}{r_3'-r_2'}, \frac{r_3'-x'}{r_3'-r_2'}, \frac{r_3'-x'}{r_3'-r_2'} + \theta_r' \frac{x'-r_2'}{r_3'-r_2'}), \text{if} x' \in (r_2', \frac{r_2'+r_3'}{2}]; \\ (\frac{r_3'-x'}{r_3'-r_2'} - \theta_l' \frac{r_3'-x'}{r_3'-r_2'}, \frac{r_3'-x'}{r_3'-r_2'}, \frac{r_3'-x'}{r_3'-r_2'} + \theta_r' \frac{r_3'-x'}{r_3'-r_2'}), \text{if} x' \in (\frac{r_2'+r_3'}{2}, r_3']. \end{cases}$$

2.3 Critical Values for RFVs

The different forms of critical values (CV) of a RFV $\tilde{\tau}'$ are given by Qin et al. (2011) as follows:

1. The optimistic CV of $\tilde{\tau}'$, denoted by $CV^*[\tilde{\tau}']$, is defined as:

$$CV^*[\tilde{\tau}'] = \sup_{\alpha' \in [0,1]} [\alpha' \wedge Pos\{\tilde{\tau}' \geq \alpha'\}].$$

2. The pessimistic CV of $\tilde{\tau}'$, denoted by $CV_*[\tilde{\tau}']$, is defined as:

$$CV_*[\tilde{\tau}'] = \sup_{\alpha' \in [0,1]} [\alpha' \wedge Nec\{\tilde{\tau}' \geq \alpha'\}]$$

3. The CV of $\tilde{\tau}'$, denoted by $CV[\tilde{\tau}']$ is defined as:

$$CV[\tilde{\tau}'] = \sup_{\alpha' \in [0,1]} [\alpha' \wedge Cr\{\tilde{\tau}' \geq \alpha'\}].$$

2.4 CV-Based Reduction Approach for T2 FVs

CV-based reduction approach is introduced by Qin et al. (2011) which reduces a T2 FV to a type-1 fuzzy variable. Let $\tilde{\tau}'$ be a T2 FV with secondary membership function $\tilde{\mu}_{\tilde{\tau}'}(y)$. In RFV $\tilde{\mu}_{\tilde{\tau}'}(y)$ is represented by the CVs, i.e. $CV^*[\tilde{\mu}_{\tilde{\tau}'}(y)], CV_*[\tilde{\mu}_{\tilde{\tau}'}(y)]$ or $CV[\tilde{\mu}_{\tilde{\tau}'}(y)]$ which are accordingly called optimistic CV reduction, pessimistic CV reduction and CV reduction approach.

Theorem 2.4.1 (Qin et al. (2011)): Suppose that $\tilde{\tau}' = (s_1', s_2', s_3'; \eta_l', \eta_r')$ be a type-2 triangular fuzzy variable. Then we have:

1. The possibility distribution of the reduction $\tilde{\tau}_1^{'}$ of $\tilde{\tau}^{'}$ using the optimistic CV reduction approach is given as follows:

$$\mu_{\tau_1^{'}}(x^{'}) = \begin{cases} \dfrac{(1+\eta_r^{'})(x^{'}-s_1^{'})}{s_2^{'}-s_1^{'}+\eta_r^{'}(x^{'}-s_1^{'})}, \text{if} x^{'} \in [s_1^{'}, \dfrac{s_1^{'}+s_2^{'}}{2}]; \\ \dfrac{(1-\eta_r^{'})x^{'}+\eta_r^{'}s_2^{'}-s_1^{'}}{s_2^{'}-s_1^{'}+\eta_r^{'}(s_2^{'}-x^{'})}, \text{if} x^{'} \in (\dfrac{s_1^{'}+s_2^{'}}{2}, s_2^{'}]; \\ \dfrac{(-1+\eta_r^{'})x^{'}-\eta_r^{'}s_2^{'}+s_3^{'}}{s_3^{'}-s_2^{'}+\eta_r^{'}(x^{'}-s_2^{'})}, \text{if} x^{'} \in (s_2^{'}, \dfrac{s_2^{'}+s_3^{'}}{2}]; \\ \dfrac{(1+\eta_r^{'})(s_3^{'}-x^{'})}{s_3^{'}-s_2^{'}+\eta_r^{'}(s_3^{'}-x^{'})}, \text{if} x^{'} \in (\dfrac{s_2^{'}+s_3^{'}}{2}, s_3^{'}]. \end{cases}$$

2. The possibility distribution of the reduction $\tilde{\tau}_2^{'}$ of $\tilde{\tau}^{'}$ using the pessimistic CV reduction approach is given as follows:

$$\mu_{\tau_2^{'}}(x^{'}) = \begin{cases} \dfrac{(x^{'}-s_1^{'})}{s_2^{'}-s_1^{'}+\eta_l^{'}(x^{'}-s_1^{'})}, \text{if} x^{'} \in [s_1^{'}, \dfrac{s_1^{'}+s_2^{'}}{2}]; \\ \dfrac{x^{'}-s_1^{'}}{s_2^{'}-s_1^{'}+\eta_l^{'}(s_2^{'}-x^{'})}, \text{if} x^{'} \in (\dfrac{s_1^{'}+s_2^{'}}{2}, s_2^{'}]; \\ \dfrac{(s_3^{'}-x^{'})}{s_3^{'}-s_2^{'}+\eta_l^{'}(x^{'}-s_2^{'})}, \text{if} x^{'} \in (s_2^{'}, \dfrac{s_2^{'}+s_3^{'}}{2}]; \\ \dfrac{(s_3^{'}-x^{'})}{s_3^{'}-s_2^{'}+\eta_l^{'}(s_3^{'}-x^{'})}, \text{if} x^{'} \in (\dfrac{s_2^{'}+s_3^{'}}{2}, s_3^{'}]. \end{cases}$$

3. The possibility distribution of the reduction $\tilde{\tau}_3^{'}$ of $\tilde{\tau}^{'}$ using the CV reduction approach is given as follows:

$$\mu_{\tau_3^{'}}(x^{'}) = \begin{cases} \dfrac{(1+\eta_r^{'})(x^{'}-s_1^{'})}{s_2^{'}-s_1^{'}+2\eta_r^{'}(x^{'}-s_1^{'})}, \text{if} x^{'} \in [s_1^{'}, \dfrac{s_1^{'}+s_2^{'}}{2}]; \\ \dfrac{(1-\eta_l^{'})x^{'}+\eta_l^{'}s_2^{'}-s_1^{'}}{s_2^{'}-s_1^{'}+2\eta_l^{'}(s_2^{'}-x^{'})}, \text{if} x^{'} \in (\dfrac{s_1^{'}+s_2^{'}}{2}, s_2^{'}]; \\ \dfrac{(-1+\eta_l^{'})x^{'}-\eta_l^{'}s_2^{'}+s_3^{'}}{s_3^{'}-s_2^{'}+2\eta_l^{'}(x^{'}-s_2^{'})}, \text{if} x^{'} \in (s_2^{'}, \dfrac{s_2^{'}+s_3^{'}}{2}]; \\ \dfrac{(1+\eta_r^{'})(s_3^{'}-x^{'})}{s_3^{'}-s_2^{'}+2\eta_r^{'}(s_3^{'}-x^{'})}, \text{if} x^{'} \in (\dfrac{s_2^{'}+s_3^{'}}{2}, s_3^{'}]. \end{cases}$$

Theorem 2.4.2 (Qin et al. (2011)) Let ξ_i be the reduction of the type-2 fuzzy variable $\tilde{\xi}_i = (r_1^i, r_2^i, r_3^i; \theta_{l,i}, \theta_{r,i})$ obtained by the CV reduction method for $i = 1,2,\ldots,n$. Suppose $\xi_1, \xi_2, \ldots\ldots\ldots, \xi_n$ are mutually independent, and $k_i \geq 0$ for $i = 1,2,\ldots,n$.

1. Given the generalized credibility level $\alpha \in (0,0.5]$, if $\alpha \in (0,0.25]$, then $\widetilde{Cr}\{\sum_{i=1}^{n} k_i\xi_i \leq t\} \geq \alpha$ is equivalent to $\sum_{i=1}^{n} \frac{(1-2\alpha+(1-4\alpha)\theta_{r,i})k_i r_1^i + 2\alpha k_i r_2^i}{1+(1-4\alpha)\theta_{r,i}} \leq t,$ and if $\alpha \in (0.25,0.5]$, then $\widetilde{Cr}\{\sum_{i=1}^{n} k_i\xi_i \leq t\} \geq \alpha$ is equivalent to $\sum_{i=1}^{n} \frac{(1-2\alpha)k_i r_1^i + (2\alpha+(4\alpha-1)\theta_{l,i})k_i r_2^i}{1+(4\alpha-1)\theta_{l,i}} \leq t$.
2. Given the generalized credibility level $\alpha \in (0.5,1]$, if $\alpha \in (0.5,0.75]$, then $\widetilde{Cr}\{\sum_{i=1}^{n} k_i\xi_i \leq t\} \geq \alpha$ is equivalent to $\sum_{i=1}^{n} \frac{(2\alpha-1)k_i r_3^i + (2(1-\alpha)+(3-4\alpha)\theta_{l,i})k_i r_2^i}{1+(3-4\alpha)\theta_{l,i}} \leq t$, and if, $\alpha \in (0.75,1]$ then $\widetilde{Cr}\{\sum_{i=1}^{n} k_i\xi_i \leq t\} \geq \alpha$ is equivalent to $\sum_{i=1}^{n} \frac{(2\alpha-1+(4\alpha-3)\theta_{r,i})k_i r_3^i + 2(1-\alpha)k_i r_2^i}{1+(4\alpha-3)\theta_{r,i}} \leq t.$

Corollary 2.4.3: Given the generalized credibility level $\alpha \in (0,0.5]$, if $\alpha \in (0,0.25]$ from (i) of the above theorem, then $\widetilde{Cr}\{\sum_{i=1}^{n} k_i\xi_i \geq t\} \geq \alpha$ i.e $\widetilde{Cr}\{\sum_{i=1}^{n} k_i\xi_i^{'} \leq t^{'}\} \geq \alpha$ is equivalent to

$$\sum_{i=1}^{n} \frac{(1-2\alpha+(1-4\alpha)\theta_{l,i})k_i(-r_3^i) + 2\alpha k_i(-r_2^i)}{1+(1-4\alpha)\theta_{l,i}} \leq t^{'} = -t,$$

$$\Rightarrow \sum_{i=1}^{n} \frac{(1-2\alpha+(1-4\alpha)\theta_{l,i})k_i r_3^i + 2\alpha k_i r_2^i}{1+(1-4\alpha)\theta_{l,i}} \geq t,$$

and if

$\alpha \in (0.25,0.5]$

then

$\widetilde{Cr}\{\sum_{i=1}^{n} k_i\xi_i \geq t\} \geq \alpha$

which implies that

$$\sum_{i=1}^{n} \frac{(1-2\alpha)k_i(-r_3^i) + (2\alpha+(4\alpha-1)\theta_{r,i})k_i(-r_2^i)}{1+(4\alpha-1)\theta_{r,i}} \leq -t$$

$$\Rightarrow \sum_{i=1}^{n} \frac{(1-2\alpha)k_i r_3^i + (2\alpha + (4\alpha - 1)\theta_{r,i})k_i r_2^i}{1 + (4\alpha - 1)\theta_{r,i}} \geq t.$$

The other values of α are similarly derived from other expressions.

2.5 Nearest Interval Approximation of Continuous Type-2 Fuzzy Variables

Kundu et al. (2015) proposed the interval approximation of type-2 triangular fuzzy variables by applying the α cut of the optimistic, pessimistic and credibilistic approximation of type-2 triangular fuzzy variables given by theorem 1. Lastly, using interval approximation method to these α cuts estimated crisp intervals are obtained which are given below:

1. Applying α cut of the optimistic CV based reduction (optimistic interval approximation): The nearest interval approximation of $\tilde{\tau}'$ is calculated as $[C_L, C_R]$ where $C_L = C_{L_1} + C_{L_2}$

$$C_{L1} = \frac{(1+\eta_r')s_1'}{\eta_r'} \ln(\frac{1+\eta_r'}{1+0.5\eta_r'}) - \frac{s_2' - s_1' - \eta_r' s_1'}{\eta_r'^2}\left[0.5\eta_r' - (1+\eta_r')\ln(\frac{1+\eta_r'}{1+0.5\eta_r'})\right], \tag{1}$$

$$C_{L2} = -\frac{s_1' - \eta_r' s_2'}{\eta_r'} \ln(1-0.5\eta_r') + \frac{s_2' - s_1' + \eta_r' s_2'}{\eta_r'^2}\left[0.5\eta_r' + (1-\eta_r')\ln(1-0.5\eta_r')\right].$$

$$C_R = C_{R_1} + C_{R_2} \tag{2}$$

$$C_{R1} = \frac{(1+\eta_r')s_3'}{\eta_r'} \ln(\frac{1+\eta_r'}{1+0.5\eta_r'}) + \frac{s_3' - s_2' + \eta_r' s_3'}{\eta_r'^2}\left[0.5\eta_r' - (1+\eta_r')\ln(\frac{1+\eta_r'}{1+0.5\eta_r'})\right]$$

$$C_{R2} = -\frac{s_3' - \eta_r' s_2'}{\eta_r'} \ln(1-0.5\eta_r') - \frac{s_3' - s_2' - \eta_r' s_2'}{\eta_r'^2}\left[0.5\eta_r' + (1-\eta_r')\ln(1-0.5\eta_r')\right].$$

2. Applying α cut of the pessimistic CV based reduction (pessimistic interval approximation): The nearest interval approximation of $\tilde{\tau}'$ is calculated as $[C_L, C_R]$ where, $C_L = C_{L_1} + C_{L_2}$

$$C_{L1} = -\frac{s_1'}{\eta_l'} \ln(1-0.5\eta_l') - \frac{s_2' - s_1' - \eta_l' s_1'}{\eta_l'^2}\left[0.5\eta_l' + \ln(1-0.5\eta_l')\right], \tag{3}$$

$$C_{L2} = \frac{s_1^{'}}{\eta_l^{'}}\ln(\frac{1+\eta_l^{'}}{1+0.5\eta_l^{'}}) + \frac{s_2^{'} - s_1^{'} + \eta_l^{'}s_2^{'}}{\eta_l^{'2}}[0.5\eta_l^{'} - \ln(\frac{1+\eta_l^{'}}{1+0.5\eta_l^{'}})].$$

$$C_R = C_{R_1} + C_{R_2} \tag{4}$$

$$C_{R1} = -\frac{s_3^{'}}{\eta_l^{'}}\ln(1-0.5\eta_l^{'}) + \frac{s_3^{'} - s_2^{'} + \eta_l^{'}s_3^{'}}{\eta_l^{'2}}\left[0.5\eta_l^{'} + \ln(1-0.5\eta_l^{'})\right],$$

$$C_{R2} = \frac{s_3^{'}}{\eta_l^{'}}\ln(\frac{1+\eta_l^{'}}{1+0.5\eta_l^{'}}) + \frac{s_3^{'} - s_2^{'} - \eta_l^{'}s_2^{'}}{\eta_l^{'2}}[0.5\eta_l^{'} - \ln(\frac{1+\eta_l^{'}}{1+0.5\eta_l^{'}})].$$

3. Applying α cut of the CV reduction (credibilistic interval approximation): The nearest interval approximation of $\tilde{\tau}^{'}$ is calculated as $[C_L, C_R]$ where, $C_L = C_{L_1} + C_{L_2}$

$$C_{L1} = \frac{(1+\eta_r^{'})s_1^{'}}{2\eta_r^{'}}\ln(1+\eta_r^{'}) - \frac{s_2^{'} - s_1^{'} - 2\eta_r^{'}s_1^{'}}{4\eta_r^{'2}}\left[\eta_r^{'} - (1+\eta_r^{'})\ln(1+\eta_r^{'})\right], \tag{5}$$

$$C_{L2} = \frac{s_1^{'} - \eta_l^{'}s_2^{'}}{2\eta_l^{'}}\ln(1+\eta_l^{'}) + \frac{s_2^{'} - s_1^{'} + 2\eta_l^{'}s_2^{'}}{4\eta_l^{'2}}\left[\eta_l^{'} - (1-\eta_l^{'})\ln(1+\eta_l^{'})\right].$$

$$C_R = C_{R_1} + C_{R_2} \tag{6}$$

$$C_{R1} = \frac{(1+\eta_r^{'})s_3^{'}}{2\eta_r^{'}}\ln(1+\eta_r^{'}) + \frac{s_3^{'} - s_2^{'} + 2\eta_r^{'}s_3^{'}}{4\eta_r^{'2}}\left[\eta_r^{'} - (1+\eta_r^{'})\ln(1+\eta_r^{'})\right],$$

$$C_{R2} = \frac{s_3^{'} - \eta_l^{'}s_2^{'}}{2\eta_l^{'}}\ln(1+\eta_l^{'}) - \frac{s_3^{'} - s_2^{'} - 2\eta_l^{'}s_2^{'}}{4\eta_l^{'2}}\left[\eta_l^{'} - (1-\eta_l^{'})\ln(1+\eta_l^{'})\right].$$

3. MAIN FOCUS OF THE CHAPTER

The main focuses of the chapter are:

1. The solid transportation problem with type-2 fuzzy variables is an emerging area. The advantage of type-2 fuzzy sets is that they are helpful in some cases where it is difficult to find the exact membership functions for fuzzy sets.
2. The authors have formulated a multi-item multi objective fixed charged solid transportation problem with budget constraints and conditions on a few brands and carriages in the sense that a few specific brands are restricted to be shipped over a few particular carriages.
3. The problem follows some transportation parameters e.g. unit transportation penalty, fixed costs, amounts, requirements, budget amount and carriage quantities as type-2 triangular fuzzy variables.

It is impossible to form a transportation plan for the future months because the values for the amount of quantity necessary at every origin, the requirement at every destination and the carriage quantity are uncertain. It is difficult to predict the exact transportation cost for a certain time period. Soft computing is the method to handle uncertainty, partial truth, and approximation to attain practicability, robustness and low solution cost.

3.1 Problem

The authors have proposed the model for the multi-item multi-objective fixed charged solid transportation problem with budget constraints and restriction on conveyances in this section.

Suppose that K different forms of conveyances are required to transport l items from m origins (or sources) $O_i (i = 1, 2, ..., m)$ to n destinations $D_j (j = 1, 2, ..., n)$ and also $t(t = 1, 2, \dots, R)$ objectives are to be minimized. In addition to that there are a few conditions on a few particular components and carriages so that a few components cannot be shipped over a few carriages. Let us denote J_k as the set of items which can be shipped through conveyance k $(k = 1, 2,.., K)$. We use notation p' $(p' = 1, 2,.., l)$ to denote the items.

The fixed charged solid transportation problem (FCSTP) is associated with two types of costs, unit transportation cost (direct cost) for shipping unit product from source i to destination j and a fixed (/additional) cost for the route (i, j). The multi-item multi-objective FCSTP with m sources, n destinations, k conveyances, direct costs, fixed cost parameters, and budget constraints in T2 FVs defined as follows:

$$\text{Min } Z_t = \sum_{p'=1}^{l} \sum_{i=1}^{m} \sum_{j=1}^{n} \sum_{k=1}^{K} d_{ijk}^{p'} c_{ijk}^{tp'} x_{ijk}^{p'} + e_{ijk}^{tp'} y_{ijk}^{p'}, t = 1, 2, ...R \tag{7}$$

subject to

$$\sum_{j=1}^{n} \sum_{k=1}^{K} d_{ijk}^{p'} x_{ijk}^{p'} \leq a_i^{p'}, i = 1, 2, ..., m; p' = 1, 2, ..., l,$$

$$\sum_{i=1}^{m} \sum_{k=1}^{K} d_{ijk}^{p'} x_{ijk}^{p'} \geq b_j^{p'}, j = 1, 2, ..., n; p' = 1, 2, ..., l,$$

$$\sum_{p'=1}^{l}\sum_{i=1}^{m}\sum_{j=1}^{n} d_{ijk}^{p'} x_{ijk}^{p'} \le f_k, k = 1,2,\ldots,K,$$

$$\sum_{i=1}^{m}\sum_{k=1}^{K} d_{ijk}^{p'}(c_{ijk}^{tp'} x_{ijk}^{p'}) + e_{ijk}^{tp'} y_{ijk}^{p'} \le B_j^{tp'}, j = 1,2,\ldots,n; p' = 1,2,\ldots.l, t = 1,2,\ldots R.$$

$$x_{ijk}^{p'} \ge 0, \forall i,j,k,p', y_{ijk}^{p'} = \begin{cases} 1, \text{if } x_{ijk}^{p'} > 0; \\ 0, \text{otherwise,} \end{cases}$$

where $d_{ijk}^{p'}$ is defined as

$$d_{ijk}^{p'} = \begin{cases} 1, \text{if } p' \in I_K \ \forall i,j,k,p'; \\ 0, \text{otherwise.} \end{cases}$$

Here, $x_{ijk}^{p'}$ is the decision variable representing the amount of p'-th item transported from source i to destination j, $e_{ijk}^{tp'}$ is the type-2 fuzzy fixed cost associated with route *(i, j)* for the objective Z_t. The unit transportation cost $c_{ijk}^{tp'}$ for the objective Z_t, total supply of p'-th item $a_i^{p'}$ at i-th origin, total demand of p'-th item $b_j^{p'}$ at j-th destination and total capacity f_k of k-th conveyance are all type-2 fuzzy variables. Here $c_{ijk}^{tp'}$ $(t = 1,2,\ldots,R)$ represent unit transportation cost so that the available type-2 fuzzy budget amount for jth destination i.e. $B_j^{tp'}$ is imposed for the objective Z_t.

4. SOLUTION AND RECOMMENDATIONS

The model (7) is then transformed into the corresponding crisp deterministic form by applying nearest interval approximation and chance constrained programming using generalized credibility in this following section.

4.1 Applying Nearest Interval Approximation

Suppose that $c_{ijk}^{tp'}$, $e_{ijk}^{tp'}$, $a_i^{p'}$, $b_j^{p'}$, f_k and $B_j^{tp'}$ are independent type-2 triangular fuzzy variables defined by:

$$c_{ijk}^{tp'} = (c_{ijk}^{tp'1}, c_{ijk}^{tp'2}, c_{ijk}^{tp'3}; \theta_{l,ijk}^{tp'}, \theta_{r,ijk}^{tp'}),\ e_{ijk}^{tp'} = (e_{ijk}^{tp'1}, e_{ijk}^{tp'2}, e_{ijk}^{tp'3}; \theta_{l,ijk}^{'tp'}, \theta_{r,ijk}^{'tp'}),\ a_i^{p'} = (a_i^{p'1}, a_i^{p'2}, a_i^{p'3};$$

$$\theta_{l,i}^{p'}, \theta_{r,i}^{p'}),\ b_j^{p'} = (b_j^{p'1}, b_j^{p'2}, b_j^{p'3}; \theta_{l,j}^{p'}, \theta_{r,j}^{p'}),\ e_k = (e_k^1, e_k^2, e_k^3; \theta_{l,k}, \theta_{r,k})$$

and

$$B_j^{tp'} = (B_j^{tp'1}, B_j^{tp'2}, B_j^{tp'3}; \theta_{l,j}^{tp'}, \theta_{r,j}^{tp'})$$

The nearest interval approximations (credibilistic interval approximation) of $c_{ijk}^{tp'}, e_{ijk}^{tp'}, a_i^{p'}, b_j^{p'}, f_k$ and $B_j^{tp'}$ are obtained using (5)-(6) and suppose these are $[c_{ijkL}^{tp'}, c_{ijkR}^{tp'}], [e_{ijkL}^{tp'}, e_{ijkR}^{tp'}], [a_{iL}^{p'}, a_{iR}^{p'}], [b_{jL}^{p'}, b_{jR}^{p'}], [f_{kL}, f_{kR}]$ and $[B_{jL}^{tp'}, B_{jR}^{tp'}]$ respectively. The nearest interval approximations of the model (7) are given below:

$$\text{Min} Z_t = \sum_{p'=1}^{l}\sum_{i=1}^{m}\sum_{j=1}^{n}\sum_{k=1}^{K} d_{ijk}^{p'}([c_{ijkL}^{tp'}, c_{ijkR}^{tp'}]x_{ijk}^{p'}) + [e_{ijkL}^{tp'}, e_{ijkR}^{tp'}]y_{ijk}^{p'}$$

subject to

$$\sum_{j=1}^{n}\sum_{k=1}^{K} d_{ijk}^{p'} x_{ijk}^{p'} \leq [a_{iL}^{p'}, a_{iR}^{p'}], i = 1,2,\ldots,m; p' = 1,2,\ldots,l,$$

$$\sum_{i=1}^{m}\sum_{k=1}^{K} d_{ijk}^{p'} x_{ijk}^{p'} \geq [b_{jL}^{p'}, b_{jR}^{p'}], j = 1,2,\ldots,n; p' = 1,2,\ldots,l,$$

$$\sum_{p'=1}^{l}\sum_{i=1}^{m}\sum_{j=1}^{n} d_{ijk}^{p'} x_{ijk}^{p'} \leq [f_{kL}, f_{kR}], k = 1,2,\ldots,K, \tag{8}$$

$$\sum_{p'=1}^{l}\sum_{i=1}^{m}\sum_{j=1}^{n}\sum_{k=1}^{K} d_{ijk}^{p'}([c_{ijkL}^{tp'}, c_{ijkR}^{tp'}]x_{ijk}^{p'}) + [e_{ijkL}^{tp'}, e_{ijkR}^{tp'}]y_{ijk}^{p'} \leq [B_{jL}^{tp'}, B_{jR}^{tp'}]$$

$$x_{ijk}^{p'} \geq 0, \forall i, j, k, p', \; y_{ijk}^{p'} = \begin{cases} 1, if\, x_{ijk}^{p'} > 0; \\ 0, \text{otherwise.} \end{cases}$$

where $d_{ijk}^{p'}$ is defined as

$$d_{ijk}^{p'} = \begin{cases} 1, if\, p' \in I_K \; \forall\, i, j, k, p'; \\ 0, \text{otherwise,} \end{cases}$$

4.1.1 Deterministic Form

Suppose the left hand side expressions of the origin, destination, carriage quantity constraints, and the budget constraints of the model (8) be denoted by $S_i^{p'}, D_j^{p'}, F_k$ and $M_j^{tp'}$ respectively. The possibility degree of fulfilment of these constraints are represented as

$$P_{S_i^{p'} \leq [a_{iL}^{p'}, a_{iR}^{p'}]} = \begin{cases} 1, S_i^{p'} \leq a_{iL}^{p'}; \\ \dfrac{a_{iR}^{p'} - S_i^{p'}}{a_{iR}^{p'} - a_{iL}^{p'}}, a_{iL}^{p'} < S_i^{p'} \leq a_{iR}^{p'}; \\ 0, S_i^{p'} > a_{iR}^{p'}. \end{cases}$$

$$P_{D_j^{p'} \geq [b_{jL}^{p'}, b_{jR}^{p'}]} = \begin{cases} 0, D_j^{p'} < b_{jL}^{p'}; \\ \dfrac{D_j^{p'} - b_{jL}^{p'}}{b_{jR}^{p'} - b_{jL}^{p'}}, \ b_{jL}^{p'} \leq D_j^{p'} < b_{jR}^{p'}; \\ 1, \ D_j^{p'} > b_{jL}^{p'}. \end{cases}$$

$$P_{F_k \leq [f_{kL}, f_{kR}]} = \begin{cases} 1, \ F_k \leq f_{kL}; \\ \dfrac{f_{kR} - F_k}{f_{kR} - f_{kL}}, f_{kL} < F_k \leq f_{kR}; \\ 0, F_k > f_{kR}. \end{cases}$$

$$P_{M_j^{tp'} \geq [B_{jL}^{tp'}, B_{jR}^{tp'}]} = \begin{cases} 1, \ M_j^{tp'} \leq B_{jR}^{tp'}; \\ \dfrac{B_{jR}^{tp'} - M_j^{tp'}}{B_{jR}^{tp'} - B_{jL}^{tp'}}, B_{jL}^{tp'} < M_j^{tp'} \leq B_{jR}^{tp'}; \\ 0, M_j^{tp'} > B_{jR}^{tp'}. \end{cases}$$

The constraints are allowed to satisfy with a few predetermined possibility degree levels $\alpha_i^{p'}, \beta_j^{p'}, \gamma_k$ and $\delta_j^{tp'} (0 < \alpha_i^{p'}, \beta_j^{p'}, \gamma_k, \delta_j^{tp'} \leq 1)$ respectively, i.e. $P_{S_i^{p'} \leq [a_{iL}^{p'}, a_{iR}^{p'}]} \geq \alpha_i^{p'}$, $P_{D_j^{p'} \geq [b_{jL}^{p'}, b_{jR}^{p'}]} \geq \beta_j^{p'}$ $P_{F_k \leq [f_{kL}, f_{kR}]} \geq \gamma_k$, $P_{M_j^{tp'} \geq [B_{jL}^{tp'}, B_{jR}^{tp'}]} \geq \delta_j^{tp'} \forall i, j, k, p'$ and then the identical deterministic inequalities of the various constraints are given below:

$$S_i^{p'} \leq a_{iR}^{p'} - \alpha_i^{p'} [a_{iR}^{p'} - a_{iL}^{p'}] \tag{9}$$

$$D_j^{p'} \geq b_{jL}^{p'} + \beta_j^{p'}[b_{jR}^{p'} - b_{jL}^{p'}], \tag{10}$$

$$F_k \leq f_{kR} - \gamma_k[f_{kR} - f_{kL}]. \tag{11}$$

$$M_j^{tp'} \leq B_{jR}^{tp'} - \delta_j^{tp'}[B_{jR}^{p'} - B_{jL}^{p'}] \tag{12}$$

The minimum objective function value (say $\underline{Z}_t$) and maximum possible objective function value (say $\overline{Z}_t$) for $[c_{ijkL}^{tp'}, c_{ijkR}^{tp'}]$, $[e_{ijkL}^{tp'}, e_{ijkR}^{tp'}]$ are obtained by solving the following two models:

$$\underline{Z}_t = \min_{c_{ijkL}^{tp'} \leq c_{ijk}^{tp'} \leq c_{ijkR}^{tp'}, e_{ijkL}^{tp'} \leq e_{ijk}^{tp'} \leq e_{ijkR}^{tp'}} \left[Min \sum_{p'=1}^{l} \sum_{i=1}^{m} \sum_{j=1}^{n} \sum_{k=1}^{K} d_{ijk}^{p'}(c_{ijk}^{tp'} x_{ijk}^{p'}) + e_{ijk}^{tp'} y_{ijk}^{p'}\right] \tag{13}$$

$$\overline{Z}_t = \max_{c_{ijkL}^{tp'} \leq c_{ijk}^{tp'} \leq c_{ijkR}^{tp'}, e_{ijkL}^{tp'} \leq e_{ijk}^{tp'} \leq e_{ijkR}^{tp'}} \left[Min \sum_{p'=1}^{l} \sum_{i=1}^{m} \sum_{j=1}^{n} \sum_{k=1}^{K} d_{ijk}^{p'}(c_{ijk}^{tp'} x_{ijk}^{p'}) + e_{ijk}^{tp'} y_{ijk}^{p'}\right] \tag{14}$$

subject to the above constraints (9)-(12) for both cases.

4.2 Chance Constrained Programming Using Generalized Credibility

Suppose that $c_{ijk}^{tp''}$, $e_{ijk}^{tp''}$, $a_i^{p''}$, $b_j^{p''}$, $f_{k''}$ and $B_j^{tp''}$ are the reduced type-1 fuzzy variables from type-2 fuzzy variables $c_{ijk}^{p'}, e_{ijk}^{p'}, a_i^{p'}, b_j^{p'}, f_{k'}$ and $B_j^{tp'}$ respectively based on CV-based reduction method. A chance-constrained programming is formulated with these reduced fuzzy parameters to solve the above problem. The uncertain constraints are allowed to be violated such that constraints must be satisfied at some chance (/confidence) level in chance-constrained programming. The usual credibility measure cannot be used if the reduced fuzzy parameters $c_{ijk}^{p''}, e_{ijk}^{p''}, a_{i'}^{p''}, b_{j'}^{p''}, f_{k''}$, and $B_j^{tp''}$ are not normalized. The following chance-constrained programming model is formulated for the above problem (7) using generalized credibility as the problem is a minimization problem:

$$\text{Min}_x(\text{Min } \tilde{f})$$

subject to

$$Cr\{\sum_{p=1}^{l} \sum_{i=1}^{m} \sum_{j=1}^{n} \sum_{k=1}^{K} d_{ijk}^{p'} c_{ijk}^{tp''} x_{ijk}^{p'} + e_{ijk}^{tp''} y_{ijk}^{p'} \leq \tilde{f}\} \geq \alpha,$$

$$Cr\{\sum_{j=1}^{n}\sum_{k=1}^{K} d_{ijk}^{p'} x_{ijk}^{p'} \leq a_i^{p''}\} \geq \alpha_i^{p'}, i = 1,2,..,m; p' = 1,2..,l,$$

$$Cr\{\sum_{i=1}^{m}\sum_{k=1}^{K} d_{ijk}^{p'} x_{ijk}^{p'} \geq b_j^{p''}\} \geq \beta_j^{p'}, j = 1,2,...,n; p' = 1,...,l,$$

$$Cr\{\sum_{p'=1}^{l}\sum_{i=1}^{m}\sum_{j=1}^{n} d_{ijk}^{p'} x_{ijk}^{p'} \leq f_{k'}\} \geq \gamma_k, k = 1,2,...,K, \tag{15}$$

$$\sum_{i=1}^{m}\sum_{k=1}^{K} d_{ijk}^{p'}(c_{ijk}^{tp''} x_{ijk}^{p'}) + e_{ijk}^{tp''} y_{ijk}^{p'} \leq B_j^{tp''} \geq \delta_j^{p'}, j = 1,2,...,n; p' = 1,2,...,l, t = 1,2,...R.$$

$$x_{ijk}^{p'} \geq 0, \forall i,j,k,p', y_{ijk}^{p'} = \begin{cases} 1, \text{if } x_{ijk}^{p'} > 0; \\ 0, \text{otherwise}, \end{cases}$$

where d_{ijk}^{p} is defined as

$$d_{ijk}^{p'} = \begin{cases} 1, \text{if } p' \in I_K \forall i,j,k,p'; \\ 0, \text{otherwise}. \end{cases}$$

where Min $\tilde{f}$ indicates the minimum possible crisp form that the objective function achieves with generalized credibility at least $\alpha(0 < \alpha \leq 1)$. In other words, α indicates that the minimization of the α critical value of the objective function. $\alpha_i^{p'}, \beta_j^{p'}, \gamma_k, \delta_j^{p'}$ $(0 < \alpha_i^{p'}, \beta_j^{p'}, \gamma_k, \delta_j^{p'} \leq 1)$ are predetermined generalized credibility levels of satisfaction of the respective constraints for all *i, j, k, p'*. The first constraint indicates that total amount of p' '-th item transported from source *i* must be less than or equal to its supply capacity at the credibility level atleast $\alpha_i^{p'}$; the second constraint indicates that total amount transported to destination *j* must satisfy its requirement at the credibility atleast $\beta_j^{p'}$ and the third constraint indicates that total amount transported through conveyance *k* must not be more than its capacity at the credibility atleast γ_k, and for the fourth constraint indicates that for the objective Z_l must satisfy its budget at the credibility atleast $\delta_j^{p'}$.

4.2.1 Crisp Equivalence

The chance constrained model formulation (15) is turned into the following crisp equivalent parametric programming problems from Theorem 3.3.2 and its corollary 3.3.3:

Case I: $0 < \alpha \leq 0.25$: The equivalent parametric programming for model (15) is

$$\text{Min}\sum_{p'=1}^{l}\sum_{i=1}^{m}\sum_{j=1}^{n}\sum_{k=1}^{K} d_{ijk}^{p'}[\frac{(1-2\alpha+(1-4\alpha)\theta_{r,ijk}^{tp'})c_{ijk}^{tp'1}x_{ijk}^{p'}+2\alpha c_{ijk}^{tp'2}x_{ijk}^{p'}}{1+(1-4\alpha)\theta_{r,ijk}^{tp'}}]+$$
$$\frac{(1-2\alpha+(1-4\alpha)\theta_{r,ijk}^{'tp'})e_{ijk}^{tp'1}y_{ijk}^{p'}+2\alpha e_{ijk}^{tp'2}y_{ijk}^{p'}}{1+(1-4\alpha)\theta_{r,ijk}^{'tp'}},$$

subject to

$$\sum_{j=1}^{n}\sum_{k=1}^{K} d_{ijk}^{p'}x_{ijk}^{p'} \le a_i^{p'}, i=1,2,...,m; p'=1,2,...,l,$$

$$\sum_{i=1}^{m}\sum_{k=1}^{K} d_{ijk}^{p'}x_{ijk}^{p'} \ge b_j^{p'}, j=1,2,...,n; p'=1,2,...,l,$$

$$\sum_{p'=1}^{l}\sum_{i=1}^{m}\sum_{j=1}^{n} d_{ijk}^{p'}x_{ijk}^{p'} \le f_k, k=1,2,...,K, \tag{16}$$

$$\sum_{i=1}^{m}\sum_{k=1}^{K} d_{ijk}^{p'}(c_{ijk}^{tp'}x_{ijk}^{p'})+e_{ijk}^{tp'}y_{ijk}^{p'} \le B_j^{tp'}, j=1,2,...,n; p'=1,2,....l, t=1,2,...R.$$

$$x_{ijk}^{p'} \ge 0, \forall i,j,k,p', y_{ijk}^{p'} = \begin{cases}1, \text{if } x_{ijk}^{p'}>0;\\ 0, \text{otherwise,}\end{cases}$$

where $d_{ijk}^{p'}$ is defined as $d_{ijk}^{p'} = \begin{cases}1, \text{if } p' \in I_K \ \forall i,j,k,p';\\ 0, \text{otherwise.}\end{cases}$ where $F_{a_i^p}, F_{b_j^p}, F_{f_k}\ F_{B_j^{p'}}$ is defined by (18), (19), (20) and (21) respectively.

Case II: $0.25 < \alpha \le 0.5$: The equivalent parametric programming for model (15) is

$$\text{Min}\sum_{p'=1}^{l}\sum_{i=1}^{m}\sum_{j=1}^{n}\sum_{k=1}^{K} d_{ijk}^{p'}[\frac{(1-2\alpha)c_{ijk}^{tp'1}x_{ijk}^{p'}+(2\alpha+(4\alpha-1)\theta_{l,ijk}^{tp'})c_{ijk}^{tp'2}x_{ijk}^{p'}}{1+(4\alpha-1)\theta_{l,ijk}^{tp'}}]+$$
$$\frac{(1-2\alpha)e_{ijk}^{tp'1}y_{ijk}^{p'}+(2\alpha+(4\alpha-1)\theta_{l,ijk}^{'tp'})e_{ijk}^{tp'2}y_{ijk}^{p'}}{1+(4\alpha-1)\theta_{l,ijk}^{'tp'}},$$

subject to

$$\sum_{j=1}^{n}\sum_{k=1}^{K} d_{ijk}^{p'} x_{ijk}^{p'} \leq a_i^{p'}, i = 1,2,...,m; p' = 1,2,...,l,$$

$$\sum_{i=1}^{m}\sum_{k=1}^{K} d_{ijk}^{p'} x_{ijk}^{p'} \geq b_j^{p'}, j = 1,2,...,n; p' = 1,2,...,l,$$

$$\sum_{p'=1}^{l}\sum_{i=1}^{m}\sum_{j=1}^{n} d_{ijk}^{p'} x_{ijk}^{p'} \leq f_k, k = 1,2,...,K, \tag{17}$$

$$\sum_{i=1}^{m}\sum_{k=1}^{K} d_{ijk}^{p'} (c_{ijk}^{tp'} x_{ijk}^{p'}) + e_{ijk}^{tp'} y_{ijk}^{p'} \leq B_j^{tp'}, j = 1,2,...,n; p' = 1,2,....l, t = 1,2,...R.$$

$$x_{ijk}^{p'} \geq 0, \forall i,j,k,p', y_{ijk}^{p'} = \begin{cases} 1, \text{if } x_{ijk}^{p'} > 0; \\ 0, \text{otherwise}, \end{cases}$$

where $d_{ijk}^{p'}$ is defined as

$$d_{ijk}^{p'} = \begin{cases} 1, \text{if } p' \in I_K \ \forall i,j,k,p'; \\ 0,\text{otherwise}. \end{cases}$$

where $F_{a_i^{p'}}, F_{b_j^{p'}}, F_{f_k}, F_{B_j^{p'}}$ is defined by (18), (19), (20) and (21) respectively.

Case III: $0.5 < \alpha \leq 0.75$: The equivalent parametric problem for the model (15) is

$$\text{Min}\sum_{p'=1}^{l}\sum_{i=1}^{m}\sum_{j=1}^{n}\sum_{k=1}^{K} d_{ijk}^{p'}[\frac{(2\alpha-1)c_{ijk}^{tp'3} x_{ijk}^{p'} + (2-2\alpha) + (3-4\alpha)\theta_{l,ijk}^{tp'})c_{ijk}^{tp'2} x_{ijk}^{p'}}{1+(3-4\alpha)\theta_{l,ijk}^{tp'}}] +$$

$$\frac{(2\alpha-1)e_{ijk}^{tp'3} y_{ijk}^{p'} + (2-2\alpha+(3-4\alpha)\theta_{l,ijk}^{'tp'})e_{ijk}^{tp'2} y_{ijk}^{p'}}{1+(3-4\alpha)\theta_{l,ijk}^{'tp'}}$$

$$\sum_{j=1}^{n}\sum_{k=1}^{K} d_{ijk}^{p'} x_{ijk}^{p'} \leq a_i^{p'}, i = 1,2,...,m; p' = 1,2,...,l,$$

$$\sum_{i=1}^{m}\sum_{k=1}^{K} d_{ijk}^{p'} x_{ijk}^{p'} \geq b_j^{p'}, j = 1,2,...,n; p' = 1,2,...,l,$$

$$\sum_{p'=1}^{l}\sum_{i=1}^{m}\sum_{j=1}^{n} d_{ijk}^{p'} x_{ijk}^{p'} \leq f_k, k = 1, 2, ..., K, \tag{18}$$

$$\sum_{i=1}^{m}\sum_{k=1}^{K} d_{ijk}^{p'} (c_{ijk}^{tp'} x_{ijk}^{p'}) + e_{ijk}^{tp'} y_{ijk}^{p'} \leq B_j^{tp'}, j = 1, 2, ..., n; p' = 1, 2, ...l, t = 1, 2, ...R.$$

$$x_{ijk}^{p'} \geq 0, \forall i, j, k, p', y_{ijk}^{p'} = \begin{cases} 1, \text{if } x_{ijk}^{p'} > 0; \\ 0, \text{otherwise,} \end{cases}$$

where $d_{ijk}^{p'}$ is defined as

$$d_{ijk}^{p'} = \begin{cases} 1, \text{if } p' \in I_K \ \forall i, j, k, p'; \\ 0, \text{otherwise.} \end{cases}$$

where $F_{a_i^{p'}}$, $F_{b_j^{p'}}$, F_{f_k} $F_{B_j^{p'}}$ is defined by (18), (19), (20) and (21) respectively.

Case IV: $0.75 < \alpha \leq 1$: The equivalent parametric problem for the model (15) is

$$\text{Min} \sum_{p'=1}^{l}\sum_{i=1}^{m}\sum_{j=1}^{n}\sum_{k=1}^{K} d_{ijk}^{p'} \left[\frac{\left\{ \left(2\alpha - 1 + \left(4\alpha - 3\right)\theta_{r,ijk}^{tp'}\right) c_{ijk}^{tp'3} + (2 - 2\alpha) c_{ijk}^{tp'2} \right\} x_{ijk}^{p'}}{1 + (4\alpha - 3)\theta_{r,ijk}^{tp'}} \right] +$$
$$\frac{(2\alpha - 1 + (4\alpha - 3)\theta_{r,ijk}^{'tp'}) e_{ijk}^{tp'3} y_{ijk}^{p'} + 2(1 - \alpha) e_{ijk}^{tp'2} y_{ijk}^{p'}}{1 + (4\alpha - 3)\theta_{r,ijk}^{'tp'}}$$

$$\sum_{j=1}^{n}\sum_{k=1}^{K} d_{ijk}^{p'} x_{ijk}^{p'} \leq a_i^{p'}, i = 1, 2, ..., m; p' = 1, 2, ..., l,$$

$$\sum_{i=1}^{m}\sum_{k=1}^{K} d_{ijk}^{p'} x_{ijk}^{p'} \geq b_j^{p'}, j = 1, 2, ..., n; p' = 1, 2, ..., l,$$

$$\sum_{p'=1}^{l}\sum_{i=1}^{m}\sum_{j=1}^{n} d_{ijk}^{p'} x_{ijk}^{p'} \leq f_k, k = 1, 2, ..., K, \tag{19}$$

$$\sum_{i=1}^{m}\sum_{k=1}^{K} d_{ijk}^{p'}(c_{ijk}^{tp'}x_{ijk}^{p'}) + e_{ijk}^{tp'}y_{ijk}^{p'} \leq B_j^{tp'}, j = 1,2,\ldots,n; p' = 1,2,\ldots.l, t = 1,2,\ldots R.$$

where $d_{ijk}^{p'}$ is defined as

$$d_{ijk}^{p'} = \begin{cases} 1, \text{if } p' \in I_K \ \forall i,j,k,p'; \\ 0, \text{otherwise.} \end{cases}$$

where, $F_{a_i^{p'}}, F_{b_j^{p'}}, F_{f_k}, F_{B_j^{p'}}$ is defined by (20), (21), (22) and (23) respectively.where

$$F_{a_i^p} = \begin{cases} \dfrac{\left(1-2\alpha_i^p+(1-4\alpha_i^p)\theta_{l,i}^p\right)a_i^{p3}+2\alpha_i^p a_i^{p2}}{1+(1-4\alpha_i^p)\theta_{l,i}^p}, \text{ if } 0<\alpha_i^p\leq 0.25; \\ \dfrac{(1-2\alpha_i^p)a_i^{p3}+\left(2\alpha_i^p+(4\alpha_i^p-1)\theta_{r,i}^p\right)a_i^{p2}}{1+(4\alpha_i^p-1)\theta_{r,i}^p}, \text{ if } 0.25<\alpha_i^p\leq 0.5; \\ \dfrac{(2\alpha_i^p-1)a_i^{p1}+\left(2(1-\alpha_i^p)+(3-4\alpha_i^p)\theta_{r,i}^p\right)a_i^{p2}}{1+(3-4\alpha_i^p)\theta_{r,i}^p}, \text{ if } 0.5<\alpha_i^p\leq 0.75; \\ \dfrac{\left(2\alpha_i^p-1+(4\alpha_i^p-3)\theta_{l,i}^p\right)a_i^{p1}+2(1-\alpha_i^p)a_i^{p2}}{1+(4\alpha_i^p-3)\theta_{l,i}^p}, \text{ if } 0.75<\alpha_i^p\leq 1 \end{cases} \tag{20}$$

$$F_{b_j^p} = \begin{cases} \dfrac{\left(1-2\beta_j^p+(1-4\beta_j^p)\theta_{r,j}^p\right)b_j^{p1}+2\beta_j^p b_j^{p2}}{1+(1-4\beta_j^p)\theta_{r,j}^p}, \text{if } 0<\beta_j^p\leq 0.25; \\ \dfrac{(1-2\beta_j^p)b_j^{p1}+\left(2\beta_j^p+(4\beta_j^p-1)\theta_{l,j}^p\right)b_j^{p2}}{1+(4\beta_j^p-1)\theta_{l,j}^p}, \text{if } 0.25<\beta_j^p\leq 0.5; \\ \dfrac{(2\beta_j^p-1)b_j^{p3}+\left(2(1-\beta_j^p)+(3-4\beta_j^p)\theta_{l,j}^p\right)b_j^{p2}}{1+(3-4\beta_j^p)\theta_{l,j}^p}, \text{if } 0.5<\beta_j^p\leq 0.75; \\ \dfrac{\left(2\beta_j^p-1+(4\beta_j^p-3)\theta_{r,j}^p\right)b_j^{p3}+2(1-\beta_j^p)b_j^{p2}}{1+(4\beta_j^p-3)\theta_{r,j}^p}, \text{if } 0.75<\beta_j^p\leq 1. \end{cases} \tag{21}$$

$$F_{f_k} = \begin{cases} \dfrac{(1-2\gamma_k+(1-4\gamma_k)\theta_{l,k})f_k^3+2\gamma_k f_k^2}{1+(1-4\gamma_k)\theta_{l,k}}, \text{if } 0<\gamma_k\le 0.25; \\ \dfrac{(1-2\gamma_k)f_k^3+(2\gamma_k+(4\gamma_k-1)\theta_{r,k})f_k^2}{1+(4\gamma_k-1)\theta_{r,k}}, \text{if } 0.25<\gamma_k\le 0.5; \\ \dfrac{(2\gamma_k-1)f_k^1+(2(1-\gamma_k)+(3-4\gamma_k)\theta_{r,k})f_k^2}{1+(3-4\gamma_k)\theta_{r,k}}, \text{if } 0.5<\gamma_k\le 0.75; \\ \dfrac{(2\gamma_k-1+(4\gamma_k-3)\theta_{l,k})f_k^1+2(1-\gamma_k)f_k^2}{1+(4\gamma_k-3)\theta_{l,k}}, \text{if } 0.75<\gamma_k\le 1. \end{cases} \tag{22}$$

$$F_{B_j^{p'}} = \begin{cases} \dfrac{\left(1-2\delta_j^{p'}+(1-4\delta_j^{p'})\theta_{l,j}^{p'}\right)B_j^{p'3}+2\delta_j^{p'}B_j^{p'2}}{1+(1-4\delta_j^{p})\theta_{l,j}^{p'}}, \text{if } 0<\delta_i^{p'}\le .25; \\ \dfrac{(1-2\delta_j^{p'})B_j^{p'3}+\left(2\delta_j^{p'}+(4\delta_j^{p'}-1)\theta_{r,j}^{p'}\right)B_j^{p'2}}{1+(4\delta_j^{p'}-1)\theta_{r,j}^{p'}}, \text{if } 0.25<\delta_j^{p'}\le .5; \\ \dfrac{(2\delta_j^{p'}-1)B_j^{p'1}+\left(2(1-\delta_j^{p'})+(3-4\delta_j^{p'})\theta_{r,j}^{p'}\right)B_j^{p'2}}{1+(3-4\delta_j^{p'})\theta_{r,j}^{p'}}, \text{if } 0.5<\delta_j^{p'}\le .75; \\ \dfrac{\left(2\delta_j^{p'}-1+(4\delta_j^{p'}-3)\theta_{l,j}^{p'}\right)B_j^{p'1}+2(1-\delta_j^{p'})B_j^{p'2}}{1+(4\delta_j^{p'}-3)\theta_{l,j}^{p'}}, \text{if } 0.75<\delta_j^{p'}\le 1. \end{cases} \tag{23}$$

4.3 Solution Procedure of Multi-Objective STP

The authors used the fuzzy programming technique to solve the crisp deterministic forms (8) and (15) of the problem (7).

4.3.1 Fuzzy Programming Technique

The following are the steps to solve the multi-objective models applying fuzzy programming technique:

Step 1: The multi-objective models are solved as a one objective model applying, every time, single objective $\overline{Z}_t$ to find the optimal solution $X^{t*}=x_{ijk}^{p'}$ of R distinct single objective model.

Step 2: The values of R objective functions at all these R optimal solutions X^{t*} are calculated and the upper and lower bound for every objective is fixed by $U_t=Max\left\{\overline{Z}_t\left(X^{1*}\right),\overline{Z}_t\left(X^{2*}\right),\ldots,\overline{Z}_t\left(X^{t*}\right)\right\}$ and $L_t=\overline{Z}_t(X^{t*})$.

Step 3: An introductory fuzzy model is defined as

Find x subject to $Z_t(x) \le L_t, t = 1,2,......,R$ and the constraints of (7)where $x = x_{ijk}^{p'}, i = 1,2,....,m; j = 1,2,....,n; k = 1,2,....,K; p = 1,2,...,l.$

Step 4: The linear membership function $\mu_t(\overline{Z_t})$ corresponding to t^{th} objective is calculated as

$$\mu_t(\overline{Z_t}) = \begin{cases} 1, if\, \overline{Z_t} \le L_t; \\ \dfrac{U_t - \overline{Z_t}}{U_t - L_t}, if\, L_t < \overline{Z_t} < U_t; \\ 0, if\, \overline{Z_t} \ge U_t, \forall t. \end{cases}$$

Step 5: The fuzzy linear programming model is expressed applying max-min operator as

Max δ subject to

$$\delta \le \mu_t\left(\overline{Z_t}\right) = \frac{U_t - \overline{Z_t}}{U_t - L_t}, \forall t \tag{24}$$

and the constraints of (7) $\delta \ge 0$ and $\delta = \min_t \{\mu_t(\overline{Z_t})\}$.

Step 6: The diminished model is worked out and the optimum solutions are obtained.

4.3.2 Numerical Model

The proposed problem is illustrated numerically in this section with hypothetical data. The proposed approach ability is solved numerically by taking one example of the model. Consider the model with sources $(i = 1,2,3)$, destinations $(j = 1,2,3)$, carriage $(k = 1,2,3,4)$ and components $(p = 1,2,3)$. Suppose that $J_1 = \{1,2\}, J_2 = \{1,2,3\}$, $J_3 = \{3\}, J_4 = \{1,2,3\}$. The supplies, demands and conveyance capacities are the following type-2 fuzzy data:

$a_1^1 = (20,22,24;0.5,0.5), a_1^2 = (24,26,28;0.9,0.5), a_1^3 = (22,24,26;0.5,0.5), a_2^1 = (24,26,28;0.7,0.5),$
$a_2^2 = (18,20,22;0.9,0.6), a_2^3 = (20,22,24;0.9,0.5), a_3^1 = (26,28,30;0.5,0.5), a_3^2 = (30,32,34;0.7,0.9),$
$a_3^3 = (22,24,26;0.7,0.5), b_1^1 = (10,11,13;0.5,0.5), b_2^1 = (14,15,16;0.9,0.9), b_3^1 = (16,17,19;0.9,0.9),$

$b_1^2 = (12,14,16;0.7,0.5), b_2^2 = (11,14,16;0.5,1), b_3^2 = (13,15,17;0.5,0.5), b_1^3 = (11,13,16;0.5,0.5),$
$b_2^3 = (10,12,14;0.5,0.5), b_3^3 = (12,14,16;0.5,1), e_1 = (33.35.38;0.9,0.9), e_2 = (45,47,49;0.9,0.5),$
$e_3 = (29,31,32;0.5,0.5), e_4 = (41,44,46;1,0.5), B_1^{11} = (115,135,165;0.5,0.5), B_2^{11} = (190,210,240;0.5,0.5),$
$B_3^{11} = (410,430,460;0.5,0.5), B_1^{21} = (300,320,350;0.5,0.5),$

Table 1. Best possible result of the model

Transportation Cost	Transportation Amount
$\underline{Z}_1 = 331.7422$	$x^1_{131} = 13.75, x^1_{211} = 11.55, x^1_{321} = 5.8, x^1_{322} = 8.6, x^2_{121} = 3.65, x^2_{114} = 15.2,$ $x^2_{134} = 6.75, x^2_{224} = 10.61, x^2_{234} = 5.65, x^3_{132} = 14.41, x^3_{312} = 8.2, x^3_{322} = 15.4,$ $x^3_{213} = 4.29, x^3_{114} = 5.06.$
$\overline{Z}_1 = 575.4114$	$x^1_{111} = 1.114472, x^1_{121} = 5.762385, x^1_{131} = 13.75, x^1_{211} = 10.43553,$ $x^1_{321} = 0.6255842, x^1_{322} = 8.012031, x^2_{131} = 3.062031, x^2_{222} = 1.766386,$ $x^2_{232} = 3.111583, x^2_{114} = 15.2, x^2_{124} = 3.987969, x^2_{224} = 8.505645,$ $x^2_{234} = 6.226386, x^3_{132} = 10.12, x^3_{312} = 14.41145, x^3_{322} = 9.188546,$ $x^3_{133} = 4.29,$ $x^3_{114} = 3.138546, x^3_{124} = 6.211454.$
$\underline{Z}_2 = 428.3$	$x^1_{111} = 7.78, x^1_{131} = 13.75, x^1_{211} = 3.7, x^1_{112} = 0.07, x^1_{322} = 14.4, x^2_{122} = 14.26,$ $x^2_{114} = 11.34, x^2_{214} = 3.86, x^2_{234} = 12.4, x^3_{322} = 3.47, x^3_{332} = 14.41,$ $x^3_{123} = 11.93,$ $x^3_{213} = 1.88, x^3_{114} = 15.67.$
$\overline{Z}_2 = 617.394$	$x^1_{131} = 13.75, x^1_{211} = 11.55, x^1_{122} = 7.85, x^1_{322} = 6.55, x^2_{122} = 14.26,$ $x^2_{114} = 11.34,$ $x^2_{134} = 6.75, x^2_{214} = 3.86, x^2_{234} = 12.4, x^3_{332} = 14.41, x^3_{213} = 17.28, x^3_{114} = 0.27,$ $x^3_{124} = 15.4.$
$\lambda = 0.3859705$ $Z_1 = [373.8916, 544.0562]$ $Z_2 = [490.9515, 688.1844]$	$x^1_{121} = 7.85, x^1_{131} = 13.75, x^1_{211} = 11.55, x^1_{322} = 6.55, x^2_{122} = 4.453625,$ $x^2_{232} = 1.829704, x^2_{114} = 15.2, x^2_{134} = 5.946375, x^2_{224} = 9.806375, x^3_{322} = 9.51,$ $x^2_{234} = 4.623921, x^3_{132} = 10.17667, x^3_{312} = 9.856671, x^3_{332} = 4.233329,$ $x^3_{123} = 5.89, x^3_{114} = 7.693329.$

$B_2^{21} = (390, 410, 440; 0.5, 0.5), B_3^{21} = (560, 580, 610; 0.5, 0.5)$.

4.3.2.1 Solution Using Nearest Interval Approximation

The credibilistic interval approximations of the triangular type-2 fuzzy parameters are calculated using (5) and (6). The transportation costs and fixed costs for this model are given in Tables 3-26. The supplies, demands, carriage capacities, and budget capacities are given as follows:

$a_1^1 = [21, 23]$, $a_1^2 = [25.0192, 26.9808]$, $a_1^3 = [23, 25]$, $a_2^1 = [25.0102, 29.9898]$, $a_2^2 = [19.0139, 20.9861]$, $a_2^3 = [21.0192, 22.9808]$, $a_3^1 = [27, 29]$, $a_3^2 = [30.991, 33.009]$, $a_3^3 = [23.0102, 24.9898]$, $b_1^1 = [10.5, 12]$, $b_1^2 = [13.0102, 14.9898]$, $b_1^3 = [12, 14.5]$, $b_2^1 = [14.5, 15.5]$, $b_2^2 = [12.4651, 15.0232]$, $b_2^3 = [11, 13]$, $b_3^1 = [16.5, 18]$, $b_3^2 = [14, 16]$, $b_3^3 = [12.9768, 15.0232]$, $e_1 = [34, 36.5]$, $e_2 = [46.0192, 47.9808]$, $e_3 = [30, 31.5]$, $e_4 = [42.5349, 44.9768]$, $B_1^{11} = [125, 150]$, $B_2^{11} = [200, 225]$, $B_3^{11} = [420, 445]$, $B_1^{21} = [310, 335]$, $B_2^{21} = [400, 425]$, $B_3^{21} = [570, 595]$

The corresponding deterministic forms of all the constraints are attained using (9)-(12) by taking $\alpha_i^{p'} = 0.7, \beta_j^{p'} = 0.7, \gamma_k^{p'} = 0.7, \delta_j^{p'} = 0.7, t = 1, 2$. We get minimum and maximum possible values of the objective function by solving (13)- (14) and the solutions are given in Table 1.

Here,
$L_{11} = 331.7422, U_{11} = 425.6453$, $L_{12} = 572.4114, U_{12} = 602.004$, $L_{21} = 428.3, U_{21} = 533.2985$, $L_{22} = 617.394$, and $U_{22} = 732.6823$ where L_{11} represents the value of the lower bound of the lower interval of the first objective function, L_{12} represents the value of the lower bound of the upper interval of the first objective function, U_{11} represents the value of the upper bound of the lower interval of the first objective function, U_{12} represents the value of the upper bound of the upper interval of first objective function, L_{21} represents the value of the lower bound of the lower interval of second objective function, L_{22} represents the value of the lower bound of second objective function, U_{21} represents the value of the upper bound of the lower interval of second objective function, and U_{22} represents the value of the upper interval of first objective function,

4.3.2.2 Solution Using Chance-Constrained Programming

The chance constrained programming model for this problem is formulated here. The fixed general credibility levels for objective function and constraints are reserved as $\alpha = 0.7, \alpha_i^p = 0.7$, $\beta_j^{p'} = 0.7, \gamma_k = 0.7, \delta_j^{p'} = 0.7, t = 1, 2, p' = 1, 2, 3, i = 1, 2, 3, j = 1, 2, 3, k = 1, 2, 3, 4$.

The corresponding deterministic form of the model using (18) is given below:

$$\text{Min} \sum_{p'=1}^{l}\sum_{i=1}^{m}\sum_{j=1}^{n}\sum_{k=1}^{K} d_{ijk}^{p'} \frac{0.4c_{ijk}^{tp'3}x_{ijk}^{p} + (0.6 + 0.2\theta_{l,ijk}^{tp'})c_{ijk}^{tp'2}x_{ijk}^{p}}{1 + 0.2\theta_{l,ijk}^{tp'}} + \frac{0.4e_{ijk}^{tp'3}y_{ijk}^{p} + (0.6 + 0.2\theta_{l,ijk}^{'tp'})e_{ijk}^{tp'2}y_{ijk}^{p'}}{1 + 0.2\theta_{l,ijk}^{'tp}}$$

subject to

$$\sum_{j=1}^{n}\sum_{k=1}^{K} d_{ijk}^{p'} x_{ijk}^{p'} \le F_{a_i^{p'}}, i = 1,2,...,m; p' = 1,2,...,l$$

$$\sum_{i=1}^{m}\sum_{k=1}^{K} d_{ijk}^{p'} x_{ijk}^{p'} \ge F_{b_j^{p'}}, j = 1,2,...,n; p' = 1,2,...,l,$$

$$\sum_{p'=1}^{l}\sum_{i=1}^{m}\sum_{j=1}^{n} d_{ijk}^{p'} x_{ijk}^{p'} \le F_{f_k}, k = 1,2,...,K\ ,$$

$$\sum_{i=1}^{m}\sum_{k=1}^{K} d_{ijk}^{p'} x_{ijk}^{p'} \le F_{B_j^{p'}}, j = 1,2,...,n; p' = 1,2,...,l,$$

$$\sum_{i=1}^{m}\sum_{k=1}^{K} d_{ijk}^{p'} (c_{ijk}^{tp'} x_{ijk}^{p'}) + e_{ijk}^{tp'} y_{ijk}^{p'} \le B_j^{tp'} \ge \delta_j^{p'}, j = 1,2,...,n; p' = 1,2,....l, t = 1,2,...R.$$

$$x_{ijk}^{p'} \ge 0$$

where $F_{a_i^{p'}}, F_{b_j^{p'}}, F_{f_k}, F_{B_j^{p'}}$ are calculated from (20), (21), (22) and (23) as follows:

$F_{a_1^1} = 21.27, F_{a_1^2} = 25.27, F_{a_1^3} = 23.27, F_{a_2^1} = 25.27, F_{a_2^2} = 19.29, F_{a_2^3} = 21.27, F_{a_3^1} = 27.27, F_{a_3^2} = 31.32,$
$F_{a_3^3} = 23.27, F_{b_1^1} = 11.73, F_{b_1^2} = 14.7, F_{b_1^3} = 14.09, F_{b_2^1} = 15.34, F_{b_2^2} = 14.73, F_{b_2^3} = 12.73, F_{b_3^1} = 17.68,$
$F_{b_3^2} = 15.73, F_{b_3^3} = 14.73, F_{f_1} = 34.32, F_{f_2} = 46.27, F_{f_3} = 30.27, F_{f_4} = 42.91, F_{B_1^{11}} = 127.73, F_{B_2^{11}} = 202.73,$
$F_{B_3^{11}} = 422.73, F_{B_1^{21}} = 312.73, F_{B_2^{21}} = 402.73, F_{B_3^{21}} = 572.73$.

Here, $L_1 = 487.9816, U_1 = 555.6238, L_2 = 607.9701, U_2 = 662.9047$ are the lower and upper bounds corresponding to the first and second objective functions respectively. The compromise optimal solution of (24) applying LINGO 16 solver, based upon GRG technique are given in the table no 2.

4.4 Discussion

The two solution methods are used to solve the model because the decision makers have the alternatives to get a best solution. It is hard to finish up which method is predominant, from the past knowledge or current circumstances a decision maker can decide any of the methods as he/she demand. The transported amount, solutions are different for the two methods because the crisp value is not the same after defuzzification of the type-2 triangular fuzzy variables. The first objective function value 511.2464, found using chance-constrained programming lies in the [373.8916,544.0562], the solution obtained using nearest

Table 2. Best possible result of the model

Transportation Cost	Transportation Amount
$\overline{Z}$ =487.9816	$x^1_{111} = 7.18$, $x^1_{131} = 14.09$, $x^1_{211} = 4.55$, $x^1_{321} = 8.5$, $x^1_{322} = 6.2$, $x^2_{122} = 2.07$, $x^2_{114} = 15.34$, $x^2_{134} = 7.86$, $x^2_{224} = 12.66$, $x^2_{234} = 4.87$, $x^3_{132} = 14.73$, $x^3_{312} = 7.54$, $x^3_{322} = 15.73$, $x^3_{213} = 7.96$, $x^3_{114} = 2.18$.
$\underline{Z}$ =607.9701	$x^1_{131} = 14.09$, $x^1_{211} = 11.73$, $x^1_{122} = 7.18$, $x^1_{322} = 7.52$, $x^2_{122} = 14.73$, $x^2_{114} = 10.54$, $x^2_{214} = 4.8$, $x^2_{234} = 12.73$, $x^3_{322} = 0.445$, $x^3_{332} = 14.73$, $x^3_{213} = 17.68$, $x^3_{124} = 14.84$.
Z_1 =511.2464, Z_2 =626.8643 δ =0.656061	$x^1_{111} = 7.18$, $x^1_{131} = 14.09$, $x^1_{211} = 4.55$, $x^1_{321} = 8.5$, $x^1_{322} = 6.2$, $x^2_{122} = 14.73$, $x^2_{114} = 10.54$, $x^2_{214} = 4.8$, $x^2_{234} = 12.73$, $x^3_{132} = 7.37262$, $x^3_{322} = 10.61$, $x^3_{332} = 7.35738$, $x^3_{213} = 7.96$, $x^3_{114} = 9.72$, $x^3_{124} = 5.12$.

interval approximation. The second objective function value 626.8643 found using chance-constrained programming lies in the [490.9515,688.1844], the solution obtained using nearest interval approximation.

5. FUTURE RESEARCH DIRECTIONS

The presented model is quite general and can also be solved by taking type-2 normal fuzzy variable, type-2 gamma fuzzy variable, etc. The model introduced in this chapter can be extended considering various constraints namely space contraints, transportation time, and deterioration of items during transportation, etc. The breakability constrainst can also be considered in the solid transportation model if some breakable items (e.g. glass, ceremic etc.) are transported. The safety factor (especially in the case of bad road conditions, landslide etc.) of the goods is another important issue in the solid transportation system. Another model can be developed by considering the safety factor (another constraint) of the goods transported from the origins to the destinations through some conveyances. The future extension of the model where all the variables are uncertain. That is why, Soft Computing is the best method to solve these cases of the uncertainites in the future model.

6. CONCLUSION

The authors have solved a multi-item solid fixed charged transportation model in budget constraints with type-2 triangular fuzzy variables for the first time ever here. A nearest interval approximation approach and chance-constrained programming are used to solve the model applying LINGO 16 solver. In many practical applications like Oil Company, Paper mill industry etc., it is genuine to suppose that the amount that can be sent on any specific road accepts a fixed charge for that road.

REFERENCES

Baidya, A., Bera, U. K., & Maiti, M. (2013). Solution of multi-item interval valued solid transportation problem with safety measure using different methods. *Opsearch*, 1-22.

Balinski, M. L. (1961). Fixed-cost transportation problems. *Naval Res. Logistics Q.*, *8*(1), 41–54. doi:10.1002/nav.3800080104

Bit, A. K., Biswal, M. P., & Alam, S. S. (1993). Fuzzy programming approach to multiobjective solid transportation problem. *Fuzzy Sets and Systems*, *57*(2), 183–194. doi:10.1016/0165-0114(93)90158-E

Chen, S. M., & Wang, C. Y. (2013). Fuzzy decision making systems based on interval type-2 fuzzy sets. *Inf. Sci.*, *242*, 1–21. doi:10.1016/j.ins.2013.04.005

Choi, B., & Rhee, F. (2009). Interval type-2 fuzzy membership function generation methods for pattern recognition. *Inform. Sci.*, *179*(13), 2102–2122. doi:10.1016/j.ins.2008.04.009

Das, A., Bera, U. K., & Manoranjan, M. (2016). A breakable multi-item multi stage solid transportation problem under budget with Gaussian type-2 fuzzy parameters. *Applied Intelligence*, *45*(3), 923–951. doi:10.1007/s10489-016-0794-y

Dereli, T., Baykasoglu, A., Altun, K., Durmusoglu, A., & Türksen, I. B. (2011). Industrial applications of type-2 fuzzy sets and systems: A concise review. *Computers in Industry*, *62*(2), 125–137. doi:10.1016/j.compind.2010.10.006

Dutta, A., & Jana, D. K. (2017). Expectations of the reductions for type-2 trapezoidal fuzzy variables and its application to a multi-objective solid transportation problem via goal programming technique. *Journal of Uncertainty Analysis and Applications*, *5*(3), 1–21.

Dutta, D., & Sen, M. (2017). A new method for solving multi-item multi-objective solid fixed charged shipment model with type-2 fuzzy variables. *Global and Stochastic Analysis*, *4*(1), 35–47.

Gen, M., Ida, K., Li, Y., & Kubota, E. (1995). Solving bicriteria solid transportation problem with fuzzy numbers by a genetic algorithm. *Computers & Industrial Engineering*, *29*(1-4), 537–541. doi:10.1016/0360-8352(95)00130-S

Gilan, S. S., Sebt, M. H., & Shahhosseini, V. (2012). Computing with words for hierarchical competency based selection of personal in construction companies. *Applied Soft Computing*, *12*(2), 860–871. doi:10.1016/j.asoc.2011.10.004

Hitchcock, F. L. (1941). The distribution of product from several sources to numerous localities. *Journal of Mathematical Physics*, *20*(1-4), 224–230. doi:10.1002/sapm1941201224

Jimenez, F., & Verdegay, J. L. (1998). Uncertain solid transportation problems. *Fuzzy Sets and Systems*, *100*(1-3), 45–57. doi:10.1016/S0165-0114(97)00164-4

Karnik, N. N., & Mendel, J. M. (1999). Applications of type-2 fuzzy logic systems to forecasting of time-series. *Inform. Sci*, *120*(1-4), 89–111. doi:10.1016/S0020-0255(99)00067-5

Kundu, P., Kar, S., & Maiti, M. (2012). Multi-objective multi-item solid transportation problem in fuzzy environment. *Applied Mathematical Modelling*, *37*(4), 2028–2038. doi:10.1016/j.apm.2012.04.026

Kundu, P., Kar, S., & Maiti, M. (2014). Multi-objective solid transportation problems with budget constraint in uncertain environment. *International Journal of Systems Science*, *45*(8), 1668–1682. doi:10.1080/00207721.2012.748944

Kundu, P., Kar, S., & Maiti, M. (2014). Fixed charge transportation problem with type-2 fuzzy variables. *Inf. Sci.*, *255*, 170–186. doi:10.1016/j.ins.2013.08.005

Kundu, P., Kar, S., & Maiti, M. (2015). Multi-item solid transportation problem with type-2 fuzzy parameters. *Applied Soft Computing*, *31*, 61–80. doi:10.1016/j.asoc.2015.02.007

Li, Y., Ida, K., & Gen, M. (1997). Improved genetic algorithm for solving multiobjective solid transportation problem with fuzzy numbers. *Computers & Industrial Engineering*, *33*(3-4), 589–592. doi:10.1016/S0360-8352(97)00199-X

Li, Y., Ida, K., Gen, M., & Kobuchi, R. (1997). Neural network approach for multicriteria solid transportation problem. *Computers & Industrial Engineering*, *33*(3-4), 465–468. doi:10.1016/S0360-8352(97)00169-1

Liu, B. (2009). *Theory and Practice of Uncertain Programming* (3rd ed.). UTLAB; http://orsc.edu.cn/liu/up.pdf doi:10.1007/978-3-540-89484-1

Liu, B., & Iwamura, K. (1998). Chance constrained programming with fuzzy parameters. *Fuzzy Sets and Systems*, *94*(2), 227–237. doi:10.1016/S0165-0114(96)00236-9

Liu, B., & Liu, Y. K. (2002). Expected value of fuzzy variable and fuzzy expected value models. *IEEE Transactions on Fuzzy Systems*, *10*(4), 445–450. doi:10.1109/TFUZZ.2002.800692

Liu, S. (2006). Fuzzy total transportation cost measures for fuzzy solid transportation problem. *Applied Mathematics and Computation*, *174*(2), 927–941. doi:10.1016/j.amc.2005.05.018

Liu, Z. Q., & Liu, Y. K. (2010). Type-2 fuzzy variables and their arithmetic. *Soft Computing*, *14*(7), 729–747. doi:10.1007/s00500-009-0461-x

Mendel, J. M. (2007). Computing with words and its relationship with fuzzistics. *Inform. Sci.*, *177*(4), 988–1006. doi:10.1016/j.ins.2006.06.008

Nagarjan, A., & Jeyaraman, K. (2010). Solution of chance constrained programming problem for multi-objective interval solid transportation problem under stochastic environment using fuzzy approach. *International Journal of Computers and Applications*, *10*(9), 19–29. doi:10.5120/1510-1690

Nahmias, S. (1978). Fuzzy variable. *Fuzzy Sets and Systems*, *1*(2), 97–101. doi:10.1016/0165-0114(78)90011-8

Ojha, A., Das, B., Mondal, S., & Maiti, M. (2009). An entropy based solid transportation problem for general fuzzy costs and time with fuzzy equality. *Mathematical and Computer Modelling*, *50*(1-2), 166–178. doi:10.1016/j.mcm.2009.04.010

Pandian, P., & Anuradha, D. (2010). A new approach for solving solid transportation problems. *Appl. Math. Sci.*, *4*, 3603–3610.

Qin, R., Liu, Y. K., & Liu, Z. Q. (2011). Methods of critical value reduction for type-2 fuzzy variables and their applications. *Journal of Computational and Applied Mathematics*, *235*(5), 1454–1481. doi:10.1016/j.cam.2010.08.031

Ramirez, C. (2011). Simulation of the bird age structured population growth based on an interval type-2 fuzzy cellular structure. *Inform. Sci*, *181*(3), 519–535. doi:10.1016/j.ins.2010.10.011

Schell, E. D. (1955). Distribution of a product by several properties. *Proceedings 2nd Symposium in Linear Programming*. DCS/Comptroller, HQ US Air Force.

Wu, D., & Tan, W. (2006). A simplified type-2 fuzzy logic controller for real-time control. *ISA Transactions*, *45*(4), 503–516. doi:10.1016/S0019-0578(07)60228-6 PMID:17063934

Yang, L., & Feng, Y. (2007). A bicriteria solid transportation problem with fixed charge under stochastic environment. *Applied Mathematical Modelling*, *31*(12), 2668–2683. doi:10.1016/j.apm.2006.10.011

Yang, L., & Liu, L. (2007). Fuzzy fixed charge solid transportation problem and algorithm. *Applied Soft Computing*, *7*(3), 879–889. doi:10.1016/j.asoc.2005.11.011

Zadeh, L. A. (1965). Fuzzy Sets. *Information and Control*, *8*(3), 338–353. doi:10.1016/S0019-9958(65)90241-X

Zadeh, L. A. (1975). The Concept of a Linguistic Variable and its Application to Approximate Reasoning-I. *Information Sciences*, *8*(3), 199–249. doi:10.1016/0020-0255(75)90036-5

Zavardehi, S. (2013). Solving a fuzzy fixed charge solid transportation problem by metaheuristics. *Mathematical and Computer Modelling*, *57*(5-6), 1543–1558. doi:10.1016/j.mcm.2012.12.031

Zhang, Q., Fan, Z., & Pan, D. (1999). A ranking approach for interval numbers in uncertain multiple attribute decision making problems. *Syst. Eng. Theory Pract*, *5*, 129–133.

Zimmermann, H.-J. (1978). Fuzzy programming and linear programming with several objective functions. *Fuzzy Sets and Systems*, *1*(1), 45–55. doi:10.1016/0165-0114(78)90031-3

APPENDIX

Table 3. c_{ijk}^{11}

i/j	1	2	3	k
1	(2,3,4;0.7,1)	(3,4,5;0.6,0.5)	(2,3,4;0.6,0.5)	
2	(1,3,4;0.6,1)	(4,5,6;0.8,0.6)	(3,5,6;0.5,0.6)	1
3	(3,4,5;0.8,0.5)	(2,4,5;0.6,0.5)	(3,5,6;0.8,1)	
1	(3,5,6;0.6,1)	(5,6,7;0.6,0.4)	(5,8,9;0.8,0.8)	
2	(7,9,10;0.3,0.9)	(2,6,7;0.6,0.3)	(3,7,8;0.8,0.7)	2
3	(1,3,4;0.3,0.8)	(1,3,4;0.8,0.2)	(2,4,5;0.9,0.6)	
1	(1,4,5;0.4,0.7)	(2,4,5;0.2,0.1)	(3,6,7;0.6,0.5)	
2	(2,4,5;0.9,0.6)	(2,6,7;0.3,1)	(3,4,5;0.9,0.4)	4
3	(3,6,7;0.5,0.5)	(3,6,7;0.6,0.9)	(5,8,9;0.9,0.3)	

Table 4. c_{ijk}^{12}

i/j	1	2	3	k
1	(4,5,7;0.6,0.9)	(3,4,8;0.7,0.9)	(5,6,8;0.6,1)	
2	(3,4,6;0.6,0.5)	(6,7,8;0.8,1)	(6,7,9;0.9,0.5)	1
3	(6,7,9;0.9,0.5)	(7,8,10;0.6,0.5)	(8,9,11;0.6,0.5)	
1	(4,5,7;0.2,0.5)	(2,3,4;0.8,0.5)	(4,5,6;0.6,0.5)	
2	(5,6,7;0.6,1)	(3,4,6;0.6,0.3)	(3,4,5;0.3,0.6)	2
3	(8,9,10;1,0.5)	(6,7,11;0.6,0.3)	(6,7,12;0.9,0.5)	
1	(1,2,3;0.6,0.5)	(3,4,5;0.6,0.8)	(1,2,3;0.8,0.5)	
2	(2,3,4;0.6,0.3)	(2,3,5;0.9,0.5)	(1,2,4;0.6,0.3)	4
3	(7,8,9;0.9,1)	(3,4,6;0.8,0.7)	(5,6,8;1,0.5)	

Table 5. c_{ijk}^{13}

i/j	1	2	3	k
1	(3,4,5;0.6,0.3)	(3,4,5;0.8,0.4)	(1,2,3;0.8,0.4)	
2	(5,6,7;0.6,0.3)	(4,6,7;0.9,0.3)	(5,6,8;0.9,0.6)	1
3	(1,2,3;0.9,1)	(1,2,3;0.9,0.5)	(2,3,4;0.9,0.6)	
1	(9,10,11;0.9,0.5)	(4,5,7;0.9,1)	(5,6,8;0.9,1)	
2	(3,4,5;0.9,0.5)	(9,10,12;0.8,0.4)	(8,9,10;0.6,0.3)	2
3	(9,10,11;0.5,1)	(10,11,12;0.4,0.8)	(11,12,13;0.3,0.6)	
1	(2,3,4;0.8,0.4)	(3,4,5;0.8,0.4)	(4,5,6;0.7,0.5)	
2	(10,11,12;0.6,0.3)	(11,12,13;0.6,0.7)	(7,13,14;0.9,0.5)	4
3	(6,7,8;1,0.5)	(7,8,9;0.9,1)	(8,9,10;0.8,0.4)	

Table 6. e_{ijk}^{11}

i/j	1	2	3	k
1	(3,4,5;0.9,1)	(2,4,5;0.5,0.7)	(5,6,7;0.6,0.7)	
2	(2,3,4;0.9,0.5)	(1,2,4;0.9,0.5)	(4,5,9;0.8,0.9)	1
3	(2,4,5;0.6,0.7)	(3,5,6;1,0.5)	(4,6,7,0.9,0.5)	
1	(3,4,6;0.5,1)	(4,5,6;0.9,0.5)	(5,6,7;0.4,0.3)	
2	(6,7,8;0.6,1)	(7,8,9;0.6,1)	(8,9,10;0.8,0.5)	2
3	(9,10,11;0.6,0.5)	(10,11,12;0.7,0.3)	(11,12,13;0.9,1)	
1	(6,7,9;0.9,1)	(7,8,10;0.6,0.9)	(8,9,11;0.6,0.9)	
2	(9,10,12;0.5,0.9)	(10,11,13;0.7,0.8)	(11,12,14;0.9,1)	4
3	(3,4,6;0.7,1)	(4,5,7;0.6,0.9)	(5,6,8;0.5,0.9)	

Table 7. e_{ijk}^{12}

i/j	1	2	3	k
1	(3,4,5;0.9,0.5)	(4,5,6;0.9,1)	(5,6,7;0.9,0.5)	
2	(6,7,8;0.7,0.5)	(7,8,9;0.8,0.5)	(8,9,10;0.8,1)	1
3	(9,10,11;0.8,0.5)	(10,11,12;0.7,0.5)	(11,12,13;0.7,1)	
1	(5,6,7;0.9,0.5)	(6,7,8;0.9,1)	(7,8,9;0.9,0.5)	
2	(8,9,10;0.8,0.5)	(9,10,11;0.8,1)	(10,11,12;0.7,0.5)	2
3	(11,12,13;0.6,1)	(12,13,14;0.6,0.5)	(13,14,15;0.7,0.5)	
1	(3,4,5;0.8,1)	(5,6,7;0.7,0.5)	(6,7,8;0.6,0.5)	
2	(2,3,4;0.7,0.5)	(3,4,5;0.8,0.5)	(4,5,6;0.9,1)	4
3	(5,6,7;0.9,1)	(6,7,8;0.6,0.5)	(7,8,9;0.9,0.5)	

Table 8. e_{ijk}^{13}

i/j	1	2	3	k
1	(1,2,3;0.7,1)	(1,2,3;0.9,0.5)	(2,3,4;1,0.5)	
2	(3,4,5;0.9,0.5)	(4,5,6;0.8,0.5)	(5,6,7;0.9,0.5)	1
3	(6,7,8;0.6,0.5)	(7,8,9;0.7,0.5)	(8,9,10;0.8,0.5)	
1	(3,4,5;0.6,0.5)	(4,5,6;0.6,1)	(5,6,7;0.6,1)	
2	(8,9,10;0.7,0.9)	(9,10,11;0.8,0.5)	(10,11,12;0.6,1)	2
3	(13,14,15;0.9,0.5)	(14,15,16;0.8,1)	(15,16,17;1,0.5)	
1	(3,4,5;0.6,1)	(4,5,6;0.9,0.5)	(5,6,7;0.9,0.5)	
2	(6,7,8;0.6,0.9)	(7,8,9;0.9,0.5)	(8,9,10;1,0.5)	4
3	(9,10,11;0.8,1)	(10,11,12;0.8,0.5)	(11,12,13;0.8,1)	

Table 9. c_{ijk}^{21}

i/j	1	2	3	k
1	(1,3,4;0.7,0.5)	(3,4,5;0.6,1)	(2,3,5;0.6,0.3)	
2	(2,3,4;0.6,0.3)	(4,5,6;0.8,0.5)	(3,5,6;0.5,0.5)	1
3	(3,4,5;0.8,0.4)	(3,4,5;0.6,0.9)	(4,5,6;0.8,0.2)	
1	(1,3,4;0.6,0.9)	(2,3,4;0.6,0.9)	(3,4,5;0.8,0.4)	
2	(2,4,11;0.3,0.7)	(3,4,9;0.6,0.9)	(3,6,10;0.8,0.4)	2
3	(2,3,5;0.3,0.9)	(2,3,5;0.8,0.5)	(3,4,6;0.9,0.1)	
1	(9,10,12;0.4,0.8)	(9,10,11;0.2,0.9)	(10,12,13;0.6,0.4)	
2	(8,10,12;0.9,0.5)	(11,12,13;0.3,0.6)	(9,12,14;0.9,0.1)	4
3	(11,12,13;.0.5,1)	(10,12,13;0.6,1)	(12,14,15;0.9,1)	

Table 10. c_{ijk}^{22}

i/j	1	2	3	k
1	(5,6,8;0.6,0.4)	(4,5,9;0.7,0.5)	(6,7,9;0.8,0.8)	
2	(4,5,7;0.6,0.4)	(7,8,9;0.8,0.6)	(7,8,10;0.9,0.9)	1
3	(7,8,10;0.9,0.4)	(8,9,11;0.6,0.7)	(9,10,12;0.6,1)	
1	(5,6,8;0.2,0.5)	(1,2,4;0.8,0.9)	(5,6,7;0.6,0.2)	
2	(6,7,8;0.6,0.6)	(4,5,7;0.6,0.8)	(4,5,6;0.3,0.5)	2
3	(9,10,11;1,0.7)	(7,8,12;0.8,1)	(7,8,13;0.9,0.1)	
1	(1,2,4;0.6,0.5)	(4,5,6;0.6,0.5)	(1,3,4;0.8,0.6)	
2	(3,4,5;0.6,0.9)	(3,4,6;0.9,0.7)	(2,3,5;0.6,1)	4
3	(8,9,10;0.9,1)	(3,5,11;0.8,1)	(6,7,9;0.8,1)	

Table 11. c_{ijk}^{23}

i/j	1	2	3	k
1	(8,9,11;0.6,1)	(7,8,9;0.8,0.5)	(5,6,7;0.8,1)	
2	(7,9,12;0.6,0.5)	(10,11,13;0.9,0.5)	(6,7,9;0.9,0.4)	1
3	(4,5,7;0.9,0.1)	(5,6,8;0.9,0.5)	(4,5,7;0.9,0.4)	
1	(10,11,12;0.9,0.1)	(4,6,8;0.9,0.5)	(4,7,9;0.9,0.5)	
2	(4,5,6;0.9,0.5)	(10,11,13;0.8,0.4)	(9,10,11;0.6,1)	2
3	(10,11,12;0.5,0.7)	(11,12,13;0.4,0.1)	(12,13,14;0.3,0.9)	
1	(2,4,5;0.6,0.8)	(3,5,6;0.6,0.2)	(3,6,7;0.5,0.8)	
2	(11,12,13;0.4,0.9)	(12,13,14;0.4,0.3)	(11,14,15;0.7,0.3)	4
3	(5,8,9;1,1)	(4,9,10;0.9,0.4)	(4,10,11;0.8,0.4)	

Table 12. e_{ijk}^{21}

i/j	1	2	3	k
1	(1,2,3;0.9,0.6)	(1,2,3;0.5,0.5)	(3,4,5;0.6,0.4)	
2	(2,3,4;0.9,0.7)	(2,3,4;0.9,0.6)	(2,3,4;0.8,0.5)	1
3	(1,2,3;0.6,0.8)	(1,3,4;1,0.7)	(2,4,5;0.9,0.6)	
1	(1,2,4;0.5,0.9)	(1,3,4;0.9,0.8)	(3,4,5;0.4,0.7)	
2	(4,5,6;0.6,0.1)	(5,6,7;0.6,0.9)	(6,7,8;0.8,0.8)	2
3	(7,8,9;0.6,0.1)	(8,9,10;0.7,1)	(9,10,11;0.9,0.9)	
1	(4,5,7;0.9,0.2)	(5,6,8;0.6,0.1)	(5,7,9;0.6,1)	
2	(7,8,10;0.5,0.3)	(7,9,11;0.7,0.2)	(9,10,12;0.9,0.1)	4
3	(1,2,4;0.7,0.4)	(2,3,5;0.6,0.3)	(3,4,6;0.5,0.2)	

Table 13. e_{ijk}^{22}

i/j	1	2	3	k
1	(1,2,3;0.9,0.1)	(2,3,4;0.9,1)	(3,4,5;0.9,0.9)	
2	(4,5,6;0.7,0.2)	(5,6,7;0.8,0.1)	(6,7,8;0.8,1)	1
3	(7,8,9;0.8,0.3)	(8,9,10;0.7,0.2)	(9,10,11;0.7,0.1)	
1	(5,6,7;0.9,0.4)	(6,7,8;0.9,0.3)	(7,8,9;0.9,0.2)	
2	(8,9,10;0.8,0.5)	(9,10,11;0.6,0.4)	(10,11,12;0.7,0.3)	2
3	(11,12,13;0.6,0.6)	(12,13,14;0.6,0.5)	(13,14,15;0.7,0.4)	
1	(1,2,3;0.8,0.7)	(3,4,5;0.7,0.6)	(4,5,6;0.6,0.5)	
2	(1,2,3;0.7,0.8)	(1,2,3;0.8,0.7)	(2,3,4;0.7,0.6)	4
3	(2,4,5;0.9,0.9)	(4,5,6;0.6,0.8)	(5,6,7;0.9,0.7)	

Table 14. e_{ijk}^{23}

i/j	1	2	3	k
1	(1,2,3;0.7,0.8)	(1,2,3;0.9,0.7)	(1,2,3;1,0.6)	
2	(1,2,3;0.9,0.9)	(2,3,4;0.8,0.8)	(2,4,5;0.9,0.7)	1
3	(4,5,6;0.6,1)	(4,6,7;0.7,0.9)	(3,7,8;0.8,0.8)	
1	(1,3,4;0.6,0.1)	(2,4,5;0.6,1)	(2,5,6;0.6,0.9)	
2	(6,7,8;0.7,0.2)	(5,8,9;0.8,0.1)	(4,9,10;0.6,1)	2
3	(11,12,13;0.9,0.3)	(9,13,14;0.8,0.2)	(11,14,15;0.1,0.1)	
1	(1,2,3;0.6,0.4)	(2,3,4;1,0.3)	(1,4,5;0.9,0.2)	
2	(4,5,6;0.6,0.5)	(3,6,7;0.9,0.4)	(3,7,8;1,0.3)	4
3	(7,8,9;0.7,0.6)	(5,9,10;0.7,0.5)	(5,10,11;0.7,0.4)	

Table 15. c_{ijk}^{11}

i/j	1	2	3	k
1	[2.4935,3.5065]	[3.5026,4.4974]	[2.5026,3.4974]	
2	[1.982,3.509]	[4.5048,5.4952]	[3.9947,5.5026]	1
3	[3.5074,4.4926]	[3.0053,4.4974]	[3.9916,5.5042]	
1	[3.982,5.509]	[5.5055,6.4945]	[6.5,8.5]	
2	[7.969,9.5155]	[4.0341,6.4915]	[5.0092,7.4977]	2
3	[1.9734,3.5133]	[2.0332,3.4834]	[3.0139,4.4931]	
1	[2.4762,4.5079]	[3.0072,4.4964]	[4.5079,6.4974]	
2	[3.0139,4.4931]	[3.9299,6.5175]	[3.5124,4.4876]	4
3	[4.5,6.5]	[4.4792,6.5069]	[6.5484,8.4845]	

Table 16. c_{ijk}^{12}

i/j	1	2	3	k
1	[4.4931,6.0139]	[3.4955,6.0179]	[5.491,7.018]	
2	[3.5026,4.9947]	[6.4958,7.5042]	[6.5096,7.9808]	1
3	[6.5096,7.9808]	[7.5026,8.9947]	[8.5026,9.9947]	
1	[4.4908,6.0184]	[3.4955,6.0179]	[5.491,7.018]	
2	[5.491,6.5059]	[3.5085,4.9829]	[3.4915,4.5085]	2
3	[8.5116,9.4884]	[6.5085,8.9659]	[6.5096,9.4521]	
1	[1.5026,2.4974]	[3.4952,4.5048]	[1.5074,2.4926]	
2	[2.5085,3.4915]	[2.5096,3.9808]	[1.5085,2.9829]	4
3	[7.498,8.502]	[3.5023,4.9954]	[5.5116,6.9768]	

Table 17. c_{ijk}^{13}

i/j	1	2	3	k
1	[3.5085,4.4915]	[3.5102,4.4898]	[1.5102,2.4898]	
2	[5.5085,6.4915]	[5.031,6.4845]	[5.5069,6.9861]	1
3	[1.498,2.502]	[1.5096,2.4904]	[2.5069,3.4931]	
1	[9.5096,10.4904]	[4.498,6.0041]	[5.498,7.0041]	
2	[3.5096,4.4904]	[9.5102,10.9795]	[8.5085,9.4915]	2
3	[9.4884,10.5116]	[10.4898,11.5102]	[11.4915,12.5085]	
1	[2.5102,3.4898]	[3.5102,4.4898]	[4.5051,5.4949]	
2	[10.5085,11.4915]	[11.4975,12.5025]	[10.0575,13.4904]	4
3	[6.5116,7.4884]	[7.498,8.502]	[8.5102,9.4898]	

Table 18. e_{ijk}^{11}

i/j	1	2	3	k
1	[3.498,4.502]	[2.9898,4.5051]	[5.4975,6.5025]	
2	[2.5096,3.4904]	[1.5096,2.9808]	[4.4978,7.0087]	1
3	[2.9951,4.5025]	[4.0232,5.4884]	[5.0192,6.4904]	
1	[3.4884,5.0232]	[4.5096,5.4904]	[5.5031,6.4969]	
2	[6.491,7.509]	[7.491,8.509]	[8.5074,9.4926]	2
3	[9.5026,10.4974]	[10.511,11.489]	[11.498,12.502]	
1	[6.498,8.0041]	[7.4931,9.0139]	[8.4931,10.0139]	
2	[9.4904,11.0192]	[10.4977,12.0046]	[11.498,13.0041]	4
3	[3.4935,5.013]	[4.4931,6.0139]	[5.4904,7.0192]	

Table 19. e_{ijk}^{12}

i/j	1	2	3	k
1	[3.5096,4.4904]	[4.498,5.502]	[5.5096,6.4904]	
2	[6.5051,7.4949]	[7.5074,8.4926]	[8.4958,9.5042]	1
3	[9.5074,10.4926]	[10.5051,11.4949]	[11.4935,12.5065]	
1	[5.5096,6.4904]	[6.498,7.502]	[7.5096,8.4904]	
2	[8.5074,9.4926]	[9.491,10.509]	[10.5051,11.4949]	2
3	[11.491,12.509]	[12.5026,13.4974]	[13.5051,14.4949]	
1	[3.4958,4.5042]	[5.5051,6.4949]	[6.5026,7.4974]	
2	[2.5051,3.4949]	[3.5074,4.4926]	[4.498,5.502]	4
3	[5.498,6.502]	[6.5026,7.4974]	[7.5096,8.4904]	

Table 20. e_{ijk}^{13}

i/j	1	2	3	k
1	[1.4935,2.5065]	[1.5096,2.4904]	[2.5116,3.4884]	
2	[3.5096,4.4904]	[4.5074,5.4926]	[5.5096,6.4904]	1
3	[6.5026,7.4974]	[7.5051,8.4949]	[8.5074,9.4926]	
1	[3.5026,4.4974]	[4.491,5.509]	[5.491,6.509]	
2	[8.4955,9.5045]	[9.5074,10.4926]	[10.491,11.509]	2
3	[13.5096,14.4904]	[14.4958,15.5042]	[15.5116,16.4884]	
1	[3.491,4.509]	[4.5096,5.4904]	[5.5096,6.4904]	
2	[6.4931,7.5069]	[7.5096,8.4904]	[8.5116,9.4884]	4
3	[9.4958,10.5042]	[10.5074,11.4926]	[11.4958,12.5042]	

Table 21. c_{ijk}^{21}

i/j	1	2	3	k
1	[2.0102,3.4949]	[3.491,4.509]	[2.5085,3.9829]	
2	[2.5085,3.4915]	[4.5074,5.4926]	[4,5.5]	1
3	[3.5102,4.4898]	[3.4931,4.5069]	[4.5166,5.4834]	
1	[1.9861,3.5069]	[2.4931,3.5069]	[3.5102,4.4898]	
2	[2.978,7.577]	[3.4931,6.5347]	[4.5307,7.959]	2
3	[2.4845,4.031]	[2.5074,3.9852]	[3.5224,4.9552]	
1	[9.4898,11.0205]	[9.4812,10.5188]	[11.0109,12.4945]	
2	[9.0192,10.9808]	[11.4915,12.5085]	[10.5672,12.9552]	4
3	[11.4884,12.5116]	[10.982,12.509]	[13.0448,14.4776]	

Table 22. c_{ijk}^{22}

i/j	1	2	3	k
1	[5.5055,6.9891]	[4.5051,6.9796]	[6.5,8]	
2	[4.5055,5.9891]	[7.5048,8.4952]	[7.5,9]	1
3	[7.5124,8.9752]	[8.4975,10.0049]	[9.491,11.018]	
1	[5.4908,7.0184]	[1.4978,3.0043]	[5.5119,6.4881]	
2	[6.5,7.5]	[4.4952,6.0095]	[4.4941,5.5059]	2
3	[9.5065,10.4935]	[7.4958,10.0169]	[7.5224,10.3879]	
1	[1.5026,2.9947]	[4.5026,5.4974]	[2.0095,3.4952]	
2	[3.4931,4.5069]	[3.5045,4.991]	[2.491,4.018]	4
3	[8.498,9.502]	[3.9916,8.0253]	[6.4958,8.0084]	

Table 23. c_{ijk}^{23}

i/j	1	2	3	k
1	[8.491,10.018]	[7.5074,8.4926]	[5.4958,6.5042]	
2	[8.0053,10.4921]	[10.5096,11.9808]	[6.5124,7.9752]	1
3	[4.5224,5.9552]	[5.5096,6.9808]	[4.5124,5.9752]	
1	[10.5224,11.4776]	[5.0192,6.9808]	[5.5287,7.9808]	
2	[4.5096,5.4904]	[10.5102,11.9795]	[9.491,10.509]	2
3	[10.4949,11.5051]	[11.51,12.49]	[12.4845,13.5155]	
1	[2.9905,4.5048]	[4.0237,5.4881]	[4.4778,6.5074]	
2	[11.4876,12.5124]	[12.5031,13.4969]	[12.533,14.489]	4
3	[6.5,8.5]	[6.5621,9.4876]	[7.0615,10.4898]	

Table 24 e_{ijk}^{21}

i/j	1	2	3	k
1	[1.5069,2.4931]	[1.5,2.5]	[3.5055,4.4945]	
2	[2.5045,3.4955]	[2.5069,3.4931]	[2.5074,3.4926]	1
3	[1.4952,2.5048]	[2.013,3.4935]	[3.0139,4.4931]	
1	[1.4904,3.0192]	[2.0043,3.4978]	[3.4921,4.5079]	
2	[4.5155,5.4845]	[5.4931,6.5069]	[6.5,7.5]	2
3	[7.5155,8.4845]	[8.4935,9.5065]	[9.5,10.5]	
1	[4.5188,5.9624]	[5.5155,6.9691]	[5.982,8.018]	
2	[7.5059,8.9882]	[8.0286,9.9714]	[9.5224,10.9552]	4
3	[1.5079,2.9841]	[2.5085,3.9829]	[3.5092,4.9816]	

Table 25. e_{ijk}^{22}

i/j	1	2	3	k
1	[1.5224,2.4776]	[2.498,3.502]	[3.5,4.5]	
2	[4.5143,5.4857]	[5.5202,6.4798]	[6.4958,7.5042]	1
3	[7.5133,8.4867]	[8.5143,9.4857]	[9.5179,10.4821]	
1	[5.5124,6.4876]	[6.5155,7.4845]	[7.5188,8.4812]	
2	[8.5074,9.4926]	[9.5055,10.4945]	[10.511,11.489]	2
3	[11.5,12.5]	[12.5026,13.4974]	[13.5079,14.4921]	
1	[1.5023,2.4977]	[3.5025,4.4975]	[4.5026,5.4974]	
2	[1.4977,2.5023]	[1.5023,2.4977]	[2.5025,3.4975]	4
3	[3,4.5]	[4.4952,5.5048]	[5.5045,6.4955]	

Table 26. e_{ijk}^{23}

i/j	1	2	3	k
1	[1.4977,2.5023]	[1.5045,2.4955]	[1.509,2.491]	
2	[1.5,2.5]	[2.5,3.5]	[3.009,4.4955]	1
3	[4.491,5.509]	[4.991,6.5045]	[5,7.5]	
1	[2.0309,3.4845]	[2.982,4.509]	[3.4792,5.5069]	
2	[6.5143,7.4857]	[6.5607,8.4798]	[6.4551,9.509]	2
3	[11.5155,12.4845]	[11.0665,13.4834]	[12.5734,14.4755]	
1	[1.5055,2.4945]	[2.5175,3.4825]	[2.5564,4.4812]	
2	[4.5026,5.4974]	[4.5372,6.4876]	[5.0701,7.4825]	4
3	[7.5025,8.4975]	[7.0204,9.4949]	[7.5397,10.4921]	

Chapter 3
Identification of Optimal Process Parameters in Electro-Discharge Machining Using ANN and PSO

Kaushik Kumar
Birla Institute of Technology, India

Paulo J. Davim
University of Aveiro, Portugal

ABSTRACT

Electrical Discharge Machining (EDM) process is a widely used machining process in several fabrication, construction and repair work applications. Considering Pulse-On Time, Pulse OFF time, Peak-Current and Gap voltage as the inputs and among all possible outputs, in the present work Material Removal Rate and Surface Roughness are considered as outputs. In order to reduce the number of experiments Design of Experiments (DOE) was undertaken using Orthogonal Array and later on the outputs were optimized using ANN and PSO. It was found that the results obtained from both the techniques were tallying with each other.

INTRODUCTION

Electrical Discharge Machining (EDM), in line with a book composed by Elman C. Jameson (Jameson 2001), happens to be non-conventional machining technique used for making the machined surface with the aid of electrical energy. The foremost vital advantage of victimising this system is that the absence of surface contact between the tool and the work piece that takes place within the presence of a dielectric medium (Paraffin oil). The Die Sinking EDM method was developed simultaneously in USA and USSR in the period of 2nd World War. During then a technique was required to process very hard materials used in military vehicles, equipment and ammunitions. Later on the method of Die Sinking EDM was developed in various countries and was utilised in numerous Defences, Automotive, Aeronautics

DOI: 10.4018/978-1-5225-3035-0.ch003

and many other industrial areas worldwide. The method is endlessly used for years to carry out varied very important experiments connected with the method of optimization. The process is used continuously from time to time to analyze the various machine parameters like Over-cut, Material removal rate, surface roughness, etc. Since multiple inputs are being used, hence optimizing these becomes essential for obtaining desired output. Out of the various parameters that may be thought of as the output of the machining operations, the material removal rate (MRR) and Surface roughness (SR) would be thought of for the current work as the factors specifically influences the standard, price of machining and therefore the machining hour rate.EDM is an essential machining process in many industries that give importance to precision and accuracy. Several researchers have studied EDM process considering different machining characteristics. Koshy et al. (1993) have studied EDM process for MRR, tool wear rate, relative electrode wear, corner reproduction accuracy and surface finish aspects using a rotating disk electrode and compared the results with a stationary electrode. It is seen that the effective flushing of the working gap improves MRR and surface finish. Zhang et al. (1997) have investigated MRR, surface roughness and diameter of discharge points in EDM on ceramics. From the experimental results, they have shown that MRR, surface roughness and also the diameter of discharge point directly varies with pulse-on time and discharge current. Lee and Li (2001) have investigated the result of machining parameters like the tool materials, tool polarity, gap voltage, peak current, pulse length, pulse interval and flushing on the machining characteristics, like MRR, surface end and relative tool wear in EDM of WC. It's ascertained that MRR usually decreases with the rise of gap voltage and surface roughness will increase with increasing peak current. Ramaswamy and Blunt (2002, 2004) have shown that the electrical energy is that the most dominant factor in modifying the surface texture, particularly the root mean square of peaks, the material volume in EDM of M300 tool steel. Puertas and Luis (2003) have shaped centre line average roughness value (Ra) and root mean sq. roughness price (Rq) in terms of current, pulse on time and Off time in EDM on soft steel (F-1110). It's seen that this intensity has the foremost influence on surface roughness and there's a strong interaction between this current intensity and also on the spark on time. Guu et al. (2003) have studied the results of different machining parameters on surface roughness in EDM of AISI D2 steel and being brought to a conclusion that surface roughness is inversely proportional to power input. Petropoulos et al. (2004) have stressed the relation between surface texture parameters and methodology parameters in EDM of Ck60 steel plates. They have thought of amplitude, spacing, hybrid, nonetheless as random methodology and type parameters that's perennial at each scale. Amorim and Weingaertner (2005) have targeting the terms of constant quantity influence of machining parameters on volumetric relative wear, MRR and surface roughness (Ra) in EDM of AISIP20 using copper tool electrodes. Puertas et al. (2005) have done a study on the influence of EDM processing parameter quality (current intensity, pulse time, duty cycle, gap voltage and dielectric flushing pressure) across 2 spacing parameters- mean spacing between peaks and therefore the range of peaks per cm in machining of siliconised or reaction-bonded carbide (SiSiC). From the results, it's seen that intensity, pulse time and duty cycle are most potent factors arousing result on the chosen responses. Guu (2005) has terminated that a lot of outstanding discharge energy leads to a lot of poor surface structure in EDM of AISI D2 steel. Yan et al. (2005) have examined scientifically the influences of the tactic parameters (dielectric sort, peak current and pulse duration) on MRR, conductor wear rate and surface roughness parameter in EDM of pure Ti metals. Keskin et al. (2006) have represented that surface roughness enhances with rise within the discharge length. Routara et al. (2007) have depicted the roughness models of EDM method for 3 dissimilar roughness parameters employing a methodology called response surface methodology (RSM) and shown that the machining parameters like pulse current and pulse on time have the best influence on the roughness parameters whereas pulse off time has no substantial result on roughness parameters. Kiyak and Cakir (2007) have studied the influences of method parameters (pulse current, pulse time and pulse pause time) on surface roughness in EDM of 40CrMnNiMo864 alloy steel and discovered that surface roughness of sample and tool are determined by factors like pulse current and pulse time. Pradhan and Biswas (2008) have developed MRR model for 3 numerous input parameters specifically pulse current, discharge time and pulse time for EDM method of AISI D2 steel using RSM. Jaharah et al. (2008) have studied the results of input parameters (peak current, pulse on-time, and pulse Off time) on surface roughness (Ra), tool wear rate (EWR) and MRR in EDM of AISI H13 alloy steel and discovered that the peak current is of bigger importance influencing MRR and surface roughness. Chiang (2008) has analysed the implications of input parameters like discharge current, pulse on time, duty cycle and gap voltage on MRR, tool wear ratio, and surface roughness employing RSM in EDM of Al2O3+TiC mixed ceramic materials and delineated that the discharge current and duty factor are most substantial factor of MRR and the discharge current and spark – on – time have significance related to statistics on both the value of the tool wear ratio and surface finish. Kuppan et al. (2008) have studied the results of input parameters (peak current, pulse on-time, duty cycle and tool speed) on MRR and depth averaged surface roughness in EDM (small deep hole drilling) of alloy 718 materials employing RSM. Habib (2009) has developed a mathematical model utilizing RSM for association of the varied machining parameters (pulse on time, peak current, average gap voltage and therefore the volume fraction percentage of SiC within the Al matrix) on MRR, EWR, gap size and therefore the surface properties. Amin et al. (2009) have studied the influences of input parameters (peak current, voltage, pulse duration and interval time) in EDM of WC employing Taguchi methodology and shown that the peak current considerably affects the EWR and surface roughness, while, the spark duration primarily impresses MRR. Sahoo et al. (2009) have inquired concerning the influence of machining input parameters (pulse current, pulse on time and pulse Off time) on the quality of surface created in EDM victimising three completely different work-piece materials viz. mild steel, brass and Tungsten Carbide. Rahman et al. (2010) studied the influences of input parameters on MRR employing RSM in EDM of Ti6Al4V and created clear that peak current and pulse on time are the substantial machining parameters influencing MRR. They also showed that MRR is directly proportional to peak current and pulse on time. Pradhan and Biswas (2010) have investigated the results of machining parameters in EDM of AISI D2 steel and indicated that the pulse current, pulse time, voltage and pulse on time have substantial consequences in dominant MRR. Effects of tool materials on the responses are studied by altogether completely different researchers. Haron et al. (2008) have studied the results of copper and carbon tool on EWR and MRR. Jahan et al. (2009) have investigated the results of tool materials (tungsten, copper metallic element and silver metallic element) and came to a conclusion that silver tungsten tool produces sander (with minimum Ra and Rmax) among the 3 electrodes in EDM of XW42 alloy steel. Choudhury et al. (2010) have used copper, brass and carbon electrodes and depicts that copper electrode offers greater MRR and brass electrode offers greater surface finish in EDM of EN 31 die steel.

NECESSITY OF OPTIMIZATION

For the applications of advanced manufacturing methodologies, technologies are developing in rapid pace. In order to achieve these advanced manufacturing criteria the industries, companies and manufacturing units need to optimize the process parameters to save money and time. Many researchers have carried out the optimization of the process parameters using many machining techniques like EDM, ECM, Wire EDM, PAC, etc. by applying various optimizations techniques like Artificial Neural Network, Particle Swarm Optimization, Genetic Algorithm, Fuzzy Logic Algorithm etc. Researchers like Asilturk et al. (2016); Korkut et al. (2004); Mahapatra et al. (2006); Vasant et al. (2016); Vasant et al. (2011) have performed optimization considering various process parameters in order to obtain best set of process parameters those can be used for advance manufacturing techniques. Considering EDM in this present work, MRR and Surface Roughness are believed as the parametric quantity where the primary aim was to obtain the optimized minimized values of Surface Roughness and optimized Maximized values of Material Removal Rate. For getting these desired machining characteristics the machining parameters like Pulse on Time (T_{On}), Pulse Off time (T_{off}), Peak-Current (I_p), Gap voltage (V_g) has been considered as the input.

OPTIMIZATION METHODS

In this work authors have tried to provide solution to the problems using Particle Swarm optimization (PSO) and Artificial Neural Network (ANN).

ARTICIAL NEURAL NETWORK (ANN)

In machine learning, artificial neural networks (ANNs) are a group of models enlivened by biological neural networks and are utilized to surmised functions that can rely on upon a substantial number of sources of info called as inputs and are for the most part obscure or not known. Artificial Neural Network has derived its origin from the human nervous system that consists of large number of neurons interconnected to it and performs various analyzing tasks in no time. They are introduced as frameworks of interconnected "neurons" which trade messages between each other. The simulated neurons take inputs and give outputs through their non-linear connections (Sitton et al., 2017). The associations or connections have weights that can be tuned, making neural nets versatile and adaptive to inputs and equipped for learning. It provides several reconciling and perceptual solutions and re-appraisal of parametric model quality in comparatively less complicated configurations when examined to different approaches of modelling. It is an optimization technique in which less number of errors is encountered and high accuracy can be achieved (Pappu and Gummadi, 2017). Artificial Neural Network modelling technique is one of the most practical, authentic and precise modelling proficiency which is able to demonstrate non-linear relations in case of machining processes (Patra et al., 2017). ANN has the potential to show better results in the fatigue loading conditions and range of spectral type and properties of the material (Durodola et al., 2017). The capacity of the ANNs to precisely estimated obscure capacities to know the unknown function. Neural systems have been utilized to unravel a wide assortment of assignments that are difficult to understand utilizing standard manage based programming. Some of the applications of ANN are for solving the Travelling Salesman's problems, Security and loan applications, Market predictions etc.

PARTICLE SWARM OPTIMIZATION (PSO)

Particle Swarm Optimization (PSO) involves computation technique that optimizes a consequence by using a series of iterations attempting to heighten an applicant arrangement with respect to a given measure of value. It is a random optimization technique which was brought forth by Kennedy and Eberhart to provide various advantages such as fast occurrence of two or more things coming together, easy principle and simple execution (Rafiraeed et al., 2017). PSO provides accurate and authentic modelling results. As a hybrid type of algorithm, it has got outstanding feature of mixing the abilities of worldwide scope or applicability and local analyzing (El-Wakeel, 2014). PSO is mostly recommended and used algorithm due to its uncomplicated concept and simple coding implementation, it takes lesser time in optimizing results when compared to other techniques (Yu-Zhen et al., 2012). One of the main reason to work with PSO is its appealing characteristics that is lesser number of parameters are to be adjusted to use it and it covers wide range of application (Katherasan et al., 2012). It takes care of a problem by having a people of candidate solutions, here named particles, and moving these particles within the inquiry space as indicated by basic numerical formulae over the particle's position and rate. Every particle's growth is influenced by its neighbourhood best known position, but on the other way it is guided toward the best-known positions in the search space, which are upgraded as comparatively better positions are found by other particles.

EXPERIMENTATION

The choice of applicable input machining conditions for EDM characteristics like material removal rate and surface finish are having a base on the analysis associated with the varied input parameters for material removal and roughness. Attempting frequent or large number of experimental runs also is not economically viable. Expertise disclosed that the sort of material yields extra influence on the EDM performance. On the opposite hand when new and advanced materials seem within the field, it's unimaginable to use available models and thus experimental investigations are continually needed.

Input Parameters Considered

Pulse-On Time (T_{On}), Pulse Off time (T_{off}), Peak-Current (I_p), Gap voltage (V_g)

Output Parameters Considered

Material Removal Rate (MRR) and Surface Roughness (SR)

Material Used

EN 19 Tool Steel as work-piece and electrolytic copper as a tool electrode (positive polarity).

Machine Used

The entire experiment was carried out on a Die sinking EDM machine (Electronica EMT-43 Machine) (Figure 1).

The work-piece on that the EDM method was disbursed was an EN19 material whereas the tool for EDM operation was Copper. Paraffin oil was hand-picked as insulator medium. After machining operation, surface roughness (Ra) was determined employing a stylus type profilometer, known as Talysurf (Taylor Hobson, 3+) (Figure 2). A traverse speed of 1 mm/sec, cut off length of 0.8 mm and an analysis length of 8 mm were set for the stylus to work. A group of three totally different readings were evaluated for varied values of surface roughness and arithmetic mean of those values was used.

DESIGN OF EXPERIMENTS (DOE)

Industrial physicists will now not afford to conduct experiment in a very trial-and-error manner, dynamically one issue at a time, the manner early scientists like Edison, Madam Curie, J.C.Bose did whereas inventing and developing things. A way more effective methodology is to use a computer-enhanced, systematic approach to experimentation, one that considers all factors at the same time. That approach is named design of experiments (DOE), and firms worldwide are assuming it as a cost-effective way to solve serious issues affecting their operations. The design of experiments technique may be a terribly very potent tool, that permits holding out the modelling and analysis of the influence of technique variables on the response variables. The response variable is associate degrees unknown perform of the method variables, which are referred to as design factors. The aim of running experiments is to characterize unknown relations and dependencies that exist inside the determined design or method, i.e. to seek out the affecting design variables and therefore the response to variations within the design variable values.

Figure 1. EDM Machine

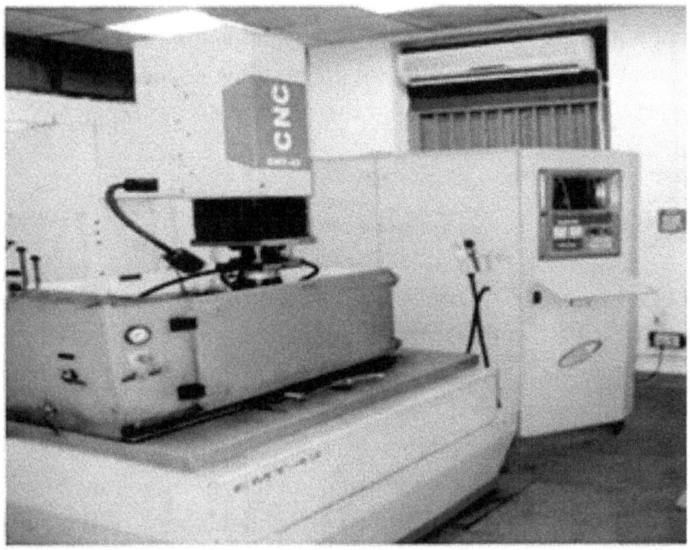

Figure 2. Stylus type profilometer(Talysurf)

A methodical scientific approach to being after the experiment must be utilized if the results are to be figured out performing minimum number of experiments. The statistical design of experiments refers to the process of planning the experiment so that relevant data that can be analyzed by statistical methods will be brought together in one place, leading in valid and objective conclusions in a logical way. Statistical methodology becomes the only logical approach for analysis when the problem involves data that may contain experimental errors. Sometimes, experiments are recurrent with a specific set of levels for all the factors to visualize the statistical technique validation and repeatability by the replicate data. This is often known as replication. to eradicate any biasness, allocation of experimental material and therefore the order of experimental runs are chosen in a random manner. This is often known as randomisation. To rearrange the experimental material into lots, or blocks, that ought to have higher homogeneity than the complete set of data is named block. So, once experiments are being undertaken this part is ought to be kept in mind. There are many totally different methodologies for style of experiments. During this work Orthogonal Array (OA) technique was used for performing the experiments.

ORTHOGONAL ARRAY

Orthogonal Arrays (OA) are special matrices used as the design matrices in the fractional factorial design for the estimation of the effect of several factors in a highly efficient way. These designs are applicable even when the factors have more than two levels and for mixed level experiments where that components do not have same number of levels. In general, when the number of process parameter enhances, large number of experiments has to be carried out for factorial design. But using orthogonal array smaller number of experiments can be performed in the specified range and the effects of process parameters can be observed quite effectively. For a two level factors 8 (L8) experiments are needed for the experimentation whereas for a three-level orthogonal arrays can be based on 9 (L9), 27 (L27) or 81(L81) experimental points. For any pair of columns in the design matrix, all combinations of the factor levels appear equal number of times.

In the present study four factors are taken with three levels, (Shown in Table 1) which have 20 degrees of freedom (4*2 + 3*2*2). Hence, L27 OA is used in the study. The three-level Orthogonal arrays (L27) for four factors viz. A, B, C and D are presented in Table 2.

EXPERIMENTATION AND MEASUREMENT OF MRR

The experiment was performed on the basis of design of experiments using L27 orthogonal array. Measuring the weight of each sample before and after each experiment, MRR was calculated experimentally. The obtained value of the MRR was employed to obtain the corresponding S/N ratio values using the

Table 1. Different variables used in the experiment and their levels

Variable	Coding	Level		
		1	2	3
Pulse On (Ton) in μs	A	200	300	400
Pulse Off (Toff) in μs	B	1800	1700	1600
Discharge Current (Ip) in A	C	8	12	16
Voltage (V) in V	D	40	60	80

Table 2. DOE using L27 Orthogonal Array

Exp. No.	A (Ton)	B (TOff)	C (Ip)	D (V)	Exp. No.	A (Ton)	B (TOff)	C (Ip)	D (V)
1	1	1	1	1	15	2	2	3	2
2	1	1	2	2	16	2	3	1	1
3	1	1	3	3	17	2	3	2	2
4	1	2	1	2	18	2	3	3	3
5	1	2	2	3	19	3	1	1	3
6	1	2	3	1	20	3	1	2	1
7	1	3	1	3	21	3	1	3	2
8	1	3	2	1	22	3	2	1	1
9	1	3	3	2	23	3	2	2	2
10	2	1	1	2	24	3	2	3	3
11	2	1	2	3	25	3	3	1	2
12	2	1	3	1	26	3	3	2	3
13	2	2	1	3	27	3	1	3	2
14	2	2	2	1					

Minitab 16 software. As higher MRR is a desired condition hence Larger the Better condition was used and results are tabulated in Table 3.

The regression equation for MRR based on the input parameters and their levels was also generated (Eqn. 1)

$$\begin{aligned} MRR &= 0.222579 + 0.000681 * Ton + 0.053207 * Toff \\ &+ 0.096158 * Ip - 0.043101 * V + 0.013372 * Ton^2 \\ &+ 0.018285 * V^2 + 0.021314 * Ip * Toff - 0.012999 * Toff * V \\ &- 0.018288\ V * Ip \end{aligned} \quad (1)$$

EXPERIMENTATION AND MEASUREMENT OF SURFACE ROUGHNESS (RA)

The machined surfaces of the work-pieces were introduced to the profilometer and surface roughness (Ra) was measured. The obtained value of the Ra was employed to obtain the corresponding S/N ratio values using the Minitab 16 software. As lower Ra is always desired hence Smaller the Better condition was used and results are tabulated in Table 4

The regression equation for Surface Roughness based on the input parameters and their levels was also generated (Eqn. 2)

Table 3. Experimental Results for MRR

Exp. No.	A (Ton)	B (T_{off})	C (Ip)	D (V)	MRR	S/N Ratio	Exp. No.	A (Ton)	B (T_{off})	C (Ip)	D (V)	MRR	S/N ratio
1	1	1	1	1	7.22	17.17	**15**	2	2	3	2	27.82	28.88
2	1	1	2	2	12.16	21.62	**16**	2	3	1	1	11.30	21.06
3	1	1	3	3	16.53	24.36	**17**	2	3	2	2	21.06	26.47
4	1	2	1	2	7.38	17.36	**18**	2	3	3	3	28.87	29.25
5	1	2	2	3	14.1	22.98	**19**	3	1	1	3	4.699	13.43
6	1	2	3	1	31	29.83	**20**	3	1	2	1	15.93	24.04
7	1	3	1	3	7.83	17.87	**21**	3	1	3	2	4.65	27.32
8	1	3	2	1	24.9	27.92	**22**	3	2	1	1	9.143	19.22
9	1	3	3	2	31.96	30.09	**23**	3	2	2	2	5.517	24.74
10	2	1	1	2	5.57	14.91	**24**	3	2	3	3	23.73	27.50
11	2	1	2	3	11.18	20.97	**25**	3	3	1	2	8.95	19.03
12	2	1	3	1	24.63	27.83	**26**	3	3	2	3	17.43	24.82
13	2	2	1	3	6.09	15.69	**27**	3	1	3	2	41.73	32.11
14	2	2	2	1	20.27	26.14							

Table 4. Experimental Results of Surface Roughness (R_a)

Exp. No.	A (Ton)	B (TOff)	C (Ip)	D (V)	Ra	S/N Ratio	Exp. No.	A (Ton)	B (TOff)	C (Ip)	D (V)	Ra	S/N ratio
1	1	1	1	1	9.41	-19.47	**15**	2	2	3	2	13.27	-22.45
2	1	1	2	2	11.6	-21.28	**16**	2	3	1	1	9.46	-19.51
3	1	1	3	3	11.65	-21.32	**17**	2	3	2	2	16.27	-24.22
4	1	2	1	2	8.49	-18.57	**18**	2	3	3	3	15.9	-24.02
5	1	2	2	3	14.43	-23.18	**19**	3	1	1	3	10.23	-20.19
6	1	2	3	1	11.41	-21.14	**20**	3	1	2	1	15.23	-23.65
7	1	3	1	3	10.53	-20.44	**21**	3	1	3	2	14.97	-23.50
8	1	3	2	1	10.71	-20.59	**22**	3	2	1	1	11.17	-20.96
9	1	3	3	2	13.77	-22.77	**23**	3	2	2	2	19.6	-25.84
10	2	1	1	2	10.67	-20.56	**24**	3	2	3	3	13.3	-22.47
11	2	1	2	3	14.63	-23.30	**25**	3	3	1	2	11.07	-20.88
12	2	1	3	1	15.8	-23.97	**26**	3	3	2	3	16.2	-24.19
13	2	2	1	3	9.87	-19.88	**27**	3	1	3	2	12.8	-21.56
14	2	2	2	1	14.57	-23.26							

$$SR = 11.0648 + 0.496 * Ton - 0.1428 * Toff + 1.1833 * Ip - 0.0392 * V - 0.2807 * Ip^2 + 0.2433 * V^2 \quad (2)$$

OPTIMIZATION OF MRR AND SR USING PARTICLE SWARM OPTIMIZATION (PSO)

For the obtained values using MiniTab16 for 27 experimental readings of Pulse-On Time (T_{On}), Pulse OFF time (T_{off}), Peak-Current (I_p), Gap voltage (V_g), MRR and SR, optimization is carried out using PSO coding for obtaining the optimum values of all the parameters listed above. The Regression Equations (1) and (2) obtained for MRR and SR in terms of Pulse-On Time (μs), Pulse Off time (μs), Peak-Current (A), Gap voltage (V).

In order to obtain the optimum values for MRR and SR, Matlab R2015b was used. Code for multivariable optimization using PSO was simulated which provided the plots which showed the optimal values for Pulse-On Time (A), Pulse Off time (B), Peak-Current (C), Gap voltage (D). For these values of Pulse-On Time (A), Pulse OFF time (B), Peak-Current (C), Gap voltage (D) the respective optimal values for MRR and SR were determined. Figure 3 and 4 shows the Maximization Plot for Material Removal Rate (MRR) and Minimization Plot for Surface Roughness (SR) respectively.

Figure 3. Maximization Plot for Material Removal Rate (MRR)

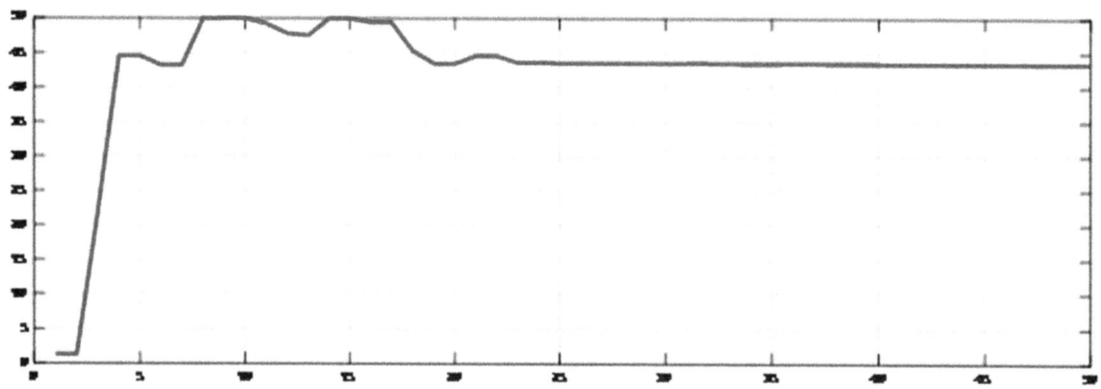

Figure 4. Minimization Plot for Surface Roughness (SR)

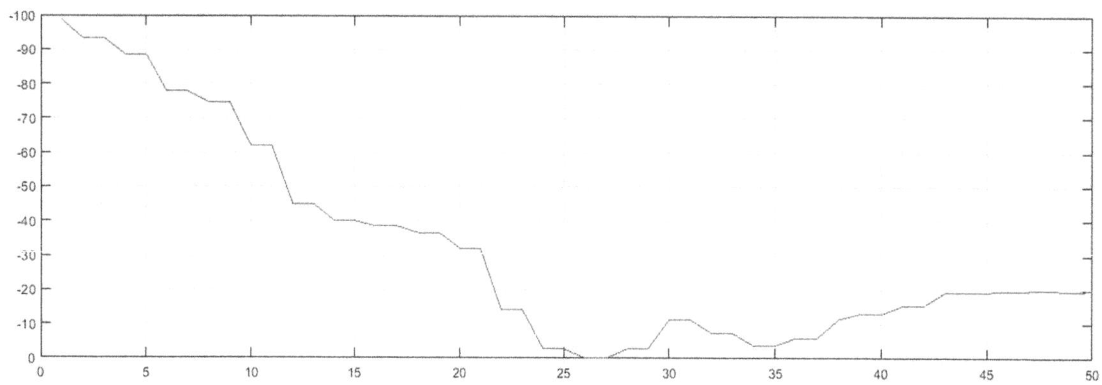

OPTIMIZATION OF MRR AND SR USING ARTIFICIAL NEURAL NETWORK

The confront study addresses with the development of artificial neural network to depict the suitable machining process parameters in order to receive optimum MRR and surface roughness characteristics in EDM process. Artificial neural network with Error Back Propagation Training Algorithms (EBPTA) has been used for the training and testing purpose of the experimental data. Initially a data set is required to train the ANN model. Then developed experimental database was used to train and simulated the ANN model to determine the predicted value. The EBPTA is based on universal delta-rule that involves dynamic weight updates system so as to reduce the mean squared error (MSE).

In this part, ANN model was used for prediction of MRR and SR. The experimental study illustrated the error for the MRR and SR properties. So, with reference to the mid-level quality characteristics supported by Artificial Neural Network the degree of non-linearity that exists in nature between the output and input variable was identified.

The multi-layer ANN network shown in Figure 5 comprises of input, output and hidden neurons layer. Multi-layer ANN architecture consists of 4-30-2 network model is used for simulation. Four neurons in the input layer corresponding to 4 inputs variable, two neurons in the output layer. One hidden layer with 30 neurons was found suitable employed in the present study. For training and testing 'nntool' is used which is available in optimization tool box in 'MATLAB' software. When maximum number of epoch and minimum MSE is achieved, training of ANN model is stopped. The mean square error (MSE) is computed using equation3

$$MSE = \frac{1}{NP}\sum_{P=1}^{NP}\sum_{k=1}^{k}\left(d_{kp} - o_{kp}\right)^2 \quad \mathrm{MSE} = \frac{1}{\mathrm{NP}}\sum_{\mathrm{p}=1}^{\mathrm{NP}}\sum_{\mathrm{k}=1}^{\mathrm{k}}\left(\mathrm{d}_{\mathrm{kp}} - \mathrm{o}_{\mathrm{kp}}\right)^2 \tag{3}$$

where

NP = number of training patterns,
dk, p = desired output for the *p*th pattern and O_{kp} is predicted output for the pth pattern.

ARTIFICIAL NEURAL NETWORK TRAINING AND TESTING

The training of ANN has been carried out for 27 input–output patterns with the help of 'nntool' available in *MATLAB* software (2012a). The factors are used for training purpose are:

- Learning rate = 0.05
- Momentum factor = 0.85
- Maximum number of epochs = 1000
- Tolerance for MSE= 0.0001

The ANN training and simulation was carried out with the help of "*training dx*" function. The variation of MSE with number of epoch is shown in Figure 6.

The training state and output regression plots (depicted in Figures 7 and 8) are generated providing the final optimal solutions and also ensures the efficacy of the system chosen.

Figure 5. Multi-layer feed forward artificial neural network

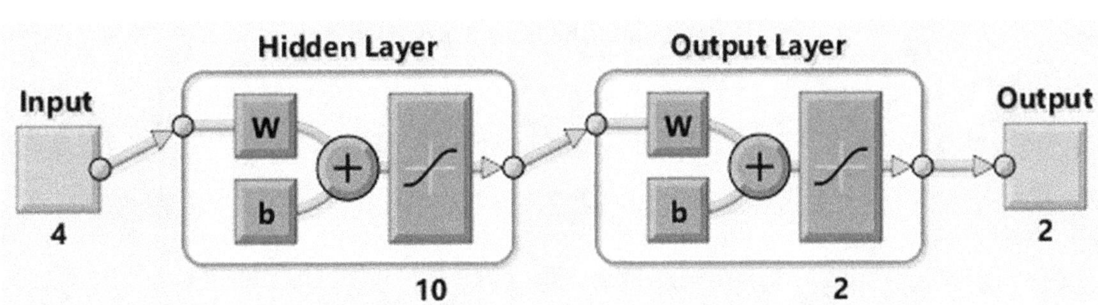

Figure 6. The variation of mean squared error (MSE)

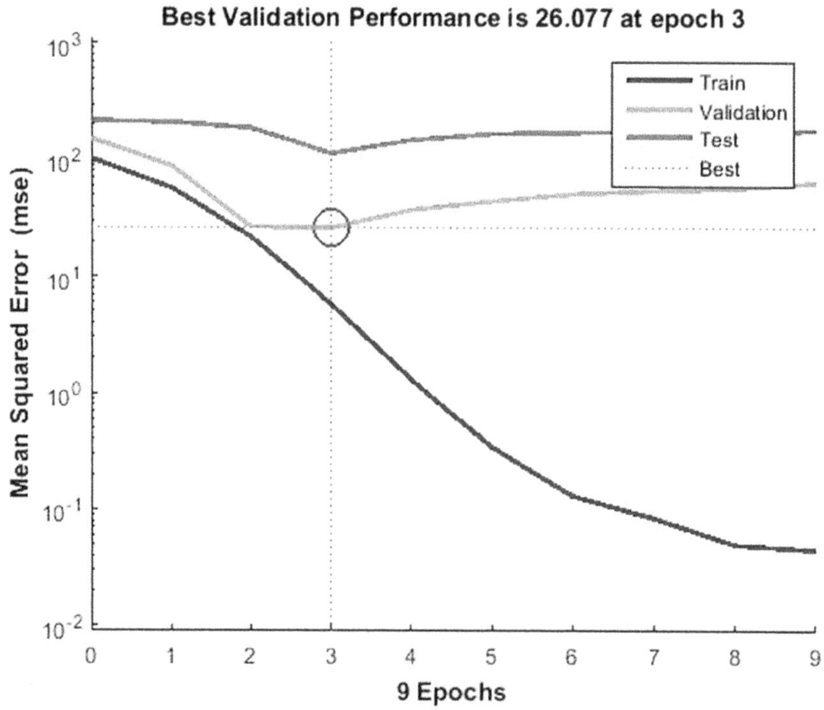

Figure 7. Correlation of the training patterns for MRR and SR with Number of epochs

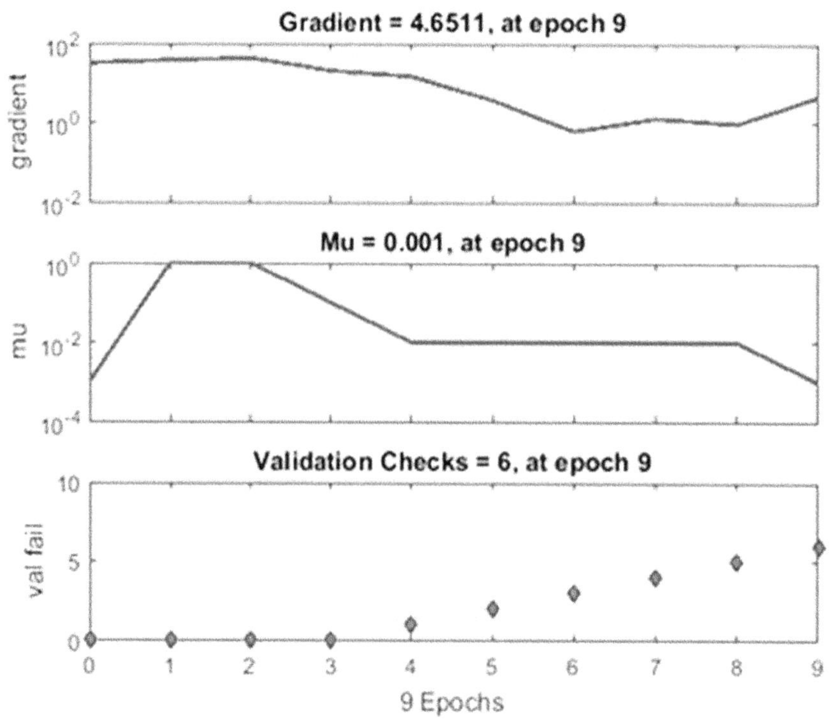

Figure 8. Regression plot for MRR and SR using ANN

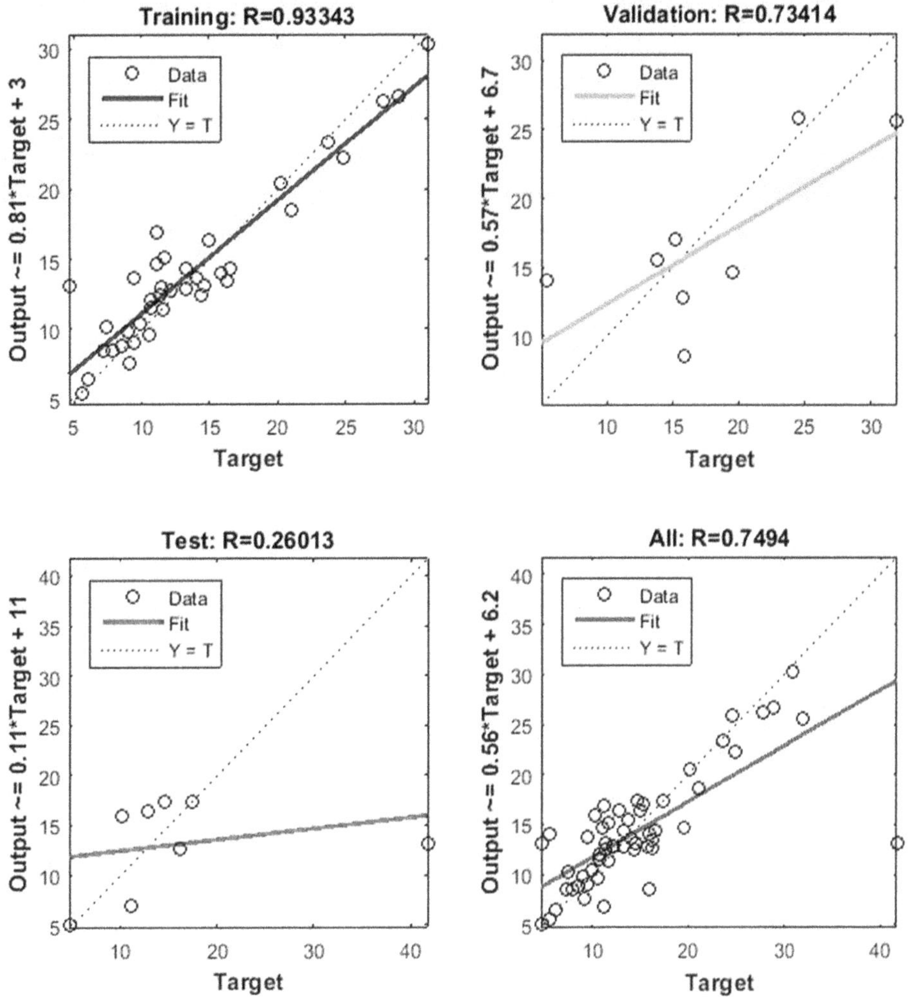

RESULTS AND DISCUSSION

Experiments were performed according to the sequence of L27 OA and the experimental results of MRR and SR are calculated. These analyses of experimental results were carried out using Minitab 16 statistical software. In this case the interaction between the particles is also taken into consideration before analysing these experimental data using software. As per ANN method the experimental data are imported into the workspace of Matlab R2015b and then the neural Network as depicted above was obtained. Further by training the input and the target values the plots for Training State, Performance, Regression, MRR and SR are obtained.

It can be noted that for Training case: R is 0.93343 with an output of 0.81*Target + 3; whereas for case of Testing R is determined as 0.26013 with an output of 0.11*Target + 11. For validation case R is found to be 0.73414 with output of 0.57*Target + 0.6.7 and the miscellaneous results case R is determined as 0.7494 and output is found to be 0.56*Target +0.02.

Further after obtaining the experimental and regression equations, ANN and PSO was used for finding the optimum values of for MRR and SR in terms of Pulse-On Time (A), Pulse Off time (B), Peak-Current (C), Gap voltage (D). Using the Multi variable code in Matlab R2015b for the regression equations for MRR and SR mentioned above in equation 1 and 2as objective function respectively it was found that the optimal values for Pulse-On Time (A), Pulse Off time (B), Peak-Current (C), Gap voltage (D) are approximately found to be400 μs, 1600 μs, 16A and 40V respectively from the minimization and maximization plots as depicted above. For these values of Pulse-On Time (A), Pulse Off time (B), Peak-Current (C), Gap voltage (D) the respective optimal values for MRR and SR are determined approximately as 0.3548 g/min and 0.004129 μm.

CONCLUSION

The main purpose of this experimentation was to optimize the machining parameters of EDM using Artificial Neural Network (ANN) and Particle Swarm Optimization (PSO) techniques for maximization of Material Removal Rate (MRR) and minimizing the surface Roughness (SR). Here in this present work Pulse-On Time (A), Pulse Off time (B), Peak-Current (C), Gap voltage (D) were conceived with equally spaced three levels within the operating range for each of the process parameters as the input parameters. From the results and discussions as stated above it can be concluded that:

For mass production of components of machines where there is a fast movement of consumer goods it is highly necessary for the Industrialists or the Manufacturers to use the optimized values of the parameters obtained from MRR. For the cases where levels are not matching enough to their desired characteristics of input parameters, using the regression equations is always an option for them for prediction of proper output.

In some cases, there are products where MRR plays least role in the product development where as SR is significant enough for the manufacturing of the product. Then the optimal values thus developed using PSO in this work can be adopted or the regression equation can be used for the cases that are having some variation in the levels of input parameters for prediction of output parameters.

For the products where both MRR and SR are to given equal amount of importance, optimized values for both MRR and SR are to be considered for predicting the output.

The results derived from the optimization techniques used here are quite agreeing to the results obtained by other researchers considering the same input, output parameters and identical levels using other techniques like Taguchi, GA, ABC etc. But it varies considerably in case some other parameters and levels are considered.

SCOPE OF FUTURE RESEARCH

This present work can also be used or applied to several other directions. This work can also be enriched by incorporating or considering several other input parameters like tool rotation, different dielectric medium, flushing pressure etc. and several other responses like tool wear, thermal stress developed etc. In order to obtain the optimized parameters other recent techniques like Hill Climbing, Grey Wolf, Stochastic Optimization, Fuzzy Optimization, Grey Optimization, Uncertain Preference, Fuzzy Multiobjective Optimization, Grey Multiobjective Optimization, Stochastic Multiobjective Optimization, Fuzzy Multilevel Programming, Grey Multilevel Programming, Stochastic Multilevel Programming, etc. can also be utilized or Hybridization of two parallel methods like Simulated annealing in Genetic Algorithm (SAGA) etc. can be employed.

ACKNOWLEDGMENT

The authors sincerely acknowledge the comments and suggestions of the reviewers that have been instrumental for improving and upgrading the paper in its final form.

REFERENCES

Amin, A. K. M. N., Lajis, M. A., & Radzi, H. C. D. M. (2009). The implementation of Taguchi method on EDM process of tungsten carbide. *European Journal of Scientific Research*, *26*, 609–617.

Amorima, F. L., & Weingaertner, W. L. (2005). The influence of generator actuation mode and process parameters on the performance of finish EDM of a tool steel. *Journal of Materials Processing Technology*, *166*(3), 411–416. doi:10.1016/j.jmatprotec.2004.08.026

Asilturk, I., Neseli, S., & Ince, M. A. (2016). Optimization of Parameters affecting surface roughness of Co28Cr6Mo medical material during CNC lathe machining by using Taguchi and RSM Methods. *Measurement*, *78*, 120–128. doi:10.1016/j.measurement.2015.09.052

Chiang, K. T. (2008). Modeling and analysis of the effects of machining parameters on the performance characteristics in the EDM process of Al2O3+TiC mixed ceramic. *International Journal of Advanced Manufacturing Technology*, *37*(5-6), 523–533. doi:10.1007/s00170-007-1002-3

Choudhary, R., Kumar, H., & Grag, R. K. (2010). Analysis and evaluation of heat affected zones in electric discharge machining of EN-31 die steel. *Indian Journal of Engineering Materials Sciences*, *17*, 91–98.

Durodola, J. F., Ramachandra, S., Li, N., & Thite, A. N. (2017). A pattern Recognition artificial neural network method for random fatigue loading life prediction. *International Journal of Fatigue*, *99*, 55–67. doi:10.1016/j.ijfatigue.2017.02.003

El-Wakeel, A. S. (2014). Design optimization of PM couplings using hybrid Particle Swarm Optimization-Simplex Method (PSO-SM) Algorithm. *Electric Power Systems Research*, *116*, 29–35. doi:10.1016/j.epsr.2014.05.003

Guu, Y. H. (2005). AFM surface imaging of AISI D2 tool steel machined by the EDM process. *Applied Surface Science*, *242*(3-4), 245–250. doi:10.1016/j.apsusc.2004.08.028

Guu, Y. H., Hocheng, H., Chou, C. Y., & Deng, C. S. (2003). Effect of electrical discharge machining on surface characteristics and machining damage of AISI D2 tool steel. *Materials Science and Engineering*, *358*(1-2), 37–43. doi:10.1016/S0921-5093(03)00272-7

Habib, S. S. (2009). Study of the parameters in electrical discharge machining through response surface methodology approach. *Applied Mathematical Modelling*, *33*(12), 4397–4407. doi:10.1016/j.apm.2009.03.021

Haron, C. H. C., Ghani, J. A., Burhanuddin, Y., Seong, Y. K., & Swee, C. Y. (2008). Copper and graphite electrodes performance in electrical-discharge machining of XW42 tool steel. *Journal of Materials Processing Technology*, *201*(1-3), 570–573. doi:10.1016/j.jmatprotec.2007.11.285

Jahan, M. P., Wong, Y. S., & Rahman, M. (2009). A study on the fine-finish die-sinking micro-EDM of tungsten carbide using different electrode materials. *Journal of Materials Processing Technology*, *209*(8), 3956–3967. doi:10.1016/j.jmatprotec.2008.09.015

Jaharah, A. G., Liang, C. G., Wahid, S. Z. M. N., Rahman, A., & Hassan, C. H. C. (2008). Performance of copper electrode in electrical discharge machining (EDM) of AISI H13 harden steel. *International Journal of Mechanical and Materials Engineering*, *3*, 25–29.

Jameson, E. C. (2001). *Electrical discharge machining*. Society of Manufacturing Engineers.

Katherasan, D., Elias, J. V., Sathiya, P., & Haq, A. N. (2012). Flux Cored Arc Welding Parameter Optimization Using Swarm Optimization Algorithm. *Procedia Engineering*, *38*, 3913–3296. doi:10.1016/j.proeng.2012.06.449

Keskin, Y., Halka, H., & Kizil, S. M. (2006). An experimental study for determination of the effects of machining parameters on surface roughness in electrical discharge machining (EDM). *International Journal of Advanced Manufacturing Technology*, *28*(11-12), 1118–1121. doi:10.1007/s00170-004-2478-8

Kiyak, M., & Cakir, O. (2007). Examination of machining parameters on surface roughness in EDM of tool steel. *Journal of Materials Processing Technology*, *191*(1-3), 141–144. doi:10.1016/j.jmatprotec.2007.03.008

Korkut, I., Kasap, M., & Seker, U. (2004). Determination of optimum cutting parameters during machining of AISI 304 austenitic stainless steel. *Materials & Design*, *25*(4), 303–305. doi:10.1016/j.matdes.2003.10.011

Koshy, P., Jain, V. K., & Lal, G. K. (1993). Experimental investigations into electrical discharge machining with a rotating disk electrode. *Precision Engineering*, *15*(1), 6–15. doi:10.1016/0141-6359(93)90273-D

Kuppan, P., Rajadurai, A., & Narayanan, S. (2008). Influence of EDM process parameters in deep hole drilling of Inconel 718. *International Journal of Advanced Manufacturing Technology*, *38*(1-2), 74–84. doi:10.1007/s00170-007-1084-y

Lee, S. H., & Li, X. P. (2001). Study of the effect of machining parameters on the machining characteristics in electrical discharge machining of tungsten carbide. *Journal of Materials Processing Technology*, *115*(3), 344–358. doi:10.1016/S0924-0136(01)00992-X

Mahapatra, S. S., Patnaik, A., & Patnaik, P. (2006). Parametric analysis and optimization of cutting parameters for turning operations. *Proceedings of the Int. Conference on Global Manufacturing and Innovation*, 1–6.

Pappu, J. S. M., & Gummadi, N. (2017). Artificial Neural Network and regression coupled genetic algorithm to optimize parameters for enhanced xylitol production by Debaryomycesnepalensis in bioreactor. *Biochemical Engineering Journal, 120*, 136–145. doi:10.1016/j.bej.2017.01.010

Patra, K., Jha, A. K., Szalay, T., Ranjan, J., & Monostori, L. (2017). Artificial Neural Network Based tool condition monitoring in micro mechanical peck drilling used thrust force signals. *Precision Engineering, 48*, 279–291. doi:10.1016/j.precisioneng.2016.12.011

Petropoulos, G., Vaxevanidis, N. M., & Pandazaras, C. (2004). Modeling of surface finish in electro-discharge machining based upon statistical multi-parameter analysis. *Journal of Materials Processing Technology, 155-156*, 1247–1251. doi:10.1016/j.jmatprotec.2004.04.189

Pradhan, M. K., & Biswas, C. K. (2008). Neuro-fuzzy model on material removal rate in electrical discharge machining in AISI D2 steel. *Proceedings of the 2nd International and 23rd All India Manufacturing Technology, Design and Research Conference, 1*, 469-474

Pradhan, M. K., & Biswas, C. K. (2010). Investigating the effect of machining parameters on EDMed components a RSM approach. *Journal of Mechanical Engineering, 7*, 47–64.

Puertas, I., & Luis, C. J. (2003). A study on the machining parameters optimization of electrical discharge machining. *Journal of Materials Processing Technology, 143-144*, 521–526. doi:10.1016/S0924-0136(03)00392-3

Puertas, I., Luis, C. J., & Villa, G. (2005). Spacing roughness parameters study on the EDM of silicon carbide. *Journal of Materials Processing Technology, 164-165*, 1590–1596. doi:10.1016/j.jmatprotec.2005.01.004

Rahman, M.M., Khan, K., A.R., K., Maleque, M.A., & Bakar, R.A. (2010). Parametric optimization in EDM of Ti-6Al-4V using copper tungsten electrode and positive polarity: A Statistical Approach. *Mathematical Methods and Techniques in Engineering and Environmental Science, 6*, 23–29.

Ramaswamy, H., & Blunt, L. (2002). 3D surface characterisation of elctropolished EDMed surface and quantitative assessment of process variables using Taguchi Methodology. *International Journal of Machine Tools & Manufacture, 42*(10), 1129–1133. doi:10.1016/S0890-6955(02)00057-3

Ramaswamy, H., & Blunt, L. (2004). Effect of EDM process parameters on 3D surface topography. *Journal of Materials Processing Technology, 148*(2), 155–164. doi:10.1016/S0924-0136(03)00652-6

Routara, B. C., Bandyopadhyay, A., & Sahoo, P. (2007). Use of desirability function approach for optimization of multiple performance characteristics of the surface roughness parameters in CNC turning. *Proceedings of the International Conference on Mechanical Engineering.*

Sahoo, P., Routara, B. C., & Bandyopadhyay, A. (2009). Roughness modelling and optimization in EDM using response surface method for different work piece materials. *International Journal of Machining and Machinability of Materials, 5*(2/3), 321–346. doi:10.1504/IJMMM.2009.023398

Sitton, D. J., Zeinali, Y., & Story, A. (2017). Rapid soil classification using artificial neural network for use in constructing compressed earth blocks. *Construction & Building Materials, 138*, 214–221. doi:10.1016/j.conbuildmat.2017.02.006

Vasant, P. M., Weber, G., & Diew, V. N. (Eds.). (2016). Handbook of Research on Modern Optimization Algorithms and Applications in Engineering and Economics. IGI Global.

Vasant, P. M., Barsoum, N., & Webb, J. (Eds.). (2011). *Innovation in Power, Control, and Optimization: Emerging Energy Technologies*. IGI Global.

Yan, B. H., Tsai, H. C., & Huang, F. Y. (2005). The effect in EDM of a dielectric of a urea solution in water on modifying the surface of titanium. *International Journal of Machine Tools & Manufacture*, *45*(2), 194–200. doi:10.1016/j.ijmachtools.2004.07.006

Yu-Zhen, Y. U., Xin-yi, R., Feng-Shan, D., & Jun-jie, S. (2012). Application of Improved PSO Algorithm in Hydraulic Pressing System Identification. *Journal of Iron and Steel Research International*, *19*(9), 29–35. doi:10.1016/S1006-706X(13)60005-9

Zhang, J. H., Lee, T. C., & Lau, W. S. (1997). Study on the electro-discharge machining of a hot pressed aluminium oxide based ceramic. *Journal of Materials Processing Technology*, *63*(1-3), 908–912. doi:10.1016/S0924-0136(96)00012-X

Chapter 4
Nature-Inspired Metaheuristic Approach for Multi-Objective Optimization During WEDM Process

Goutam Kumar Bose
Haldia Institute of Technology, India

Pritam Pain
JLD Engineering and Management College, India

ABSTRACT

In this research paper Wire-Electric Discharge Machining (WEDM) is applied to machine AISI-D3 material in order to measure the performance of multi-objective responses like high material removal rate and low roughness. This contradictory objective is accomplished by the control parameters like Pulse on Time (Ton), Pulse off Time (Toff), Wire Feed (W/Feed) and Wire Tension (W/Ten) employing brass wire. Here the orthogonal array is used to developed 625 parametric combinations. The optimization of the contradictory responses is carried out in a metaheuristic environment. Artificial Neural Network is employed to train and validate the experimental result. Primarily the individual responses are optimized by employing Firefly algorithm (FA). This is followed by a multi-objective optimization through Genetic algorithm (GA) approach. As the results obtained through GA infer a domain of solutions, therefore Grey Relation Analysis (GRA) is applied where the weights are considered through Fuzzy set theory to ascertain the best parametric combination amongst the set of feasible alternatives.

INTRODUCTION

Our advanced lifestyle has rapidly improved by the quick development of various technology. Thus, the possibility to achieve an effective solution for real-world based problems is very high. These days, it is also visible that mathematical dynamics of the related solution approaches are commonly based on nature dynamics. From this point of view, nature has an incredible role on approaching the solutions and design mathematical structure for developing effective scientific, computational methods or techniques.

DOI: 10.4018/978-1-5225-3035-0.ch004

Also, the connection in between nature and science has established a new research approach to solve real life problem. Many nature-inspired methods and algorithm that can be employed in the context of intelligent solution approach like cooperative characteristics by the swarms like a firefly, bat, bee etc. Due to their great possibility on developing effective algorithmic solution way, nature-inspired techniques have a huge effect on Artificial Intelligence based studies. The main advantages of this methods are that it can also be used with another Artificial Intelligence algorithm to form hybrid methodologies.

Alternatively, it is also possible to classify such nature-inspired techniques based on some mechanisms inspired by the theory of Evaluation. Genetic Algorithm is one of this evolutionary technique which is based on the theory of survival of the fittest by Darwin.

Traditional machining served the necessities of industries over the eras. But new advanced work materials, as well as the geometric design of products and components are putting a lot of pressure on the capabilities of traditional machining processes to manufacture the components with desired tolerance economically. This has led to the growth and founding of Non-Traditional Machining processes in industries as efficient and economic alternatives to conventional ones. Wire Electrical Discharge Machining (WEDM) is one such non-conventional, thermo-electric machining process which has an important role in high-precision and high-performance manufacturing industries due to its capability of accurate and efficient machining of parts with varying hardness or complex shapes and sharp edges of the workpiece. In WEDM material removal takes place by controlled erosion of the electrically conducting material. The material removal process governed by a series of discrete sparks between an anode workpiece which is immersed in a dielectric liquid medium and continuously feeding cathode wire through the workpiece controlled by a microprocessor. This electric discharge melts and vaporizes small amounts of the work material, and the removed material is flushed away by the flowing dielectric liquid. As the tool wire and the workpiece does not make direct contact during machining, so there is no machining stress, chatter and vibration. Therefore, this machining process can be utilized in machining of conductive material having low machinability and high temperature strength resistance.

In order to achieve high quality, high process safety, minimal manufacturing cost and lowest possible machining time the manufacturing process parameters have to choose in an optimized way. The photograph of WEDM setup is illustrated in Figure 1.

Figure 1. Machining setup of WEDM

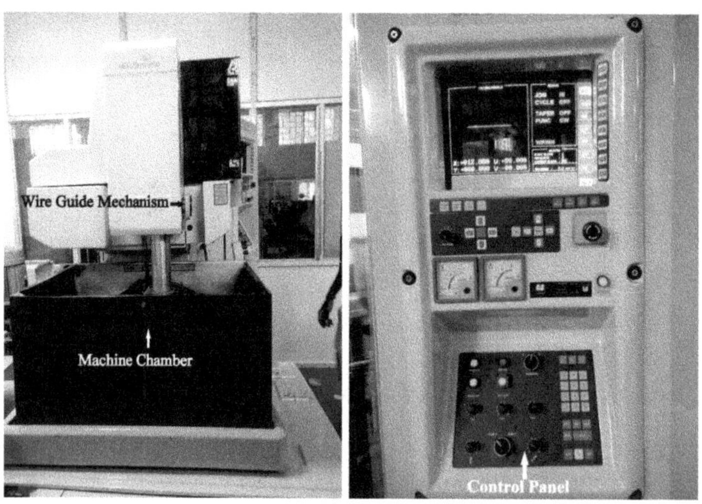

A brief literature review of the past research work is presented here. Liao et al. (2014) have studied on the association amid machining parameters and machining characteristic using a neural network and finally, genetic algorithm was used in order to find the optimal machining parameters combination for different materials.

Bobbili et al. (2013) present the impact of machine control parameters on surface roughness and MRR of high strength armor steel using WEDM. Zhang et al. (2014) studied the effects of high temperature and massive electrical discharges on tool steel (SKD11) during medium-speed WEDM and found the optimal parameters on MRR and 3D surface quality by using integrated response surface methodology along with Non-Dominated Sorting Genetic Algorithm-II. Patil and Brahmankar (2010) has investigated the effect of thermo-physical properties of the workpiece, pulse-on-time and average gap voltage on MRR in WEDM process for silicon carbide particulate reinforced aluminum matrix composites workpiece and find out significant role of coefficient of thermal expansion in this process. Mukherjee et al. (2012) applied population-based non-traditional optimization algorithm for single and multi-objective optimization of two different WEDM processes, and they observed that biogeography-based optimization algorithm outperforms the others. Ming et al. (2015) have examined on augmenting the cutting parameters in WEDM process using integrated artificial neural network (ANN) along with wolf pack algorithm based on the strategy of the leader (LWPA) and they mathematically constructed the effects of cutting parameters on machining time, machining cost and surface roughness. Kirkpatrick et al. (1983) stated that there is a profound and useful relation between statistical mechanics and combinatorial optimization. Kuriachen et al. (2015) revealed that capacitance is the predominant factor that influences the micro-WEDM process. Based on experimental observations they predict the process characteristics and also employed particle swarm optimization (PSO) algorithm to improve the performance. Pant et al. (2017) studied the newly developed PSO Algorithm for both controlled and unrestricted nonlinear condition and optimized the reliability of a composite system. Kumar et al. (2017) discussed several metaheuristic techniques and finally, Cuckoo Search Algorithm is used to solve complex life support system and bridge system. Ali et al. (2014) computed a hybridized Firefly algorithm and make the algorithm more flexible. Miguel and Miguel (2012) studied two metaheuristic algorithms, namely Harmony Search (HS) and Firefly Algorithm (FA) and demonstrate the effectiveness of both algorithms in engineering problem. Arora and Singh (2013) used FA to solve complex optimization problems. Kurikose and Shanmugham (2005) used a multiple-regression model to characterize a relationship between input and output parameters and a multi-objective optimization process based on a Non-Dominated Sorting Genetic Algorithm (NSGA) is used to optimize WEDM process. Somashekhar et al. (2010) proposed the improvement of modeling and optimization of micro-electric discharge machining process. A feed forward ANN is employed to analyze the MRR and then GA is applied to determine the optimum process parameter for machining. Garg and Mittal (2014) explain the experimental result of dejong function by using Genetic Algorithm. Rao & Krishna (2014) has analyzed to optimize the effect pulse-on-time, pulse-off-time, and wire tension on MRR, surface roughness, and wire wear ratio in WEDM process on Al7075/SiCp work-piece using the NSGA-II and obtain the set of Pareto-optimal solutions.

The objective of the current research work is to study the characteristic features of the WEDM process while machining AISI -D3 as analysis through Taguchi design based 625 experimental run with various process control parametric combinations like Pulse on Time (Ton), Pulse off Time (Toff), Wire Feed (W/Feed) and Wire Tension (W/Ten) on Material Removal Rate (MRR) and Surface Roughness (Ra). ANN is applied to identify and learn correlated patterns amongst input data sets and corresponding responses values. After training, validation and testing of the experimental data through ANN the outcomes can be predicted. Regression equations for MRR and Ra are individually developed in terms of the control parameters from the experimental data to established effect of variables on the outputs. The individual responses as indicated by their regression equations are optimized by FA technique. The contradictory objectives arising from the different responses are simultaneously optimized through GA. In order to determine the best parametric combination from the output results of GA a Multi-Criteria Decision Making (MCDM) technique popularly known as Grey Relational Analysis (GRA) is applied.

PLANNING FOR EXPERIMENTATION

In this experimentation, CNC Wirecut EDM (Maxicut-e, Electronica Make) is used. The Dielectric used is DM-Water having pH 7 (Specific gravity of 1 at 23°C and Viscosity of 0.001 Ns/m^2 at 20°C). External lateral flushing is done with a pressure of 0.8 kgf/cm^2. Experiments are organized with positive polarity of the electrode. The pulsed discharge current is employed in various steps in positive mode. The WEDM setup comprises of a dielectric reservoir, power generator, pump and dielectric flow system, and control unit, working tank, X-Y table accommodating the working table, wire holding nozzle, workpiece, flashing system as shown in Figure 2.

The servo control unit is arranged to sustain the pre-determined gap. It detects the gap voltage and compares it with the current value and the change in voltage is then used to regulate the movement of the servo motor to amend the gap. The work material AISI -D3having dimension of 250x12.6x12.6 mm is used. The percentage composition of various elements in AISI-D3 is C: 2-2.35%, Mn: 0.60%, Cr: 11-13.5%, Si: 0.60%, Ni: 0.30%, W: 1%, V: 1%, Cu: 0.25%, P: 0.03%, S: 0.03%. A constant Brass wire diameter of 0.25mm is maintained during experimentation. The length of cut is 12.66mm and angle of cut is vertical. The workpiece is at the center of the table. Servo reference voltage is taken as 35V. The injection pressure set point is 3.75kg/cm^2. Die electric temperature is 22-25 °C. Now, the MRR during experimentation is evaluated using Equation 1, which is shown below:

Figure 2. Machining Chamber

$$MRR = \text{Fm X } B \text{ X t } \left(mm^3 / min\right) \quad (1)$$

where

Fm = machining feed per minute (mm/min),
B = (2Wg + Wd) (mm)
Wg = wire gap (mm)
Wd = wire diameter (mm)
t = thickness of the job (mm)

The surface roughness of the cavity surface is articulated as Ra expressed in μmm, is measured using stylus type profilometer, Perthometer-M1 of Mahr Gmbh make.

During the experiment, levels of the factors are varied simultaneously to save time and money, and also allows for the study of relations between the factors. Initially, L625 orthogonal array is used for the experimentation. The control parameters are altered at five levels in six hundred and twenty-five experimental runs. Several other factors which may influence the output result like Flushing pressure, Lift time etc., are kept uniform during experimentation. Table 1 exhibits the different combinations of control parameters during the machining process.

ARTIFICIAL NEURAL NETWORK

An Artificial neural network (ANN) is fundamentally representing a biological neural network. Nodes are interconnected simple processes in a neural network. Each neuron receives a set of inputs (Xi, i=1, 2…., n-1, n) from associate neurons and produces the output signals (Oi, i= 1to K) at the output nodes. Each input Xi is multiplied by the associated weight Wi value in synapses and then transfer the newly developed signal to the hidden layer. Equation 2 illustrates the net input to each neuron in the hidden layer.

$$I_i = \sum_{i=1}^{n} W_i X_i \quad (2)$$

Generally, no output will be produced until the activation level of the node go beyond a threshold value. The output result of a neuron is usually described by a sigmoid function presented in equation 3.

$$O_i = f(\mathrm{I}_i) = \frac{1}{1 + e^{-I_i}} \quad (3)$$

Figure 3 shows the schematic diagram of an artificial neuron.

The most common neural network architectures are

- Feedforward neural networks
- Feedback neural networks

Table 1. Parametric settings and responses for experimental run

Exp. No.	Control Parameter				Responses		Exp. No.	Control Parameter				Responses	
	Ton (μSec)	Tof (μSec)	W/Fee d (m/min)	W/ Ten (gms)	MRR (mm3/ Sec)	Ra (μmm)		Ton (μSec)	Toff (μSec)	W/ Feed (m/ min)	W/ Ten (gms)	MRR (mm3/ Sec)	Ra (μmm)
1	10	30	5	6	0.45	25.66	314	6	30	7	10	1.16	28.71
2	6	20	3	4	0.53	12.98	315	8	30	5	10	0.69	27.88
3	14	35	3	6	0.07	33.44	316	6	30	3	6	2.44	28.93
4	10	25	7	10	1.36	18.33	317	6	25	3	6	1.19	20.15
5	8	20	4	4	0.12	12.03	318	6	40	3	4	6.50	52.11
6	14	20	7	10	3.26	10.90	319	10	35	4	2	2.71	35.64
7	10	40	7	2	3.77	45.75	320	8	20	7	2	0.39	10.86
8	14	25	6	6	2.37	15.18	321	6	20	6	6	0.05	12.76
9	6	40	4	10	5.07	52.31	322	10	25	6	8	0.93	17.87
10	8	30	7	2	1.48	25.44	323	10	40	7	6	3.01	46.45
11	14	35	5	4	0.03	32.04	324	12	30	4	4	0.04	24.42
12	10	40	4	2	4.56	47.70	325	14	25	3	10	2.40	17.10
13	10	40	6	4	3.65	46.64	326	14	20	7	4	2.63	8.87
14	14	25	7	2	2.07	13.84	327	6	20	6	2	0.31	11.89
15	12	35	3	4	1.55	34.80	328	10	25	6	4	0.43	16.86
16	6	40	4	4	6.27	51.52	329	10	30	3	4	1.15	26.20
17	10	40	7	10	2.21	47.46	330	8	30	5	8	1.00	27.38
18	10	40	6	8	2.86	47.38	331	12	35	7	6	0.27	33.10
19	14	40	7	6	0.55	42.58	332	12	40	7	2	2.59	43.79
20	6	20	4	6	0.20	13.09	333	8	20	6	8	0.80	12.60
21	8	40	7	2	4.89	47.69	334	6	20	7	10	0.56	13.91
22	8	35	7	6	2.34	36.36	335	6	40	7	2	5.93	49.60
23	14	35	5	6	0.39	32.40	336	12	40	4	6	2.58	46.20
24	14	35	3	4	0.46	33.20	337	6	20	5	8	0.12	13.41
25	10	35	6	8	1.21	35.62	338	8	35	3	4	3.54	37.93
26	14	35	5	10	1.15	33.34	339	6	35	6	2	4.13	37.74
27	10	25	4	6	0.34	17.91	340	10	40	3	8	3.56	48.87
28	10	20	7	8	1.57	11.74	341	14	40	3	6	1.60	44.93
29	6	20	4	4	0.40	12.73	342	8	20	7	8	0.94	12.50
30	8	35	4	4	3.33	37.40	343	14	25	4	2	1.45	15.15
31	14	35	6	8	1.00	32.39	344	14	25	6	8	2.64	15.76
32	6	40	5	4	6.04	50.96	345	6	35	7	2	3.91	37.24
33	14	25	5	2	1.65	14.70	346	14	25	6	2	1.86	14.26
34	6	25	6	4	0.97	18.95	347	14	30	5	2	0.85	22.27

continued on following page

Table 1. Continued

Exp. No.	Control Parameter				Responses		Exp. No.	Control Parameter				Responses	
	Ton (µSec)	Tof (µSec)	W/Fee d (m/min)	W/ Ten (gms)	MRR (mm3/ Sec)	Ra (µmm)		Ton (µSec)	Toff (µSec)	W/ Feed (m/ min)	W/ Ten (gms)	MRR (mm3/ Sec)	Ra (µmm)
35	8	35	6	4	2.89	36.41	348	10	20	7	2	1.00	10.04
36	6	35	4	4	4.23	38.95	349	8	40	4	8	4.44	50.17
37	12	35	6	8	0.14	34.02	350	12	40	4	10	1.72	46.90
38	10	40	5	8	3.10	47.86	351	8	35	3	2	3.89	37.81
39	10	35	5	6	1.78	35.62	352	12	30	3	10	0.76	25.94
40	10	40	5	4	3.91	47.23	353	10	35	7	6	1.34	34.74
41	12	25	5	2	0.79	15.79	354	6	25	4	4	1.29	19.56
42	12	35	3	6	1.18	35.03	355	8	35	5	4	3.11	36.90
43	6	25	5	6	0.89	19.62	356	10	25	7	2	0.38	16.10
44	14	20	6	4	2.45	9.15	357	10	20	6	8	1.43	11.86
45	8	25	7	8	0.34	18.70	358	8	40	6	10	3.58	49.71
46	8	40	7	6	4.14	48.35	359	14	30	5	6	1.48	22.99
47	10	35	4	10	1.26	36.85	360	6	25	7	6	0.58	19.15
48	12	40	5	2	3.14	45.11	361	6	30	6	10	1.33	28.90
49	12	35	5	2	1.44	33.42	362	8	30	6	2	1.69	25.88
50	10	40	4	4	4.16	47.84	363	14	20	6	6	2.66	9.69
51	12	25	7	2	1.19	14.98	364	8	25	5	10	0.29	19.63
52	6	40	3	8	5.69	52.45	365	10	25	3	2	0.36	17.72
53	8	40	4	4	5.25	49.69	366	12	40	6	6	2.07	45.06
54	12	20	5	2	1.33	9.85	367	14	25	6	4	2.11	14.68
55	14	20	4	4	2.11	9.76	368	6	30	4	2	2.86	28.10
56	12	25	6	2	0.99	15.38	369	12	35	4	6	0.96	34.51
57	10	40	4	8	3.33	48.36	370	8	25	3	6	0.54	19.20
58	10	20	6	4	1.02	10.77	371	14	35	6	6	0.63	31.90
59	14	40	7	10	0.27	43.65	372	8	30	7	6	0.93	26.28
60	10	25	7	6	0.85	17.06	373	12	20	3	4	1.22	10.85
61	10	20	5	4	0.87	11.02	374	12	40	4	2	3.40	45.80
62	8	20	6	2	0.23	11.13	375	10	30	4	2	1.25	25.51
63	12	25	4	8	1.41	17.34	376	8	25	7	2	0.35	17.19
64	12	40	6	8	1.66	45.49	377	14	25	5	4	1.91	15.06
65	8	20	4	2	0.08	11.72	378	6	20	3	8	0.11	13.67
66	10	25	6	6	0.68	17.33	379	6	35	7	6	3.27	37.96
67	14	25	3	2	1.26	15.63	380	12	20	3	6	1.45	11.21
68	10	30	7	4	0.34	24.47	381	6	30	4	4	2.57	28.30

continued on following page

Table 1. Continued

Exp. No.	Control Parameter				Responses		Exp. No.	Control Parameter				Responses	
	Ton (µSec)	Tof (µSec)	W/Fee d (m/min)	W/ Ten (gms)	MRR (mm3/ Sec)	Ra (µmm)		Ton (µSec)	Toff (µSec)	W/ Feed (m/ min)	W/ Ten (gms)	MRR (mm3/ Sec)	Ra (µmm)
69	6	20	3	10	0.11	14.12	382	12	20	4	4	1.38	10.54
70	8	35	6	6	2.56	36.77	383	6	35	7	4	3.60	37.56
71	10	25	5	8	0.77	18.09	384	10	35	6	10	0.85	36.15
72	10	20	3	10	1.25	12.83	385	10	30	6	2	0.83	24.53
73	8	40	7	4	4.52	47.98	386	14	25	3	6	1.81	16.21
74	14	30	5	10	2.14	24.02	387	12	35	4	8	0.58	34.87
75	10	35	5	2	2.48	35.06	388	10	25	3	8	0.45	18.60
76	14	20	3	10	2.67	11.45	389	12	20	7	8	2.27	10.95
77	6	25	4	8	0.79	20.27	390	8	25	4	4	0.64	18.57
78	14	20	5	6	2.50	9.93	391	6	20	6	4	0.13	12.29
79	14	25	7	4	2.31	14.32	392	10	30	3	8	0.50	26.78
80	6	35	3	6	4.08	39.63	393	14	40	7	4	0.95	42.15
81	6	40	4	6	5.88	51.71	394	6	25	4	10	0.52	20.74
82	6	30	4	10	1.65	29.35	395	14	40	7	2	1.34	41.81
83	6	25	5	10	0.39	20.59	396	10	20	5	6	1.07	11.47
84	8	25	6	2	0.54	17.55	397	10	20	4	6	0.93	11.69
85	6	30	7	2	2.27	26.79	398	6	20	7	4	0.01	12.10
86	12	25	6	10	2.02	17.46	399	12	25	5	10	1.85	17.65
87	14	25	5	6	2.18	15.51	400	10	30	5	4	0.75	25.29
88	10	20	7	10	1.78	12.45	401	10	35	5	4	2.14	35.30
89	14	30	4	8	1.60	23.84	402	14	30	6	8	2.01	23.12
90	12	35	3	10	0.41	35.70	403	10	35	7	8	0.99	35.25
91	10	25	5	6	0.51	17.61	404	10	35	7	10	0.64	35.83
92	12	40	6	4	2.47	44.71	405	6	30	7	6	1.73	27.60
93	14	40	5	6	1.08	43.72	406	14	30	6	2	1.08	21.75
94	14	30	4	6	1.27	23.42	407	8	35	7	2	2.99	35.61
95	6	35	3	4	4.43	39.46	408	14	35	3	8	0.31	33.76
96	12	20	5	10	2.20	11.88	409	14	35	5	2	0.33	31.76
97	6	40	7	6	5.19	50.23	410	10	30	7	8	0.25	25.50
98	8	25	5	6	0.23	18.62	411	6	25	7	4	0.81	18.67
99	14	20	5	4	2.28	9.45	412	8	30	5	4	1.60	26.61
100	12	40	7	8	1.41	45.01	413	12	20	6	10	2.34	11.78
101	14	30	5	8	1.80	23.47	414	6	35	5	2	4.35	38.26
102	6	25	6	8	0.50	19.88	415	12	20	5	6	1.75	10.72

continued on following page

Table 1. Continued

Exp. No.	Control Parameter				Responses		Exp. No.	Control Parameter				Responses	
	Ton (μSec)	Tof (μSec)	W/Fee d (m/min)	W/ Ten (gms)	MRR (mm3/ Sec)	Ra (μmm)		Ton (μSec)	Toff (μSec)	W/ Feed (m/ min)	W/ Ten (gms)	MRR (mm3/ Sec)	Ra (μmm)
103	8	30	5	2	1.89	26.34	416	8	25	3	10	0.01	19.99
104	14	20	6	2	2.25	8.68	417	8	20	5	10	0.88	13.31
105	8	25	7	10	0.59	19.35	418	12	30	7	6	0.90	23.58
106	12	20	4	10	2.06	12.01	419	12	40	6	2	2.87	44.44
107	8	40	3	2	5.89	50.23	420	12	30	5	2	0.13	23.65
108	12	25	3	10	1.53	18.09	421	12	20	6	2	1.51	9.52
109	12	30	4	8	0.61	25.15	422	12	25	4	4	0.86	16.52
110	8	30	6	10	0.52	27.64	423	8	20	5	4	0.27	11.77
111	10	30	5	10	0.18	26.62	424	12	20	7	6	2.07	10.30
112	8	25	4	6	0.39	18.90	425	8	25	5	4	0.48	18.23
113	6	30	5	2	2.67	27.65	426	8	25	4	8	0.13	19.32
114	14	35	3	2	0.83	33.03	427	8	20	3	4	0.02	12.30
115	6	20	5	10	0.33	13.98	428	10	35	3	8	1.84	36.87
116	10	30	4	10	0.01	26.89	429	14	20	5	8	2.72	10.50
117	10	25	7	8	1.10	17.66	430	10	35	4	8	1.63	36.43
118	12	25	5	4	1.05	16.14	431	14	30	4	4	0.94	23.08
119	6	40	5	2	6.42	50.79	432	14	30	3	4	0.73	23.58
120	8	25	4	2	0.89	18.31	433	8	40	7	10	3.35	49.32
121	10	35	6	2	2.24	34.50	434	10	30	3	2	1.46	26.02
122	8	25	5	8	0.03	19.09	435	12	35	7	8	0.09	33.62
123	14	35	7	2	0.18	30.57	436	12	35	5	6	0.73	34.02
124	12	20	7	2	1.68	9.21	437	6	30	6	6	1.92	27.90
125	12	30	7	2	0.32	22.67	438	14	40	5	4	1.50	43.40
126	8	30	4	2	2.09	26.82	439	8	25	3	2	1.06	18.72
127	14	20	4	6	2.34	10.19	440	10	30	6	6	0.25	25.29
128	6	30	6	4	2.20	27.52	441	8	30	4	4	1.79	27.03
129	8	30	7	8	0.64	26.82	442	14	30	6	6	1.69	22.58
130	10	20	6	10	1.64	12.52	443	8	20	3	2	0.22	12.05
131	6	35	6	10	2.79	39.26	444	6	30	6	2	2.47	27.21
132	14	20	7	2	2.43	8.35	445	8	35	4	2	3.67	37.23
133	6	35	5	10	2.98	39.55	446	10	25	7	4	0.62	16.54
134	6	40	5	8	5.26	51.51	447	6	30	5	8	1.80	28.63
135	10	25	3	10	0.74	19.05	448	14	25	4	10	2.57	16.85
136	10	40	6	10	2.45	47.86	449	8	40	3	6	5.08	50.45

continued on following page

Table 1. Continued

Exp. No.	Control Parameter				Responses		Exp. No.	Control Parameter				Responses	
	Ton (µSec)	Tof (µSec)	W/Fee d (m/min)	W/ Ten (gms)	MRR (mm3/ Sec)	Ra (µmm)		Ton (µSec)	Toff (µSec)	W/ Feed (m/ min)	W/ Ten (gms)	MRR (mm3/ Sec)	Ra (µmm)
137	10	30	7	6	0.05	24.94	450	10	30	3	6	0.83	26.45
138	6	35	3	10	3.35	40.20	451	10	30	6	4	0.55	24.87
139	14	40	5	2	1.90	43.17	452	10	20	3	4	0.57	11.59
140	10	35	5	8	1.42	36.02	453	14	20	3	4	1.94	10.10
141	8	20	7	10	1.14	13.20	454	12	20	5	4	1.54	10.25
142	6	25	3	2	1.69	19.71	455	6	30	3	2	3.04	28.58
143	8	25	7	4	0.13	17.62	456	12	30	5	4	0.17	23.96
144	12	40	3	8	2.39	47.05	457	10	35	4	4	2.36	35.83
145	8	30	3	10	1.02	28.40	458	6	20	7	8	0.37	13.23
146	6	30	7	4	2.01	27.16	459	10	35	4	6	2.00	36.09
147	12	35	6	2	1.19	32.84	460	6	25	6	10	0.25	20.46
148	6	30	7	8	1.45	28.12	461	10	25	4	4	0.08	17.56
149	12	25	7	4	1.43	15.44	462	10	35	5	10	1.06	36.49
150	8	25	6	10	0.44	19.48	463	12	20	7	10	2.48	11.69
151	10	35	7	2	2.00	33.95	464	8	35	6	2	3.22	36.13
152	10	35	3	10	1.46	37.23	465	12	40	5	8	1.91	45.99
153	10	30	5	8	0.14	26.10	466	6	40	4	2	6.66	51.42
154	10	25	5	2	0.01	16.87	467	8	25	4	10	0.15	19.80
155	8	30	3	4	1.98	27.47	468	6	30	5	10	1.49	29.11
156	14	40	6	2	1.62	42.48	469	10	30	4	8	0.32	26.43
157	12	40	3	10	1.95	47.38	470	12	35	6	4	0.85	33.16
158	14	35	4	2	0.58	32.39	471	8	40	4	6	4.85	49.90
159	12	40	7	6	1.81	44.53	472	14	25	4	6	1.99	15.85
160	6	20	6	8	0.24	13.31	473	14	30	7	8	2.22	22.78
161	8	35	5	10	2.05	38.03	474	10	40	7	4	3.40	46.06
162	10	20	4	2	0.51	10.98	475	14	30	3	10	1.75	24.67
163	14	30	3	8	1.40	24.23	476	14	40	5	10	0.23	44.57
164	10	25	4	8	0.61	18.34	477	6	25	4	6	1.04	19.87
165	14	30	5	4	1.16	22.59	478	6	20	5	2	0.45	12.16
166	8	25	6	4	0.30	17.92	479	14	30	4	10	1.94	24.34
167	14	30	7	4	1.61	21.68	480	12	35	4	10	0.20	35.31
168	8	40	3	4	5.49	50.30	481	8	20	6	10	1.01	13.24
169	8	30	6	6	1.12	26.61	482	8	30	4	8	1.18	27.69
170	10	30	5	2	1.04	25.01	483	10	30	4	4	0.95	25.74

continued on following page

Table 1. Continued

Exp. No.	Control Parameter				Responses		Exp. No.	Control Parameter				Responses	
	Ton (μSec)	Tof (μSec)	W/Fee d (m/min)	W/ Ten (gms)	MRR (mm3/ Sec)	Ra (μmm)		Ton (μSec)	Toff (μSec)	W/ Feed (m/ min)	W/ Ten (gms)	MRR (mm3/ Sec)	Ra (μmm)
171	8	35	5	8	2.41	37.58	484	8	35	5	2	3.45	36.67
172	6	25	5	4	1.13	19.24	485	10	30	4	6	0.64	26.05
173	14	35	4	8	0.54	33.29	486	8	35	4	6	2.98	37.65
174	10	35	7	4	1.67	34.31	487	8	35	3	10	2.44	38.73
175	6	40	3	10	5.27	52.73	488	10	40	3	4	4.40	48.47
176	14	35	7	8	1.23	31.97	489	14	25	6	10	2.91	16.41
177	6	40	7	4	5.57	49.88	490	6	35	5	4	4.02	38.47
178	8	20	3	6	0.20	12.62	491	12	40	7	10	1.01	45.57
179	12	40	4	4	2.99	45.96	492	14	25	7	10	3.09	16.22
180	14	25	4	4	1.72	15.46	493	12	25	7	6	1.67	15.98
181	10	25	6	10	1.19	18.48	494	6	30	3	4	2.75	28.72
182	6	25	5	2	1.37	18.95	495	8	35	3	6	3.18	38.12
183	6	40	4	8	5.48	51.97	496	12	25	5	8	1.58	17.07
184	12	40	5	4	2.74	45.33	497	8	40	3	8	4.66	50.67
185	12	20	6	6	1.91	10.50	498	12	40	5	10	1.49	46.44
186	8	30	3	2	2.28	27.31	499	8	35	7	10	1.65	37.42
187	6	35	5	8	3.34	39.12	500	6	40	6	10	4.64	51.53
188	8	40	6	2	5.14	48.29	501	6	25	7	2	1.02	18.26
189	12	20	7	4	1.87	9.71	502	14	40	3	4	2.03	44.73
190	8	20	5	8	0.67	12.72	503	10	20	3	8	1.02	12.34
191	6	35	3	2	4.78	39.36	504	12	25	3	8	1.24	17.62
192	10	20	7	6	1.37	11.10	505	6	30	3	8	2.13	29.23
193	12	30	5	6	0.48	24.34	506	6	40	6	4	5.81	50.41
194	8	40	5	6	4.61	49.36	507	10	30	3	10	0.17	27.18
195	12	25	5	6	1.31	16.57	508	14	35	6	4	0.27	31.49
196	10	40	6	2	4.04	46.38	509	14	40	6	4	1.22	42.77
197	8	25	3	8	0.27	19.56	510	10	40	5	2	4.30	47.03
198	12	20	3	2	1.00	10.57	511	14	40	4	8	0.91	44.64
199	8	20	4	8	0.54	12.86	512	14	20	7	8	3.04	10.15
200	12	25	3	4	0.68	16.92	513	8	35	3	8	2.81	38.39
201	14	25	3	8	2.10	16.62	514	14	35	5	8	0.76	32.83
202	14	25	7	8	2.82	15.51	515	6	20	4	2	0.59	12.45
203	12	25	4	10	1.69	17.86	516	6	20	3	2	0.73	12.75
204	10	35	6	6	1.56	35.17	517	14	40	4	2	2.18	43.88

continued on following page

Table 1. Continued

Exp. No.	Control Parameter				Responses		Exp. No.	Control Parameter				Responses	
	Ton (µSec)	Tof (µSec)	W/Fee d (m/min)	W/ Ten (gms)	MRR (mm3/ Sec)	Ra (µmm)		Ton (µSec)	Toff (µSec)	W/ Feed (m/ min)	W/ Ten (gms)	MRR (mm3/ Sec)	Ra (µmm)
205	6	25	6	2	1.20	18.59	518	14	20	7	6	2.83	9.47
206	12	30	6	2	0.09	23.15	519	6	20	7	6	0.19	12.63
207	14	40	6	8	0.40	43.58	520	12	20	5	8	1.97	11.26
208	12	20	4	6	1.60	10.95	521	10	40	4	6	3.75	48.06
209	8	40	4	10	4.02	50.53	522	12	35	6	6	0.50	33.55
210	6	35	4	8	3.53	39.49	523	8	20	3	10	0.65	13.49
211	10	40	3	2	4.81	48.38	524	6	40	5	10	4.86	51.91
212	6	30	6	8	1.63	28.36	525	12	20	3	8	1.68	11.64
213	12	30	3	2	0.56	24.71	526	8	25	6	6	0.07	18.36
214	10	20	3	2	0.36	11.32	527	10	25	6	2	0.19	16.47
215	8	20	3	8	0.42	13.02	528	6	35	6	4	3.81	38.00
216	10	35	3	6	2.21	36.58	529	6	20	5	4	0.27	12.50
217	14	35	4	4	0.22	32.61	530	10	20	5	2	0.67	10.65
218	12	25	4	6	1.13	16.89	531	14	40	4	4	1.76	44.06
219	12	30	6	6	0.69	23.95	532	8	35	7	4	2.67	35.94
220	6	25	3	10	0.66	20.90	533	14	30	7	6	1.91	22.19
221	10	25	4	10	0.88	18.84	534	8	35	6	8	2.21	37.21
222	8	35	4	10	2.25	38.37	535	14	25	3	4	1.53	15.88
223	6	25	6	6	0.74	19.38	536	12	25	6	4	1.24	15.78
224	6	20	3	6	0.33	13.29	537	14	35	6	10	1.37	32.95
225	6	30	5	4	2.39	27.90	538	6	40	6	8	5.04	51.08
226	8	40	3	10	4.23	50.97	539	10	20	5	10	1.51	12.61
227	8	30	3	8	1.35	28.01	540	14	40	5	8	0.66	44.10
228	12	30	3	6	0.08	25.17	541	8	20	5	2	0.07	11.42
229	12	35	6	10	0.23	34.57	542	12	30	4	2	0.35	24.17
230	8	20	6	4	0.41	11.54	543	6	20	6	10	0.44	13.94
231	14	30	4	2	0.63	22.81	544	10	25	5	4	0.25	17.20
232	14	30	3	6	1.06	23.87	545	14	35	4	6	0.16	32.91
233	10	20	3	6	0.79	11.93	546	12	25	6	6	1.49	16.27
234	8	30	7	4	1.21	25.82	547	10	25	5	10	1.04	18.65
235	6	35	5	6	3.68	38.75	548	12	35	4	4	1.32	34.23
236	6	30	4	6	2.27	28.57	549	6	40	7	10	4.43	51.16
237	8	40	6	6	4.38	48.85	550	10	30	6	8	0.05	25.79
238	10	20	4	8	1.15	12.16	551	8	25	6	8	0.18	18.89

continued on following page

Table 1. Continued

Exp. No.	Control Parameter				Responses		Exp. No.	Control Parameter				Responses	
	Ton (µSec)	Tof (µSec)	W/Fee d (m/min)	W/ Ten (gms)	MRR (mm3/ Sec)	Ra (µmm)		Ton (µSec)	Toff (µSec)	W/ Feed (m/ min)	W/ Ten (gms)	MRR (mm3/ Sec)	Ra (µmm)
239	14	40	7	8	0.14	43.08	552	6	30	4	8	1.97	28.92
240	12	35	5	10	0.01	34.93	553	14	30	6	10	2.33	23.72
241	8	40	5	2	5.40	48.92	554	8	40	7	8	3.75	48.80
242	10	30	6	10	0.36	26.36	555	14	20	3	8	2.42	10.92
243	14	25	5	10	2.74	16.62	556	12	25	6	8	1.75	16.82
244	10	35	6	4	1.91	34.80	557	8	30	4	6	1.49	27.32
245	10	20	5	8	1.29	12.00	558	10	40	5	10	2.68	48.28
246	12	40	3	4	3.25	46.61	559	10	30	7	10	0.55	26.13
247	12	30	7	10	1.51	24.80	560	6	35	3	8	3.72	39.88
248	10	35	3	4	2.58	36.38	561	6	35	6	8	3.14	38.76
249	12	30	5	8	0.80	24.80	562	14	30	7	2	1.32	21.25
250	10	25	3	4	0.10	17.94	563	10	40	3	10	3.13	49.19
251	12	30	7	8	1.20	24.15	564	12	35	5	8	0.36	34.44
252	6	35	6	6	3.48	38.35	565	14	20	4	8	2.57	10.70
253	8	25	5	2	0.72	17.92	566	12	20	6	8	2.12	11.10
254	10	20	7	4	1.18	10.53	567	6	25	3	8	0.93	20.49
255	6	25	7	8	0.35	19.72	568	6	25	4	2	1.53	19.32
256	14	20	3	2	1.72	9.80	569	6	20	7	2	0.16	11.64
257	6	40	6	2	6.18	50.18	570	10	40	4	10	2.91	48.73
258	12	35	3	8	0.80	35.33	571	6	35	4	6	3.88	39.18
259	6	25	7	10	0.11	20.35	572	10	25	3	6	0.17	18.23
260	8	30	6	8	0.82	27.09	573	12	30	3	4	0.24	24.90
261	12	30	6	8	1.00	24.46	574	10	20	4	4	0.72	11.30
262	10	20	6	6	1.22	11.28	575	6	40	6	6	5.43	50.70
263	8	35	7	8	2.00	36.85	576	12	40	3	2	3.66	46.51
264	8	30	3	6	1.67	27.70	577	8	40	5	8	4.21	49.70
265	14	40	3	2	2.45	44.61	578	6	20	5	6	0.08	12.92
266	8	20	4	6	0.33	12.40	579	12	35	3	2	1.92	34.65
267	8	40	6	8	3.98	49.24	580	8	40	4	2	5.64	49.57
268	14	35	7	6	0.87	31.43	581	14	40	3	8	1.16	45.20
269	12	30	6	10	1.32	25.05	582	12	30	6	4	0.38	23.51
270	8	30	6	4	1.41	26.21	583	14	20	5	10	2.96	11.14
271	12	30	4	10	0.94	25.63	584	14	25	7	6	2.56	14.88
272	6	35	7	8	2.94	38.43	585	14	30	6	4	1.38	22.13

continued on following page

Table 1. Continued

Exp. No.	Control Parameter				Responses		Exp. No.	Control Parameter				Responses	
	Ton (µSec)	Tof (µSec)	W/Fee d (m/min)	W/ Ten (gms)	MRR (mm3/ Sec)	Ra (µmm)		Ton (µSec)	Toff (µSec)	W/ Feed (m/ min)	W/ Ten (gms)	MRR (mm3/ Sec)	Ra (µmm)
273	8	40	5	10	3.80	50.11	586	6	30	5	6	2.10	28.23
274	12	35	7	4	0.61	32.65	587	14	40	4	6	1.34	44.31
275	12	40	4	8	2.15	46.51	588	12	40	3	6	2.82	46.79
276	10	35	3	2	2.94	36.24	589	8	30	4	10	0.86	28.13
277	12	30	4	6	0.28	24.75	590	6	25	3	4	1.45	19.89
278	14	30	3	2	0.41	23.37	591	14	25	4	8	2.28	16.31
279	6	40	3	2	6.90	52.06	592	8	20	4	10	0.76	13.39
280	12	40	5	6	2.33	45.62	593	12	20	3	10	1.93	12.15
281	6	40	3	6	6.10	52.24	594	14	20	5	2	2.07	9.03
282	8	20	6	6	0.60	12.03	595	14	20	6	8	2.88	10.31
283	14	40	4	10	0.47	45.05	596	10	20	6	2	0.83	10.34
284	10	40	7	8	2.62	46.92	597	10	40	5	6	3.51	47.50
285	14	30	7	10	2.54	23.45	598	12	30	3	8	0.42	25.52
286	6	20	4	8	0.00	13.53	599	8	20	7	6	0.75	11.87
287	12	30	5	10	1.13	25.33	600	8	40	6	4	4.76	48.53
288	14	20	3	6	2.18	10.47	601	10	40	3	6	3.98	48.63
289	14	25	5	8	2.46	16.03	602	8	40	5	4	5.01	49.10
290	10	30	7	2	0.62	24.07	603	12	35	7	10	0.44	34.22
291	6	40	5	6	5.66	51.20	604	6	40	7	8	4.81	50.66
292	8	35	6	10	1.86	37.72	605	14	40	6	6	0.82	43.14
293	14	35	6	2	0.08	31.16	606	12	40	6	10	1.25	45.99
294	14	35	7	4	0.52	30.96	607	6	25	5	8	0.65	20.06
295	12	25	4	2	0.60	16.23	608	14	35	4	10	0.93	33.74
296	14	20	6	10	3.11	11.01	609	14	40	3	10	0.71	45.56
297	14	35	7	10	1.60	32.59	610	12	40	7	4	2.21	44.12
298	10	25	4	2	0.18	17.28	611	8	25	7	6	0.10	18.12
299	12	35	5	4	1.09	33.68	612	12	25	3	2	0.41	16.68
300	10	40	6	6	3.26	46.97	613	6	35	4	10	3.17	39.87
301	12	20	4	2	1.17	10.20	614	8	25	3	4	0.81	18.93
302	14	35	3	10	0.71	34.16	615	6	35	7	10	2.60	38.98
303	8	35	5	6	2.77	37.20	616	14	20	4	2	1.89	9.41
304	8	35	4	8	2.62	37.97	617	12	35	4	2	1.68	34.03
305	12	25	7	10	2.19	17.29	618	12	30	7	4	0.60	23.09
306	6	20	4	10	0.22	14.04	619	8	20	7	4	0.56	11.33

continued on following page

Table 1. Continued

Exp. No.	Control Parameter				Responses		Exp. No.	Control Parameter				Responses	
	Ton (μSec)	Tof (μSec)	W/Fee d (m/min)	W/ Ten (gms)	MRR (mm3/ Sec)	Ra (μmm)		Ton (μSec)	Toff (μSec)	W/ Feed (m/ min)	W/ Ten (gms)	MRR (mm3/ Sec)	Ra (μmm)
307	12	20	4	8	1.83	11.44	620	6	30	3	10	1.81	29.60
308	6	35	4	2	4.57	38.80	621	12	20	6	4	1.70	9.97
309	12	25	7	8	1.93	16.60	622	8	20	5	6	0.46	12.21
310	14	40	6	10	0.02	44.10	623	8	30	7	10	0.34	27.43
311	12	25	3	6	0.96	17.23	624	14	20	4	10	2.81	11.28
312	12	35	7	2	0.95	32.27	625	8	30	5	6	1.31	26.96
313	10	20	4	10	1.38	12.71							

Figure 3. Schematic diagram of an artificial neural

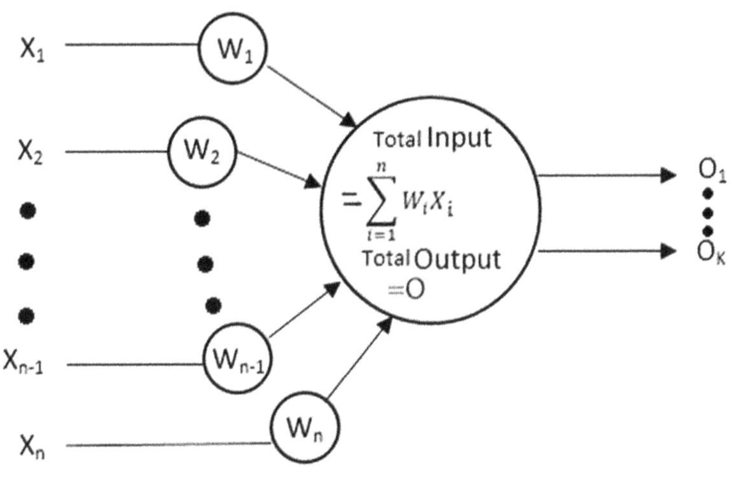

- Self-organizing neural networks

Figure 4 shows the architecture of a neural network.

Problems involving nonlinear and complex data are easily processed even if the data are imprecise and noisy. The technique is ideally suited for modeling a complex process like WEDM.

Analysis of The Result

The statistical design is carried out by the software Matlab R2015a. The algorithms considered in Matlab for calculation have followed the details: the data has been separated using the divider and function, Levenberg-Marquardt for training, Mean Squared Error (MSE) for performance. The simulink model for ANN while considering four control parameters to optimize the responses is shown if Figure 5. In

Figure 4. Architecture of a neural network

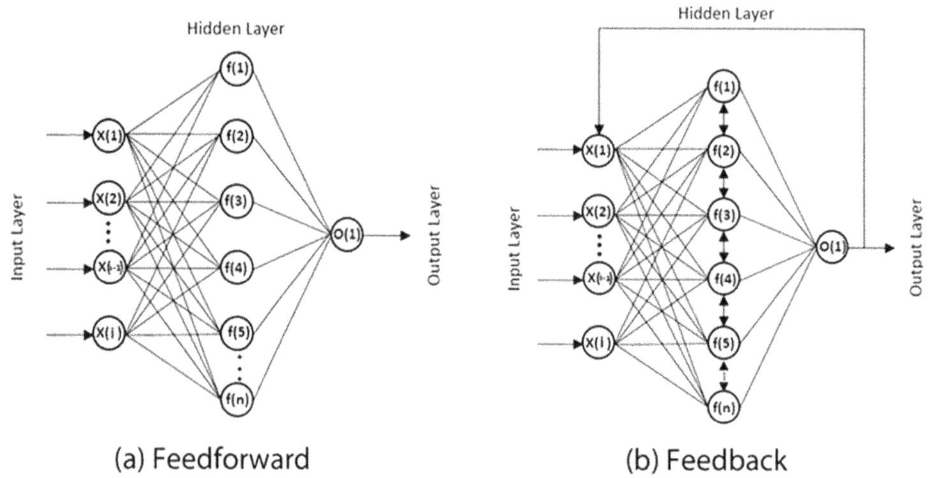

Figure 5. Simulink diagram

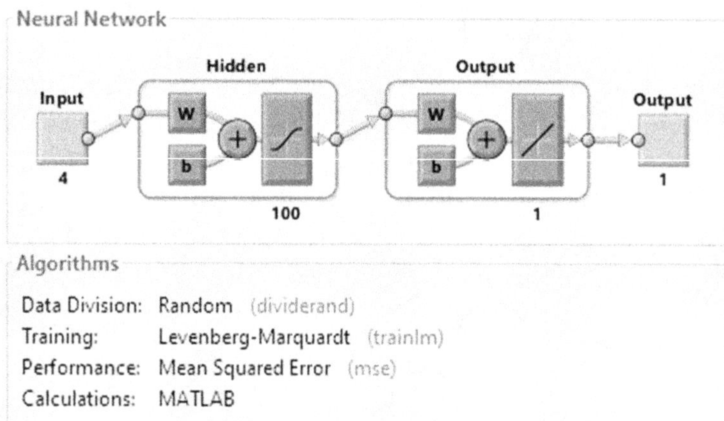

the planned ANN model 625 experimental run are conducted for training. After giving the weight, the network is ready for training.

The backpropagation algorithm is used to calculate the gradient and the Jacobian. MSE between the network output and target value is applied in ANN's model. Epoch is the presentation of the set of trainings (input and/or target) vectors to a network and the design of new weights and biases. First, the maximum number of 1000 epoch was set to terminate the training for each case.

A regression plot shows the relationship between the outputs of the network and the targets. The four plots represent the training, validation, testing and overall data. The dashed line in each plot signifies the perfect result – outputs = targets. The continuous line signifies the best fit linear regression line in between outputs and targets. If R = 1, this indicates that there is an exact linear relationship between outputs and respective targets. If R is nearly zero, then there is no linear relationship between outputs and targets.

Here outputs are separately analyzed subject on the four process control parameters. During ANN analysis, 100 hidden layers were considered.

Analysis to Maximize MRR

Here during computation, 70% of the data are considered for training, 15% of the data are considered for validation and 15% of the data are considered for testing. After 30iterations, it was terminated since the MSE is achieved. If the magnitude of the gradient is less than 1.00x10^{7}, the training will stop. It is observed that the gradient here is 1.24x10^{-3}. If the number of validation check reaches 6, the training will

Figure 6. Training performance progress for MRR

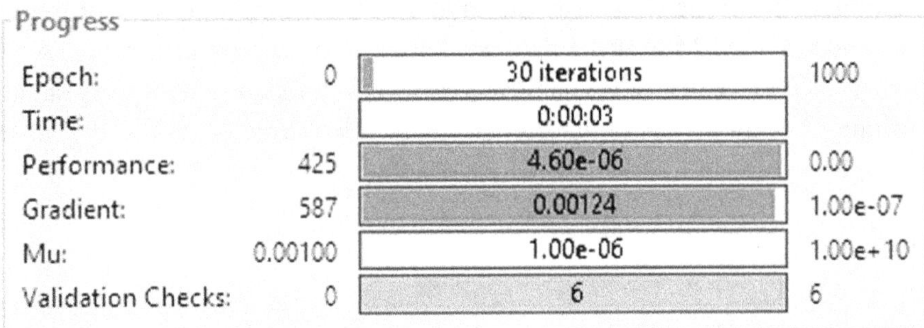

Figure 7. Performance plot for MRR

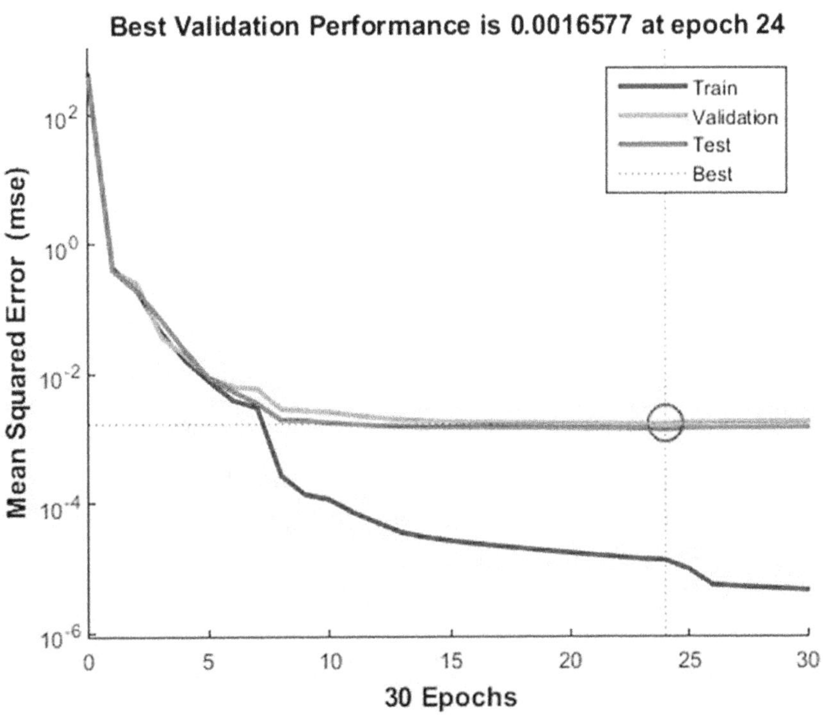

stop. Here the number of successive iterations performed for validation checks is 6. Figure 6 represents the neural network training performance progress.

Best validation performance is 1.6577x10-3 at epoch 24, which is a low prediction error measured with MSE shown in Figure 7.

Figure 7 highlights the best validation performance with respect to the iteration. The training was prolonged for 30 iterations till it got terminated. From the figure, no major problems with the training are indicated. The validation and test curves follow a similar trend.

Here in Figure 8 in case of training R = 1, in case of validation R = 0.99946 and in case of testing R=0.99967. Hence overall R = 0.99989. Therefore, the training data indicates a good fit as the validation and test results both show R values that greater than 0.9.

Analysis to Minimize Ra

Here while computing, 70% of the data are taking into account for training, 15% of the data are taking into account for validation and 15% of the data are taking into account for testing. After 8iterations, it was terminated science there was in the achieved MSE. If the magnitude of the gradient is less than $1.00x10^{-7}$, the training will stop. It is observed that the gradient here is 1.93×10^{-4}. If the number of

Figure 8. Regression plot of MRR

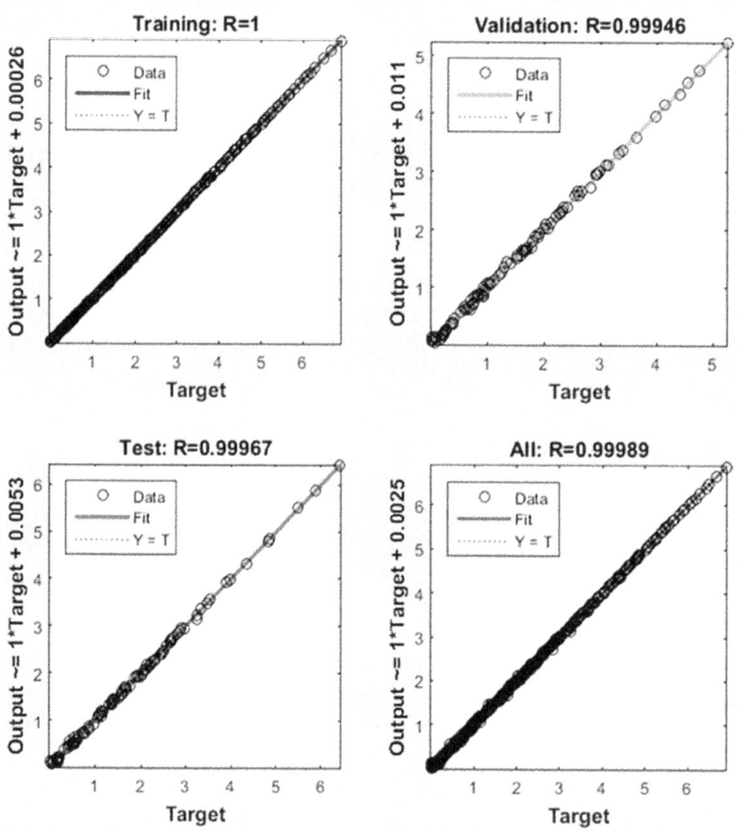

validation check reaches 6, the training will stop. Here the number of successive iterations performed for validation checks is 6. Figure 9 represents the neural network training performance progress.

Best validation performance is 6.0721×10^{-3} at epoch 2, which is a low prediction error measured with MSE shown in Figure 10.

The Figure 10 shows that the best validation performance with respect to the iteration. After 8iterations, the training stopped. No major problems with the training are observed from the figure. All the curves almost follow similar trends.

Figure 9. Training performance progress for Ra

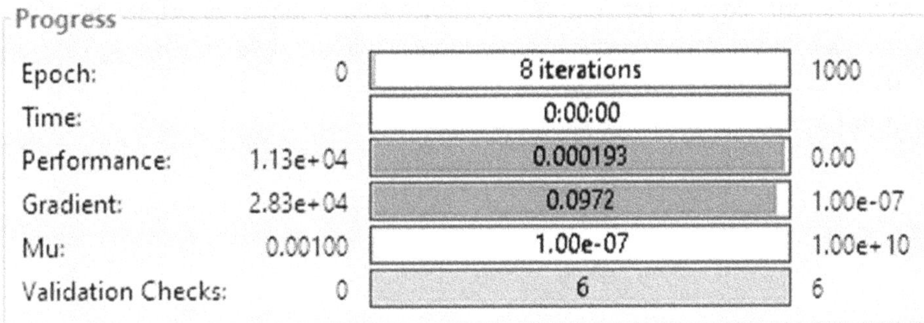

Figure 10. Performance plot for Ra

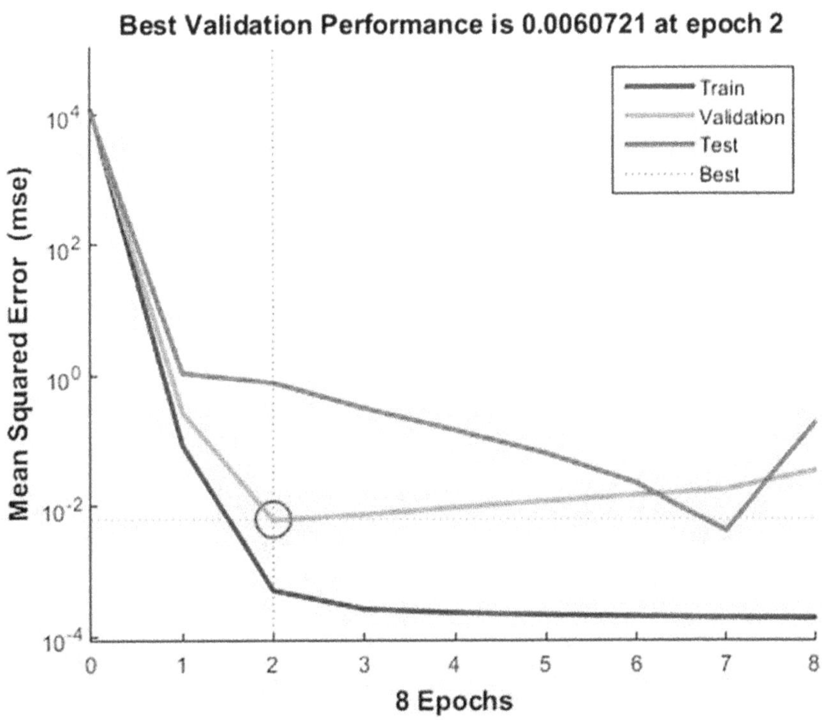

Here in Figure 11 in case of training R = 1, in case of validation R = 0.99998 and in case of testing R = 0.99772. Hence overall R = 0.99966. Therefore, the training data indicates a good fit as the validation and test results both show R values that greater than 0.9.

FIREFLY ALGORITHM

Firefly Algorithm (FA) is a nature inspired metaheuristic optimization algorithm based on the flashing behavior of fireflies. FA is one of the Swarm Intelligence (SI) based algorithm which is developed by mimicking the swarm intelligence characteristics of firefly. Here in the algorithm, it is assumed that all fireflies are unisexual, so any individual firefly will be attracted by other fireflies. The attractiveness is proportional to the brightness of their light, so the brighter one attracts the less bright one and the less bright one moves toward the brighter one. If two fireflies have the same brightness, then they move randomly. As the gap between them increases, the intensity of the light decreases (Yang & He, 2013). Brightness is related to objective function.

Variation of the attractiveness can be calculated as

$$\beta = \beta_0 e^{-\gamma d^2} \tag{4}$$

Figure 11. Regression plot of Ra

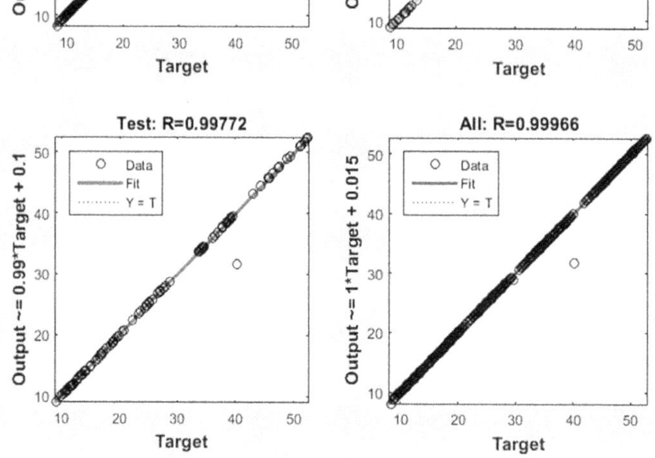

where

β = Attractiveness at any distance 'd'
β_0 = attractiveness at distance 0
γ = Absorption coefficient

FA has few major advantages as compared to other algorithms, that it can automatically subdivide and it can also deal with multimodality. First, this algorithm is created on attraction where attractiveness decreases with increase in distance. As a result, the entire population can automatically subdivide into subgroups. Then each group can swarm around local optimum. Second, this newly developed subdivision allows the fireflies to find all the optimum simultaneously if the size of the population is passably greater than the number of modes.

Planning for Analysis

Here the same set of input parameters as considered for experimentation is used for the individual optimization of output responses while using Firefly Algorithm (FA). The statistical design is carried out by the software Matlab R2015a. In order to optimize the objective functions as developed from the experimental responses the different boundary conditions are as follows:

$$6 \le T_{on} \le 14 \tag{5}$$

$$20 \le T_{off} \le 40 \tag{6}$$

$$3 \le W / Feed \le 7 \tag{7}$$

$$2 \le W / Ten \le 10 \tag{8}$$

Here a maximum number of function evaluations is set to 20000 number. The values of the parameters are Randomness $(\alpha) = 0.25$, attractiveness at distance zero $(\beta_0) = 0.2$ and absorption coefficient $(\gamma) = 1$.

Analysis to Maximize MRR

A regression equation to maximize MRR is developed as shown below.

$$MRR = 0.337 - 0.1209T_{on} + 0.10505T_{off} - 0.0485W / Feed - 0.0446W / Ten \tag{9}$$

Now a global optimum solution is obtained from the developed regression equation by applying FA in Matlab environment. Maximum MRR is obtained as 3.5789 mm³/Sec while the parametric combination of the control parameters is Ton6 µSec, Toff40 µSec, W/Feed3 m/min and W/Ten2 gms.

Analysis to Maximize Ra

A regression equation to minimize Ra is shown below.

$$Ra = -19.182 - 0.6542T_{on} + 1.80150T_{off} - 0.3765W\,/\,Feed + 0.2011W\,/\,Ten \tag{10}$$

Here a global optimum solution is determined through this regression equation by using FA in Matlab environment. Minimum Ra is obtained as 5.4559 µmm while the parametric combination of the control parameters is Ton 14 µSec, Toff 20 µSec, W/Feed 7 m/min and W/Ten 2 gms.

GENETIC ALGORITHM

While optimizing a complex design which is an assortment of continuous discrete variables, irregular and nonconvex design space, by employing standard non-linear programming techniques, for these types of problems will be ineffective. Genetic algorithm (GA) on other hand is well designed for solving such problems, and it is efficient in finding the global optimum solution with maximum probability. GA is based on the theory of survival of the fittest by Darwin.

Using the penalty function, the first transformation alters the original constrained function into an unconstraint function, as:

$$\mathrm{X} = \text{variable} \tag{11}$$

$$\text{Minimize } f(\mathrm{X}) + R\sum_{i=1}^{n} \Phi\left(g_i(\mathrm{X})\right) \tag{12}$$

$$\text{Subjected to } x_j^{(l)} \le x_j \le x_j^{(u)},\ j = 1,2,\ldots,n \tag{13}$$

where Φ is a penalty function defined as:

$$\Phi = \left(Z\right) = \left\langle Z\right\rangle^2 \tag{14}$$

where

$$\langle Z \rangle = \begin{cases} Z & if\ Z > 0 \\ 0 & if\ Z \leq 0 \end{cases} \tag{15}$$

R is the penalty parameter constant.

Maximizing the fitness function, F(X), the second transformation function is minimized as:

$$F(X) = F_{\max} - \left(f(\mathrm{X}) + R\sum_{i=1}^{n} \Phi\left(g_i(\mathrm{X})\right) \right) = F_{\max} - f'(X) \tag{16}$$

where

$$F_{\max} > f'(X) \tag{17}$$

F(X) denotes the fitness function.

It generally works with sequences of typically binary numbers (0/1) repressing the problem variables. A binary sequence can consider as a biological chromosome.

Figure 12 shows the flowchart of GA.

Analysis of the Result for Multi-Objective Global Optimization

Finally, multi-objective optimization is carried out from the regression equations (equations 9 and 10) by applying GA. The objective functions are optimized such that two contradictory objectives i.e. maximum MRR and lower Ra are satisfied simultaneously. The control parameters are subjected to constraint as follows in equations 5-8.

The following conditions are contemplated during the computation:

- Population
    - **Population Type:** Double vector
    - **Population Size:** 50 (As the input variables are less than 5)
    - **Creation Function:** Constraint dependent
    - **Initial Population:** NULL
    - **Initial Scores:** NULL
- Selection
    - **Selection Function:** Tournament
    - **Tournament Size:** 2
- Reproduction
    - **Crossover Fraction:** 0.8
- Mutation
    - **Mutation Function:** Constant dependent
- Crossover
    - **Crossover Function:** Intermediate

Figure 12. Flowchart of GA

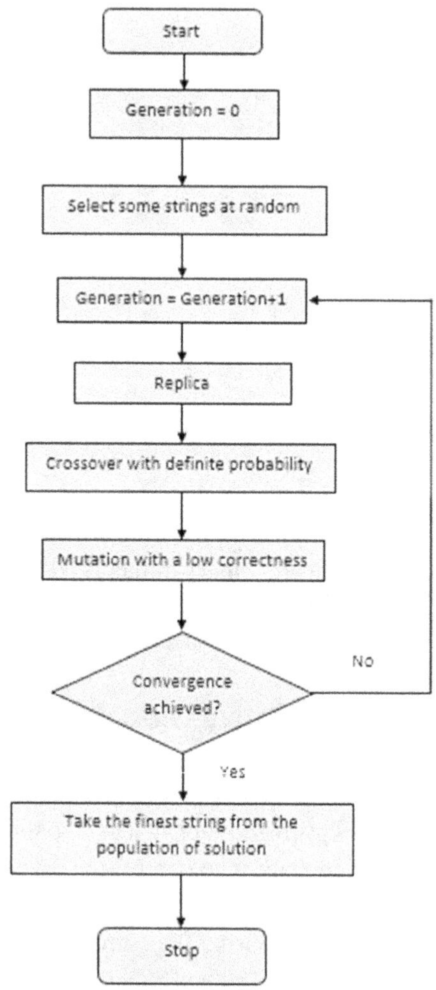

    - **Ratio:** 1
- Migration
    - **Direction:** Forward
    - **Fraction:** 0.2
    - **Interval:** 20
- Multiobjective problem settings
    - **Distance Measure Function:** @ distance crowding
    - **Pareto Front Population Fraction:** 0.35
- Hybrid function
    - **Hybrid Function:** None

- Stopping Criteria
    - **Generations:** 100 x number of variables
    - **Time Limit:** Infinity
    - **Fitness Limit:** -Infinity
    - **Stall Generations:** 100
    - **Stall Time Limit:** Infinity
    - **Function Tolerance:** 1x10-4
    - **Constrain Tolerance:** 1x10-3
- Display to command window
    - **Level of Display:** Off
- User function evaluation
    - **Evaluation Fitness Function:** In serial

Here the total number of iteration required for optimization is 145 and which gives 17 combinations for the control parameters along with responses. The corresponding output is illustrated in Table 2. Optimization terminated as the average change in the spread of Pareto solutions has been reached to its tolerance value. As the default setting of the genetic algorithm is to minimize the objective function, the negative sign is neglected for the maximizing MRR. Figure 13, represents the different plot functions of GA where four conflicting objectives are optimized simultaneously. From the Figure 13 it is evident that the response parameters vary with in a range. For MRR the range is 0.3237 mm^3/Sec to 3.5363 mm^3/Sec and for Ra, it varies between 5.6050 μmm to 48.1370 μmm. Therefore, in order to arrive at an optimal or near optimal parametric combination which will consecutively satisfy contradictory nature of the responses Fuzzy Gray Relational Analysis is conducted.

Figure 13. Plot functions for GA

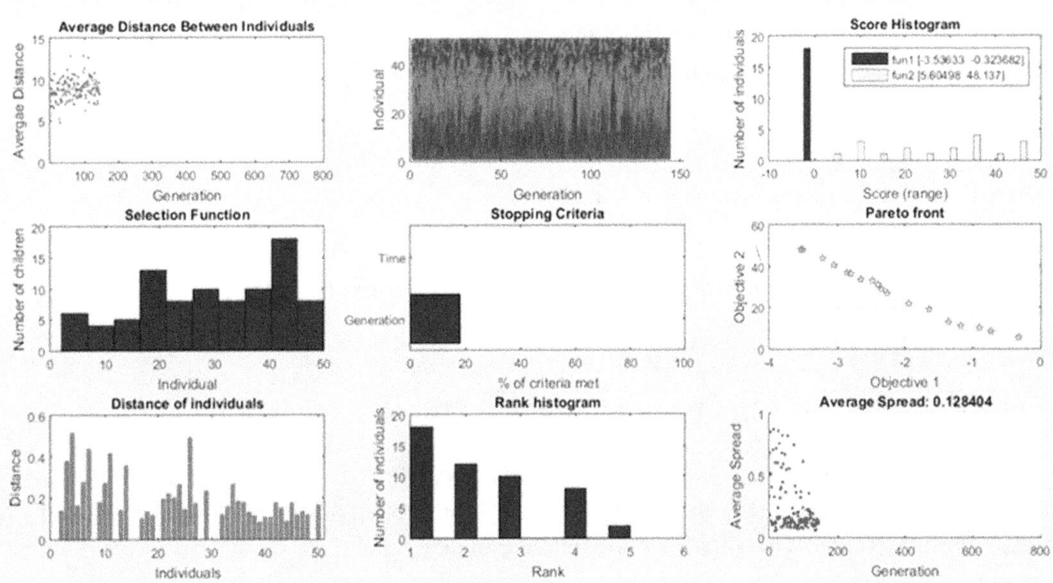

Table 2. Parametric combination with responses

Exp. No.	Control Parameter				Responses		Exp. No.	Control Parameter				Responses	
	Ton (μSec)	Toff (μSec)	W/ Feed (m/ min)	W/Ten (gms)	MRR (mm^3/ Sec)	Ra (μmm)		Ton (μSec)	Toff (μSec)	W/ Feed (m/ min)	W/ Ten (gms)	MRR (mm^3/ Sec)	Ra (μmm)
1	8	20	5	2	1.170	10.988	10	8	31	4	2	2.392	31.162
2	7	28	3	2	2.259	26.829	11	7	21	4	2	1.352	12.966
3	6	33	3	2	2.813	36.184	12	8	25	5	2	1.648	18.942
4	6	32	4	2	2.654	33.544	13	6	25	4	3	1.945	21.618
5	7	30	4	3	2.346	28.796	14	8	33	5	2	2.474	33.270
6	6	36	4	2	3.058	40.522	15	11	21	4	2	0.895	9.854
7	12	20	5	2	0.730	8.296	16	7	38	4	3	3.217	43.870
8	14	20	7	2	0.324	5.605	17	6	34	3	2	2.864	36.805
9	6	40	3	3	3.536	48.137							

MULTI OBJECTIVE MODEL USING GREY RELATION ANALYSIS

Grey Relation Analysis (GRA) can use to optimize contradictory responses simultaneously. The grey system theory was proposed by Deng in 1989's and nowadays it is commonly used for analyzing experimental design in which the model is unspecified or the information is incomplete indicating a combination of known and unknown information.

GRA build the relationship between desired (best/ ideal) with actual experimental data. Grey grade is calculated by averaging the grey coefficient of each response. The estimated grey relational grade ranges from 0 to 1, where higher order values indicate its closeness towards the idle solution. Finally, amongst all the combinations of process variables considered one that acquires the highest grey relational grade is considered to be the best parametric combination.

If higher value is for the better performance such as MRR then it is normalized as per equation,

$$X_{ij} = \frac{Y_{ij} - Min[Y_{ij,} i = 1,2,....n]}{Max[Y_{ij,} i = 1,2,....n] - Min[Y_{ij,} i = 1,2,....n]} \tag{18}$$

If lower value for the better performance such as Ra then it is expressed as,

$$X_{ij} = \frac{Max[Y_{ij,} i = 1,2,....n] - Y_{ij}}{Max[Y_{ij,} i = 1,2,....n] - Min[Y_{ij,} i = 1,2,....n]} \tag{19}$$

The grey relational co-efficient is use to express the relationship between actual normalized experimental data with reference. This relation co-efficient can be evaluate as;

$$Y(X_{oj}, X_{ij}) = \frac{\nabla_{min} + \varsigma\nabla_{max}}{\nabla_{ij} + \varsigma\nabla_{max}} [i = 1,2,....n \ \& \ j = 1,2,...m] \tag{20}$$

where,

$$\nabla_{ij} = \left|X_{oj} - X_{ij}\right|, \nabla_{min} = Min\left[\nabla_{ij}, i = 1,2,....n \ \& \ j = 1,2,...m\right]$$

and

$$\nabla_{max} = Max[\nabla_{ij}, i = 1,2,....n \ \& \ j = 1,2,...m]$$

ς = Distinguished co-efficient (range 0-1)

Generally, the distinguished co-efficient can be adjusted to fit the partial requirements. MRR and Ra both are given equal weights that assume a value of 0.5. The expression is as follows:

$$\text{“}\left(X_0, X_i\right) = \frac{1}{m}\sum_{i=1}^{m} Y\left(X_{0j}, X_{ij}\right) \tag{21}$$

where, m = Number of response parameter.

To optimize both the response at a time by the single combination of process variables, GRA is performed. The combination with the rank 1 full fills multiple response parameters optimization.

Fuzzy Set Theory

Fuzzy set theory is the best powerful tool to solve the uncertainty in decision-making problem. Uses of linguistic assessments of weights of the conditions in the problem are more useful approach instead of numerical values (Keufmann & Gupta, 1991). A decision matrix can be transformed into a fuzzy decision matrix and a weighted normalized fuzzy decision matrix while considering the decision makers' fuzzy ratings. In a cosmos of conversation X, a fuzzy set $\hat{d}$ is characterized by a membership function $\mu_{\hat{d}}\left(x\right)$ i.e., a degree of membership of x in $\hat{d}$ which maps each element x in X to a real number in the interval [0-1]. Here triangular fuzzy number (TFN), $\hat{d}$ can be defined as a triplet (d_1, d_2, d_3) and the membership function is defined (Dubois & Prade, 1979) as shown by equation 22.

$$\mu_{\tilde{d}}(x) = \begin{cases} 0, & x \leq d_1 \\ \dfrac{x - d_1}{d_2 - d_1}, & d_1 \leq x \leq d_2 \\ \dfrac{d_3 - x}{d_3 - d_2}, & d_2 \leq x \leq d_3 \\ 0, & x > d_3 \end{cases} \tag{22}$$

The translation method of fuzzy number into the non-fuzzy number, that is, a crisp value is identified as defuzzification. In this paper 'centroid of area' technique for determining Best Non-Fuzzy Performance (BNP) value is applied.

$$BNP = \frac{\left[\left(d_3 - d_1\right) - \left(d_2 - d_1\right)\right]}{3} + d_1 \tag{23}$$

Choosing the linguistic ratings for criteria:

The importance weights of various criteria are considered as linguistic variables. These linguistic variables can be shown in positive triangular fuzzy numbers in table 3.

Multi Criteria Decision Making (MCDM) Analysis

In the present research work the two responses i.e. MRR and Ra have got a different level of importance. Here the emphasis is given on MRR rather than on Ra.

Using fuzzy set theory, the criteria weights are computed for respective criteria as tabulated in Table 4.

The highest value of the grey relation grade gives the ideal relation between the reference and the comparative sequence. The higher value of the grey relation grade gives the optimized machining condition. Here 17 experimental combinations which obtain as the output of the GA is subjected to evaluation. In this MCDM analysis, the MRR is considered to be maximum i.e. larger is better and other responses are considered the minimum, i.e. smaller is better.

Table 3. Linguistic terms for criteria

Linguistic Terms	Fuzzy Number
Very High (VH)	(0.9,1.0,1.0)
High (H)	(0.7,0.9,1.0)
Moderate High (MH)	(0.5,0.7,0.9)
Moderate (M)	(0.3,0.5,0.7)
Moderate Low (ML)	(0.1,0.3,0.5)
Low (L)	(0.0,0.1,0.3)
Very Low (VL)	(0.0,0.0, 0.1)

Table 4. Weight criteria for deference responses

Criteria	Linguistic Terms	Fuzzy Number	BNP
MRR	VH	0.9,1.0,1.0	0.844
Ra	ML	0.1,0.3,0.5	0.156

Optimization of the Parameters

The given weights for MRR and Ra are 84.4% and 15.6% respectively as calculated by using Entropy method. Table 5 represents the grey relation coefficient and grades corresponding to parametric settings and responses for the material in Table 2.

Table 5 exhibits the results of grey relation coefficient, grey relation grade, and their ranks. The results show that experiment number 8 has the highest grey relational grade value. As described above, this experiment setup fulfills multiple response parameter optimizations. Therefore, the experimental run of 8 which have parametric combination Ton 14 μSec, Toff 20 μSec, W/Feed 7 m/min and W/Ten 2 gms is the best among another experimental setup for having high MRR and low Ra.

Table 5. Grey relation co-efficient along with grades and ranks

Exp. No.	Response		Grey Co-Efficient		Grey Grade	Rank
	MRR (mm^3/Sec)	Ra (μmm)	MRR (mm^3/Sec)	Ra (μmm)		
1	1.1703	10.9880	0.5340	0.5521	0.5431	5
2	2.2585	26.8286	0.6797	0.2382	0.4589	14
3	2.8129	36.1839	0.7894	0.1783	0.4838	10
4	2.6542	33.5439	0.7545	0.1919	0.4732	11
5	2.3462	28.7958	0.6950	0.2225	0.4587	15
6	3.0577	40.5223	0.8500	0.1597	0.5048	8
7	0.7296	8.2963	0.4914	0.7114	0.6014	2
8	**0.3237**	**5.6050**	**0.4577**	**1.0000**	**0.7289**	**1**
9	3.5363	48.1370	1.0000	0.1349	0.5675	3
10	2.3922	31.1621	0.7033	0.2061	0.4547	17
11	1.3515	12.9663	0.5538	0.4741	0.5139	7
12	1.6476	18.9423	0.5894	0.3322	0.4608	13
13	1.9448	21.6178	0.6301	0.2930	0.4616	12
14	2.4744	33.2699	0.7186	0.1934	0.4560	16
15	0.8945	9.8535	0.5065	0.6096	0.5581	4
16	3.2172	43.8697	0.8947	0.1478	0.5212	6
17	2.8642	36.8053	0.8014	0.1754	0.4884	9

Table 6. Results of machining performance using the initial and optimal machining parameters

Settings Levels	Predicted Result	Experimental Result
MRR	0.3237	0.5473
Ra	5.6050	9.0162
Grey Grade	0.7289	0.5681

Improvement of the grey relation grade: 0.1608

The confirmation experiment performed with the above optimal combination results in grey relational grade MRR and Ra obtained 0.5473 mm³/Sec and 9.0162 μmm respectively. It is observed that MRR and Ra improve significantly by using optimal machining variables combinations. Table 6 shows the validation results while machining at optimizing condition.

CONCLUSION

Nature Inspired intelligent schemes has proven their efficiency in solving complex optimization problems. These techniques can easily optimize a problem whose objective functions are contradictory in nature along with a wide range of possible solutions. Artificial Neural Network can easily train and validate the experimental data.

Nature inspired techniques like Firefly Algorithm can find out the ideal parametric combination to confirm the maximum MRR and minimum Ra individually. Where Genetic Algorithm gives a feasible range of the control parameters where the contradictory objective function can be achieved simultaneously. Finally, the results obtained through Genetic Algorithm are hybridized with a Fuzzy set theory to obtain a single parametric combination of the process control parameters which satisfies these two contradictory objectives function simultaneously.

The experimental study indicates that while machining AISI- D3 using WEDM process the responses are dependent on Pulse on Time, Pulse off Time, Wire Feed and Wire Tension. While analyzing the response data individually applying, ANN considering four control parameters in order to achieve maximum MRR and minimum Ra the training, validation and testing data indicates that the values of R, are almost 1.

The individual responses are optimized applying FA where maximum MRR is obtained as 3.5789 mm³/Sec while the parametric combination of the control parameters is Ton 6 μSec, Toff 40 μSec, W/Feed 3 m/min and W/Ten 2 gms. In case of Ra, minimum Ra is obtained as 5.4559 μmm while the parametric combination of the control parameters is Ton 14 μSec, Toff 20 μSec, W/Feed 7 m/min and W/Ten 2 gms.

A multi-objective response is developed which is optimized using GA. Since the objectives of the responses are contradictory in nature, therefore, using GA the optimum values of responses are obtained within a range. In case of MRR it varies between 0.3237 mm³/Sec to 3.5363 mm³/sec and for Ra, it varies in between 5.6050 μmm to 48.1370 μmm.

The grey analysis establishes the ranks of output for different variables combinations and establishes optimal combinations for a complex process like WEDM process. The experimental investigation approach for evaluating the optimum WEDM parametric combination during machining is Ton 14 μSec, Toff 20 μSec, W/Feed 7 m/min and W/Ten 2 gms.

Therefore, the experimental analysis for estimating the optimum WEDM parametric combination during machining of materials can act as valuable and an effective guideline for manufacturing of products of similar material.

REFERENCES

Ali, N., Othman, M. A., Husain, M. N., & Misran, M. H. (2014). A Review of Firefly Algorithm. *ARPN Journal of Engineering and Applies Science*, *9*(10), 1732–1736.

Arora, S., & Sing, S. (2013). The Firefly Optimization Algorithm: Convergence Analysis and Parameter Selection. *International Journal of Computers and Applications*, *69*(3).

Bobbili, R., Madhu, V., & Gogia, A. K. (2013). Effect of Wire-EDM Machining Parameters on Surface Roughness and Material Removal Rate of High Strength Armor Steel. *Materials and Manufacturing Processes*, *28*(4), 364–368. doi:10.1080/10426914.2012.736661

Deng, J. (1989). Introduction to Grey System. *Journal of Grey System*, *1*(1), 1–24.

Dubois, D., & Prade, H. (1979). Operations in a Fuzzy-Valued Logic. *Information and Control*, *43*(2), 224–240. doi:10.1016/S0019-9958(79)90730-7

Garg, R., & Mittal, S. (2014). Optimization by Genetic Algorithm. *International Journal of Advanced Research in Computer Science and Software Engineering*, *4*(4), 587–589.

Keufmann, A., & Gupta, M. M. (1985). *Introduction to Fuzzy Arithmetic: Theory and Applications*. New York: Van Nostrand Reinhold.

Kirkpatrick, S., Gelatt, C. D., & Vecchi, M. P. (1983). Optimization by Simulated Annealing. *Science*, *220*(4598), 671–680. doi:10.1126/science.220.4598.671 PMID:17813860

Kumar, A., Pant, S., & Ram, M. (2016). *System Reliability Optimization Using Gray Wolf Optimizer Algorithm. Quality and Reliability Engineering International*. John Wiley & Sons Ltd.

Kuriachen, B., Somashekhar, K. P., & Mathew, J. (2015). Multi response optimization of micro- wire electrical discharge machining process. *International Journal of Advanced Manufacturing Technology*, *76*(1-4), 91–104. doi:10.1007/s00170-014-6005-2

Kurikose, S., & Shanmugham, M. S. (2005). Multi objective optimization of wire EDM process by non-dominated sorting genetic algorithms. *Journal of Materials Processing Technology*, *170*, 133–141. doi:10.1016/j.jmatprotec.2005.04.105

Liao, Y. S., Chuang, T. J., & Yu, Y. P. (2014). Study of machining parameters optimization for different materials in WEDM. *International Journal of Advanced Manufacturing Technology*, *70*(9-12), 2051–2058. doi:10.1007/s00170-013-5458-z

Miguel, L. F. F., & Miguel, L. F. F. (2012). Novel Metaheuristic Algorithms Applied to Optimization of Structures. *WSEAS Transaction on Applied and Theoretical Mechanics*, *7*(3), 210–220.

Ming, W., Hou, J., Zhang, Z., Huang, H., Xu, Z., Zhang, G., & Huang, Y. (2015). Integrated ANN-LWPA for cutting parameter optimization in WEDM. *International Journal of Advanced Manufacturing Technology*, 1–18.

Mukherjee, R., Chakraborty, S., & Samanta, S. (2012). Selection of wire electrical discharge machining process parameters using non-traditional optimization algorithms. *Applied Soft Computing*, *12*(8), 2506–2516. doi:10.1016/j.asoc.2012.03.053

Pant, S., Kumar, A., & Singh, S. B. (2017). *Reliability Optimization of Complex Systems Using Cuckoo Search Algorithm. In Mathematical Concepts and Applications in Mechanical Engineering and Mechatronics*. Hershey, PA: IGI Global.

Patil, N. G., & Brahmankar, P. K. (2010). Determination of material removal rate in wire electro-discharge machining of metal matrix composites using dimensional analysis. *International Journal of Advanced Manufacturing Technology*, *51*(5), 599–610. doi:10.1007/s00170-010-2633-3

Rao, B. T., & Krishna, A. G. (2014). Selection of optimal process parameters in WEDM while machining Al7075/SiCp metal matrix composites. *International Journal of Advanced Manufacturing Technology*, *73*(1-4), 299–314. doi:10.1007/s00170-014-5780-0

Somashekhar, K. P., Ramachandran, N., & Mathew, J. (2010). Optimization of Material Removal Rate in Micro-EDM Using Artificial Neural Network and Genetic Algorithms. *Materials and Manufacturing Processes*, *25*(6), 467–475. doi:10.1080/10426910903365760

Yang, X. S., & He, X. (2013). Firefly Algorithm: Recent Advancement and Applications. *International Journal Swarm Intelligence*, *1*(1), 36–50. doi:10.1504/IJSI.2013.055801

Zhang, G., Zhang, Z., Ming, W., Guo, J., Huang, Y., & Shao, X. (2014). The multi-objective optimization of medium-speed WEDM process parameters for machining SKD11 steel by the hybrid method of RSM and NSGA-II. *International Journal of Advanced Manufacturing Technology*, *70*(9-12), 2097–2109. doi:10.1007/s00170-013-5427-6

Chapter 5
Optimization of Process Parameters in Plasma Arc Cutting Applying Genetic Algorithm and Fuzzy Logic

Nehal Dash
Birla Institute of Technology, India

Apurba Kumar Roy
Birla Institute of Technology, India

Sanghamitra Debta
Birla Institute of Technology, India

Kaushik Kumar
Birla Institute of Technology, India

ABSTRACT

Plasma Arc Cutting (PAC) process is a widely used machining process in several fabrication, construction and repair work applications. Considering gas pressure, arc current and torch height as the inputs and among all possible outputs, in the present work Material Removal Rate and Surface Roughness would be considered as factors that determines the quality, machining time and machining cost. In order to reduce the number of experiments Design of Experiments (DOE) would be carried out. In later stages applications of Genetic Algorithm (GA) and Fuzzy Logic would be used for Optimization of process parameters in Plasma Arc Cutting (PAC). The output obtained would be minimized and maximized for Surface Roughness and Material Removal Rate respectively using Genetic Algorithm (GA) and Fuzzy Logic.

DOI: 10.4018/978-1-5225-3035-0.ch005

INTRODUCTION

Presently Plasma Arc Cutting (PAC) process is a machining process that has wide variety of applications in various fabrication unit e.g. automotive repair shops, various fabrication shops, construction sector etc. In this technique a jet in accelerated state (accelerated jet) of hot plasma is used as the means for cutting through electrically conductive materials. Materials those can be cut using this method include steel, brass, copper, aluminium etc. When the cutting object gets in contact with the electrode in torch an arc discharge generates the heat. This generated heat is utilised for cutting operations. The arc discharge which is generated forms the working gas into the plasma with high temperature. When a gas is heated to very high temperature, then the molecules gain kinetic energy which makes them collide with each other producing electrically neutral state called plasma. This plasma in high temperature state is blown through the nozzle with a very high speed and the cutting material is fused to be cut. This method is used for cutting operations from large scale industries with CNC applications to small fabrication shops because of its high operating speed and low cost precision cutting characteristics.

Since the requirements of customers have become complex enough in the day to day life, the quality of the product to be developed and its quantity have become the primary motive. The systems and the technology nowadays are made in a way to meet the complex requirements of customers. Hence in this case the reconfigurable system designing plays a vital role in maintaining the high level performance by changing the functional requirements. This reconfigurable designing also considers many factors those affects the operations like that of time, cost and the quantity of production.

Plasma Arc Cutting (PAC) is one of the most dependable non conventional machining processes that work on the principle of thermal cutting. PAC has been a really useful technique in the cutting operations of stainless steel, high hardness metals and metals with high melting points. Many metals or alloys which are difficult enough for machining are generally machined using PAC. It acts as a replacement/alternate method to oxy-fuel process. In 1960s when Plasma welding process was considered as an effective method for joining process then PAC came out of this very technique. It was found to be an efficient method to cut sheet metal and plate in 1980s. Later CNC technology was incorporated into plasma cutting machines in 1990s. By incorporating the CNC technology it was found that PAC became even more flexible and many complex shapes were also easily cut using this technique. The only limitation with CNC plasma cutting machines is that they were limited to only some cutting patterns in two axes of X and Y.

Presently Plasma Arc Cutting (PAC) process is a widely used machining process in various fabrication unit e.g. automotive repair shops, various fabrication shops, construction sector etc. In this technique a jet in accelerated state (accelerated jet) of hot plasma is used as the means for cutting through electrically conductive materials. Materials those can be cut using PAC method include steel, brass, copper, aluminium etc. When the cutting object gets in contact with the electrode in torch an arc discharge generates the heat. This generated heat is utilised for cutting operations. The arc discharge which is generated forms the working gas into the plasma with high temperature. When a gas is heated to very high temperature, then the molecules gain kinetic energy which makes them collide with each other producing electrically neutral state called plasma. This plasma in high temperature state is blown through the nozzle with a very high speed and the cutting material is fused to be cut. This method is used for cutting operations from large scale industries with CNC applications to small fabrication shops because of its high operating speed and low cost precision cutting characteristics.

The principle behind the arc that is formed between the electrode and the work-piece is done by a using fine bore and a copper nozzle as depicted in Figure 1. This increases temperature and velocity of plasma out of the nozzle. Generally the temperature of the plasma is about 2000°C with a velocity nearly equal to the velocity of sound. While performing cutting operations the gas flow is increased and the plasma jet that penetrates into the deep into the materials cuts through the material and removes the molten material. An arc occurs between the negative electrode and the work-piece and the energy dissipated by this arc in turn leads to the ionization of cutting gas. Due to the increase in the enthalpy of the gas it turns into plasma state. Now the plasma and the heated gas in forced to flow from the internal geometry of torch to the bore that in narrow enough, which gains momentum and gets superheated. Now coming out of the nozzle the supersonic jet plume strikes the work-piece thereby melting the metal and expelling the molten material through kerf. Then a secondary shielding gas comes into picture which covers the cutting zone and constrains the plasma jet and cools the torch (Bini et al. 2008).

OPERATIONS OF PLASMA ARC CUTTING (PAC)

Plasma cutting is one of the widely used separation methods. The reason behind it is it can be flexibly used for cutting 2D and 3D cutting tasks and also it helps in the cutting operations of electrically conductive materials. Some of the most important cutting operations carried out by PAC are Bevel Cutting, Hole Cutting, Cutting of gratings, plasma gouging etc. Some of the advantages of using PAC for cutting operations is that it reduces the risk of double arcing, it undergoes cutting operations at a higher speeds and it reduces the top edge rounding possibilities. The most common input parameters those are considered before performing any operations in PAC are Gas pressure, Arc current and Torch Height. In order to avoid the mishandlings and mishaps these factors are needed to be kept in mind.

Figure 1. Schematic setup of Plasma Arc Cutting Process

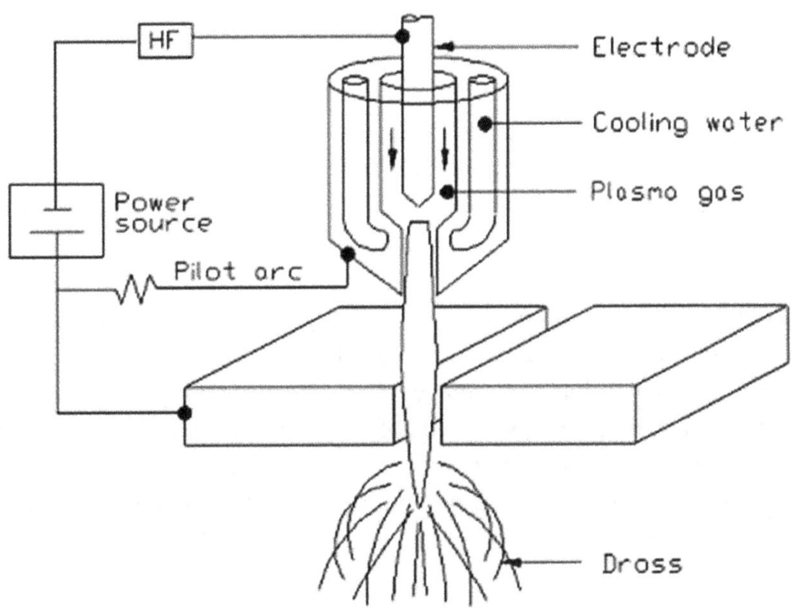

NECESSITY OF OPTIMIZATION

For the applications of advanced manufacturing methodologies, technologies are developing in rapid pace. In order to achieve these advanced manufacturing criteria the industries, companies and manufacturing units need to optimize the process parameters to save money and time. Many researchers have performed the optimization of the process parameters using many machining techniques like CNC lathe, PAC, EDM etc. by applying various optimizations techniques like Artificial Neural Network, Particle Swarm Optimization, Genetic Algorithm, Fuzzy Logic Algorithm etc. Considering PAC in this present work, MRR and Surface Roughness always are considered as the parameters where the primary aim was to obtain the optimized minimized values of Surface Roughness and optimized Maximized values of Material Removal Rate. For getting these desired machining characteristics the machining parameters like Gas pressure, Arc current and Torch height are considered as the input and are finalized in advance.

BACKGROUND OF WORK AND RELATED WORKS

Primarily researchers have concentrated on the area of heat affected zone in PAC (Yang, 2001; Vasil'ev, 2003; Gullu and Atici, 2006; Zajac and Pfeifer, 2006; Kadirgama et al., 2010; Ismail and Taha, 2011), kerf (Ramakrishnan et al., 2000; Teulet et al., 2006; Bini et al., 2007; Wang et al., 2011) and depth of cut (Gane et al., 1994; Xu et al., 2002; Gariboldi and Previtali, 2004; Yang, 2007; Bahram, 2009). Surface Roughness characteristics, MRR and conicity are some other major areas where researchers have also emphasized in PAC.

Some researchers like Ismail and Taha (2011) have tried to optimized arc current, scanning velocity and the carbon content for steel in order to obtain better plasma arc surface hardening and roughness of ASSAB 618 and ASSAB DF3 steels by applying Taguchi technique. It was also observed that the most important parameter that affects the process is arc current. Hence, they came out with optimum synchronization of process parameters for better hardness and Surface Roughness. An ANN model considering cutting current, plate thickness and cutting speed as three input neurons was developed by Radovanovic and Madic (2011) which help in prediction of the output neuron surface roughness Rz values for undergoing PAC operations in steel, aluminum and nickel. Considering AISI 4140 steel Ozek et al. (2012) applied Fuzzy logic algorithm in order to predict surface roughness in PAC process. The most important three machining parameters those were used were plasma arc current, cutting speed and thickness of the material. Salonitis and Vatousianos (2012) listed the effect of various machining parameters like cutting speed, cutting current, plasma gas pressure and torch height on surface roughness, Heat affected zone and conicity of the cutting geometry in plasma arc cutting of S235 mild steel and also developed a regression model for it. They obtained the conclusion that surface roughness and conicity were mostly affected by the cutting height. Bhuvenesh et al. (2012) have optimized the air pressure, cutting current, cutting speed and arc gap by considering the multiple performance characteristics taking MRR and Surface Roughness into consideration using Taguchi technique in PAC of AISI 1017 steel. Maity and Bagal (2015) studies PAC process on AISI 316 stainless steel and optimized the parameters those affect the process like feed rate, current, voltage and torch height considering multiple responses like kerf, chamfer, dross, surface roughness and MRR using RSM and grey relational analysis coupled together with PAC.

From the literature study, it can be concluded that several researchers have considered one or two surface roughness parameters as the responses. But it is known that for any machined surface the roughness parameters are generally considered by some machining parameters like amplitude parameters, spacing parameters and hybrid parameters. In this present work surface roughness and MRR are considered together in the operation of PAC for obtaining the optimized values of MRR and Surface Roughness using Genetic Algorithm and Fuzzy Logic technique.

OPTIMIZATION METHODS

In this work authors have tried to provide solution to problems using Genetic Algorithm and Fuzzy Logic.

GENETIC ALGORITHM

Genetic Algorithm is considered as stochastic global search method that mimics the metaphor of natural biological evolution. It operates on a population solution which uses the principle of survival of the fittest to produce better. In every generation, a new set of approximations are considered which selects the individuals based on their levels of fitness in the domain of the problem and later breeding them together using the operators borrowed from natural genetics. This process leads to the evolution of populations of individuals those are better suited to their environment than the individuals that they were created from, just as in natural adaptations.

Genetic Algorithm is meta-heuristic method which provides us with solutions to various optimization problems and search problems. It uses the operators that are inspired by biological phenomenon like mutation crossover and selection. Here the candidate solutions which are also known as the creatures or phenotypes have some properties or characteristics generally known as genotypes get mutated. These mutated solutions are depicted in binary strings. The individuals get generated in random fashion from a population in iterative manner and this process is called as Generation. Now fitness evaluation is carried out for every individual in the each and every generation. Here fitness refers to the objective function in the optimization problem. From the population taken into consideration the more fit individuals are selected and modification is carried out in their genome in order to obtain better and new population. If the desired level of fitness is reached then this algorithm terminates. For a genetic algorithm to be carried out successfully, it requires the solution domain to be in form of genetic representation and a fitness function which is used for evaluation the solution domain. Some of the applications of genetic algorithm are computer automated design, automated designs of mechanical and industrial equipment, automated designs of mechatronics systems etc.

FUZZY LOGIC

Fuzzy logic is a concept, in which the truth values can be taken as real numbers between 0 and 1, considering it to be fuzzy. According to Boolean logic these truth values may only be 0 or 1. Generally fuzzy logic is generally used for handling truths partial in nature and its values vary from a range of either completely true or completing false. But specific functions might be used if the variables are linguistic in nature. In a way, fuzzy concepts are vague which lack a fixed and precise meaning. Generally, it is understood as a concept which is "to an extent applicable" in various situations. Some of the real-time applications of fuzzy logic concepts are the traffic light operation, washing machine, air conditioners etc. Fuzzy logic at times appears very exotic and intimidating to those who are unfamiliar to it. In other words, it is both an old and new technique because although the modern and methodical science of fuzzy logic is still young, the concepts of fuzzy logic reach right down to our bones.

There are a lot of advantages of using Fuzzy Logic technique. Some of them are: It is quite easy to understand. It is one of the most flexible techniques. Non-linear functions can also be modelled using this technique. It can be blended with conventional control techniques and is based on natural language.

EXPERIMENTATION

Input Process Parameters to be Considered

- Gas pressure
- Arc Current
- Torch Height

Plasma arc cutting involves several input parameters to be considered during machining process. In this work, the synchronization factors such as gas pressure (bar), arc current (amp), torch height (mm) are considered. Three equally spaced levels of each process parameter were selected (Table 1). By undergoing through the literature survey, the control factors were chosen and some preliminary investigations. Other factors, which can be expected to have an effect on the measures of performance, are considered to be constant.

OUTPUT CONSIDERED

1. MRR

It is the rate at which the material is removed from the work-piece. The weight of the work-piece material is recorded before carrying out the machining process and is compared with the weight after machining (using Equation (1) below).

Table 1. Different variables used in the experiment and their levels (L_{27} OA)

Variable	Coding	Level		
		1	2	3
Gas pressure (bar)	A	5	6	7
Arc current (Amp)	B	180	190	200
Torch height (mm)	C	2	4	6

$$\text{MRR (gm/min)} = \frac{Wi - Wf}{t} \quad (1)$$

where,

W_i = Initial weight of work piece material (gms)
W_f = Final weight of work piece material (gms)
t = Time taken for machining in minutes

For various industrial applications MRR is most important parameter. These applications desire a larger MRR, hence it was optimized based on larger for better condition.

2. Surface Roughness

Surface roughness is usually defined as the deviation of a surface from an ideal level and is defined according to international standard (ISO 4287, 1997). Surface roughness regularly abbreviated to roughness, is a part of surface texture. It is evaluated by the deviations I the direction of vector of a real surface from its optimal frame. With vast deviations, the surface is considered to be rough; whereas if they are found to be small, the surface is smooth. Roughness is regularly thought to be the high-frequency, short-wavelength part of a deliberately measured surface. Practically speaking it is frequently important to know both the amplitude and frequency to guarantee that a surface is fit for a purpose.

Always smaller SR is desired for most of the machining operations. Hence it is optimized on basis of smaller for better condition. The Arithmetic Mean Surface Roughness Ra is depicted in Figure 2.

MATERIAL USED

EN31 steel, high carbon alloy steel with high degree of hardness, compressive strength and abrasion resistance, would be considered as work-piece. Some of its applications are automobile axle, bearing, spindle bearing rolls, dies etc.

Figure 2. Arithmetic Mean Surface Roughness Ra

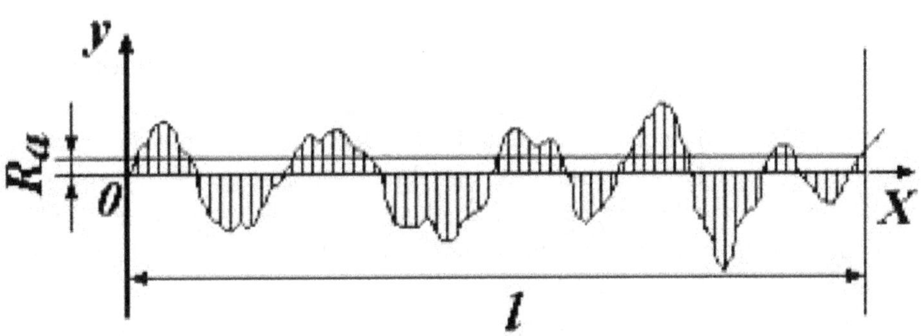

MACHINE AND INSTRUMENT

The experimentation will be carried out using a CNC plasma arc cutting (PAC) (EPP-450, 380V and 50/60 Hz) with PT-36 (Torch) supplied by ESAB. Special electrodes which were made from water cooled copper with insertion of metals like hafnium were used. Here air was used as the cutting gas. The computer controlled machines were mounted with mechanical torches. In between the torch tip and the work-piece a standoff was maintained. A rectangular block of 80 mm X 15 mm X 10 mm of EN31 steel was selected as work-piece for experimentation.

TORCH DESIGN

In single flow torch the air for cutting is restricted to small amperage and thin gauge sheet metal cutting operations. Shielding is not required for cooling the torch as the small amperage output needed for the thin gauge sheet metal cutting is quite low. There is a flow of gas or air for the plasma cutting and shielding gas flow for torch cooling in the dual flow torch. For cutting thicker materials it is widely used since it requires higher amperages. In this present work, a dual flow torch PT-36 is considered.

DESIGN OF EXPERIMENTS

Based on the operations performed on the CNC-PAC, relevant outputs are obtained using different levels of the considered inputs. Design of Experiments (DOE) would be undertaken to minimize the number of experiments.

With a primary objective of finding the factors those influence the process and the output of the process, Design of Experiments (DOE) is considered as the best process. For the better processing of the inputs in order to optimize the output it act as the most useful method which is systematic enough to carry out. For carrying out the DOE process some terms that are used are controllable and non-controllable input factors, responses, blocking, replication, hypothesis testing, interaction etc. Input parameters/factors that can be changed over time complexity are called Controllable factors, whereas the factors those cannot be modified in an experiment are called as Uncontrollable factors. These factors are needed to be recognized in advance to understand the fact how these factors are going to affect the response. For reducing the cost of design by incorporating the speed process of design and avoiding the late changes in design methodology, reducing the material for product development and labour complexity experimental design methods are taken into considerations. Another important side of DOE is that it also reduces the variation in the manufacturing process which in turn reduces the manufacturing cost of the product. It also reduces the rework, scrap and needs of inspections.

This work always required a correct Design of Experiments. For this Taguchi method of DOE was followed. Since it was considered as a methodology for formulating the engineering problems using statistical models, this methodology specified the methods or procedures particularly in case of hypothesis testing. For acquiring the best process parameters, it was always necessary to perform the required number of experiments. But this statistical method allowed the researchers to minimize the number of experiments to be carried out without significantly affecting the variation in results. Quality control and process control methods also used statistics as tool for managing conformance to various specifications of manufacturing processes and products. The main aim of this DOE is to examine the data and suggest and help in planning of future experiments. In various real-time engineering applications, the goal is more often to optimize a process or product, hence incorporating this no doubt reduced the time of experimentation and also cost of experimentation. Some other fields of applications of DOE are Time and methods engineering which uses statistics for studying repetitive operations in manufacturing systems, Reliability engineering for checking the reliability of the system or product by performance of its intended functions, Probabilistic design which involves use of probability in product development process, System identification methods also uses statistical methods for building mathematical models which includes the optimal design of experiments for efficiently generating the informative data for fitting of these mathematical models.

The three input process parameters are Gas Pressure (A), Arc Current (B) and Torch Height (C) were considered in the study with three equally spaced variables in an operating range. In this present work, research work experimentation was planned following Taguchi's design of Experiments method. Taguchi (1990) suggested that total degrees of freedom (DOF) of selected orthogonal array must be greater than or equal to total DOF of standard orthogonal array (OA). Here total DOF for four factor and their interactions in 20 (4x2+3x4) and standard OA for four factors with three levels are L9 with 8 DOF and L27 with 26 DOF. Thus, L27 OA has been selected for experimental work and tabulated below (Table 2).

OPTIMIZATION OF MRR AND SR USING GENETIC ALGORITHM

Considering the Gas Pressure (A), Arc Current (B) and Torch Height (C) as the inputs Surface roughness and Material Removal Rate are determined as output. L27 orthogonal array (OA) design is utilized

Table 2. DOE using L27 Orthogonal Array

Exp. No.	A	B	C	MRR (gm /min)	Ra (μm)	Exp. No.	A	B	C	MRR (gm /min)	Ra (μm)
1	5	180	2	1.850	1.848	15	6	190	6	1.426	1.812
2	5	180	4	1.333	2.018	16	6	200	2	2.210	1.620
3	5	180	6	1.425	1.966	17	6	200	4	1.760	1.784
4	5	190	2	1.001	3.742	18	6	200	6	2.145	1.452
5	5	190	4	1.671	1.998	19	7	180	2	1.895	1.780
6	5	190	6	1.044	3.110	20	7	180	4	2.065	1.806
7	5	200	2	1.703	1.985	21	7	180	6	1.616	1.910
8	5	200	4	1.175	1.703	22	7	190	2	1.423	1.926
9	5	200	6	1.500	1.840	23	7	190	4	1.898	1.686
10	6	180	2	1.632	3.000	24	7	190	6	1.696	1.564
11	6	180	4	1.780	1.868	25	7	200	2	1.483	1.833
12	6	180	6	2.793	1.522	26	7	200	4	1.980	1.710
13	6	190	2	1.858	1.864	27	7	200	6	1.470	1.690
14	6	190	4	1.732	1.896						

and the data obtained from respective experiments are utilized for analysing the SR and MRR, where SR and MRR were considered as the responses. Since for various practical applications MRR has to be maximum whereas SR has to be minimum hence for analysis the maximization and minimization equations are applied.

The following flow diagram (Figure 3) describes the sequential procedure of the optimization using GA technique. This sequence was followed for obtaining the optimized values for MRR and SR that is the maximized value of MRR and minimized value of SR.

As stated in the flowchart (Figure 3), at first the input and the output data are noted in the workspace of Matlab R2015b and these data were noted into the workspace of network window. The values thus imported were trained and the plots for the MRR and SR were found as below in Figure 4 and Figure 5 respectively. The optimum values found for MRR and SR are 1.858 gm/min and 1.864 μm respectively.

OPTIMIZATION OF MRR AND SR USING FUZZY LOGIC

For the obtained values using MiniTab16 for 27 experimental readings of Gas Pressure, Arc Current, Torch Height, MRR and SR, optimization is carried out using Fuzzy Logic for obtaining the optimum values of all the parameters listed above.

The flow diagram below (Figure 6) shows the steps followed in optimization of method MRR and SR by Fuzzy Logic Algorithm.

As indicated in Flow Chart (Figure 6), initially the Fuzzy Logic Network was created for the considering the inputs for the operations in PAC. The network obtained using Matlab R2015b is as shown below in Figure 7.

Figure 3. Step by step procedure followed in GA optimization technique

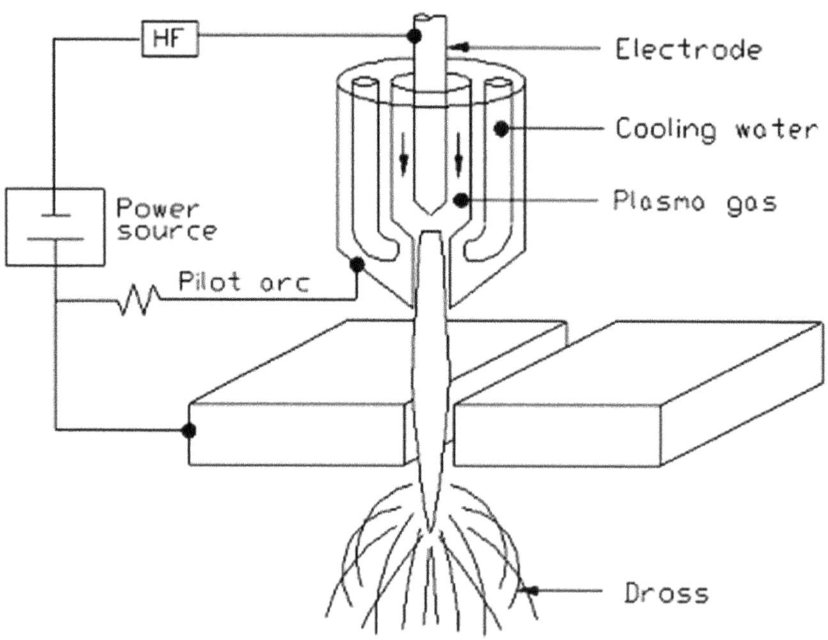

Figure 4. Maximization Plot for Material Removal Rate (MRR)

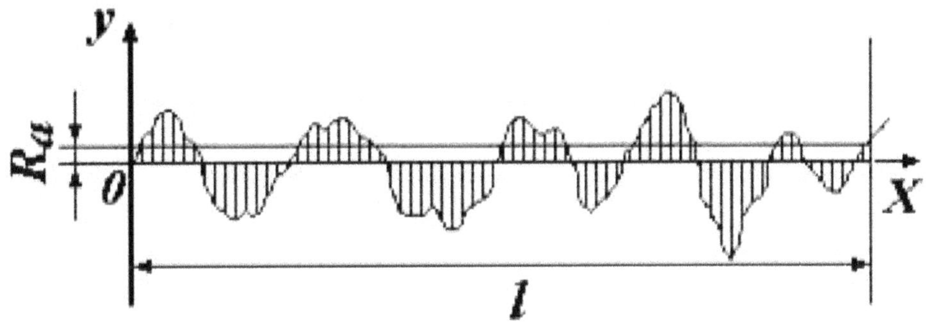

Figure 5. Minimization Plot for Surface Roughness (SR)

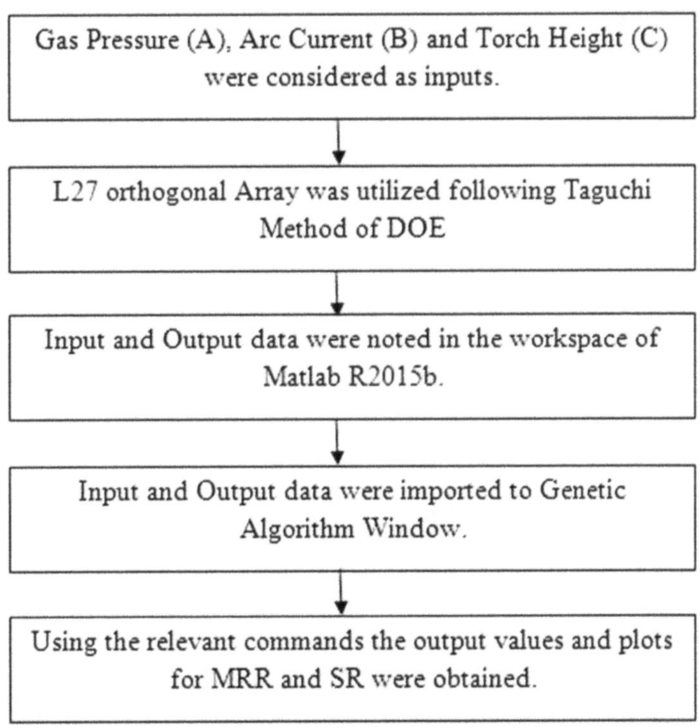

Considering the levels of the input parameters from Table 1, the values are trained and the optimum values of the input parameters were obtained utilising Matlab R2015b. The optimum values for Gas Pressure (A), Arc Current (B) and Torch Height (C) approximately were found to be 6 bar, 190 Amp and 4 mm respectively. For these values of Gas Pressure (A), Arc Current (B) and Torch Height (C), the respective optimal values for MRR and SR are determined approximately as 1.858 gm/min and 1.864 μm. respectively. Figure 8, Figure 9 and Figure 10 shows the optimum values plot for the input parameters Gas Pressure (A), Arc Current (B) and Torch Height (C) respectively.

Figure 6. Steps followed in Fuzzy Logic technique

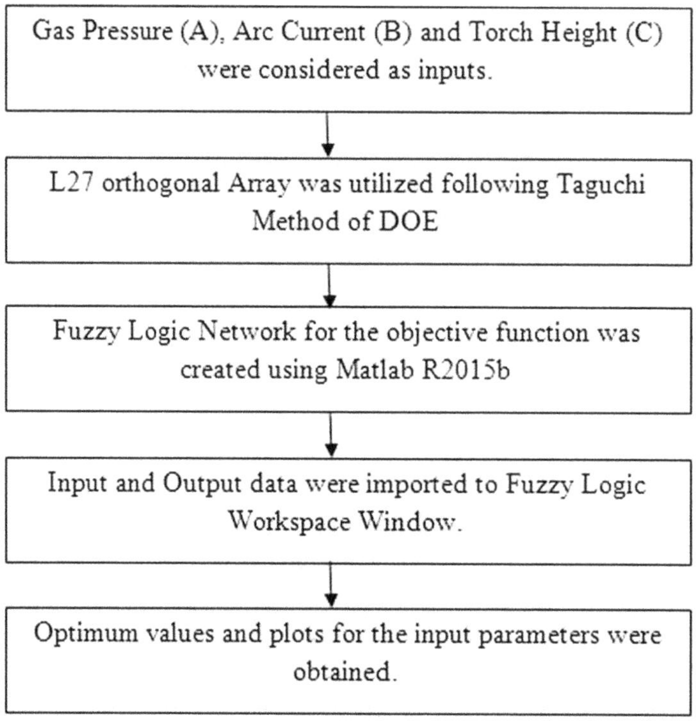

Figure 7. Fuzzy Logic Network

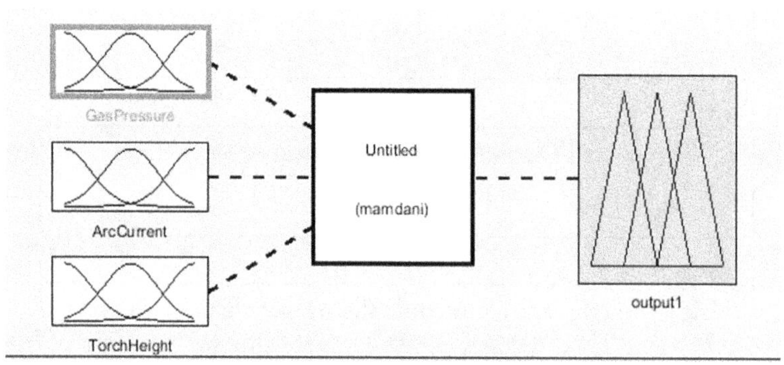

RESULTS AND DISCUSSION

Experiments were performed according to the sequence of L27 OA and the experimental results of MRR and SR were calculated. The analysis of experimental results was carried out using Minitab 16 statistical software. In this case the interaction between the particles is also taken into consideration before analysing these experimental data using software. As per GA method the experimental data are imported into the workspace of Matlab R2015b and plots for the optimized values of MRR and Surface Roughness were obtained.

Figure 8. Plot for Optimum Value of Gas Pressure (Fuzzy Logic Algorithm)

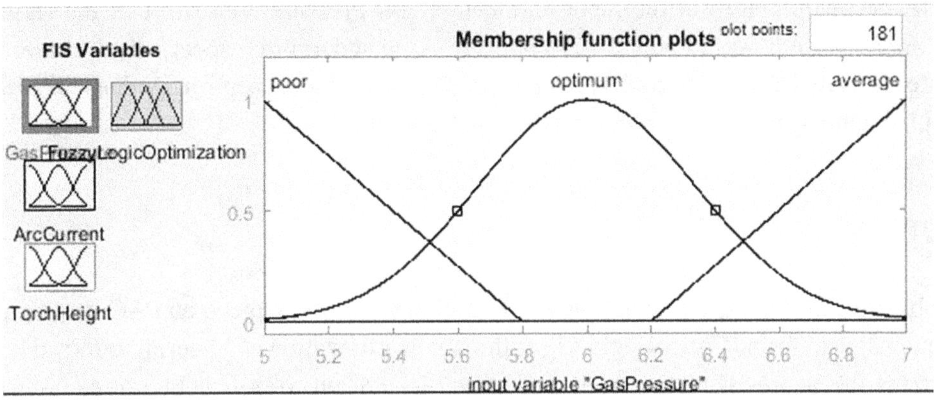

Figure 9. Plot for Optimum Value of Arc Current (Fuzzy Logic Algorithm)

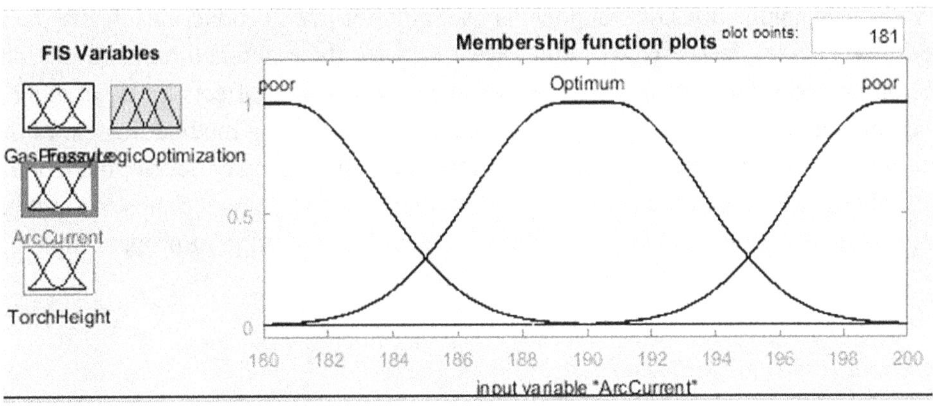

Figure 10. Plot for Optimum Value of Torch Height (Fuzzy Logic Algorithm)

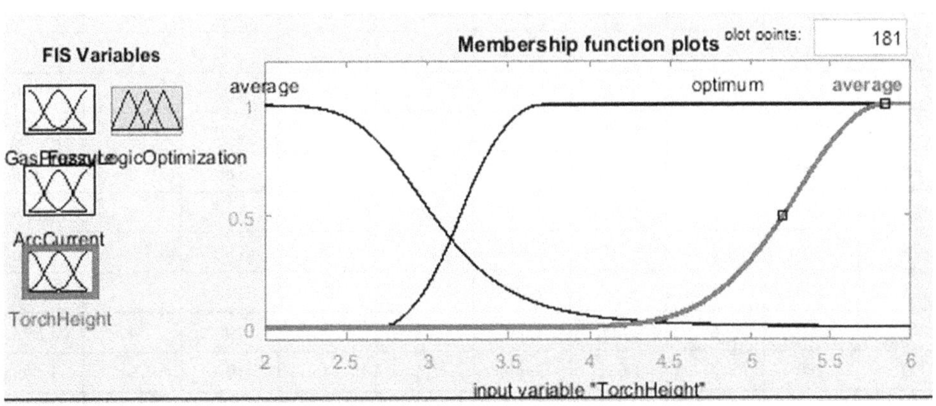

Also using the Fuzzy Logic Algorithm, the Fuzzy Logic Network was initially created using Matlab R2015b. The optimum values of the input parameters Gas Pressure (A), Arc Current (B) and Torch Height (C) approximately were found to be 6 bar, 190 Amp and 4 mm respectively. For these values of Gas Pressure (A), Arc Current (B) and Torch Height (C) the respective optimal values for MRR and SR are determined approximately as 1.858 gm/min and 1.864 μm.

CONCLUSION

The main objective of this experimentation was to optimize the work-piece on PAC machining process using Genetic Algorithm and Fuzzy Logic Algorithm for maximization of Material Removal Rate (MRR) and minimizing the surface Roughness (SR). Here in this present work Gas Pressure (A), Arc Current (B) and Torch Height (C) were considered with equally spaced three levels within the operating range for each of the process parameters as the input parameters. From the results and discussions as stated above it can be concluded that:

For any industrial applications of components generally for mass productions where the movement of consumer goods is very fast, it is very much necessary for the manufacturers and the industrialists to obtain the optimized values and use them. In some case MRR only affects the machining process of product development and SR has very less influence on the machining methodologies. In those cases, the only optimized value for MRR has to be considered for the machining or development of the product. For the cases where levels are not matching enough to their desired characteristics of input parameters, using the regression equations is always an option for them for prediction of proper output

Table 3. DOE and results for MRR and SR

Exp. No.	A	B	C	MRR (gm /min)	Ra (μm)	Exp. No.	A	B	C	MRR (gm /min)	Ra (μm)
1	5	180	2	1.850	1.848	15	6	190	6	1.426	1.812
2	5	180	4	1.333	2.018	16	6	200	2	2.210	1.620
3	5	180	6	1.425	1.966	17	6	200	4	1.760	1.784
4	5	190	2	1.001	3.742	18	6	200	6	2.145	1.452
5	5	190	4	1.671	1.998	19	7	180	2	1.895	1.780
6	5	190	6	1.044	3.110	20	7	180	4	2.065	1.806
7	5	200	2	1.703	1.985	21	7	180	6	1.616	1.910
8	5	200	4	1.175	1.703	22	7	190	2	1.423	1.926
9	5	200	6	1.500	1.840	23	7	190	4	1.898	1.686
10	6	180	2	1.632	3.000	24	7	190	6	1.696	1.564
11	6	180	4	1.780	1.868	25	7	200	2	1.483	1.833
12	6	180	6	2.793	1.522	26	7	200	4	1.980	1.710
13	6	190	2	1.858	1.864	27	7	200	6	1.470	1.690
14	6	190	4	1.732	1.896						

In some products, very little influence of MRR is found on the development of the product and SR plays a vital role in product development. In that case the optimal values obtained using these optimized techniques can be adopted for better designing and machining of the products.

For the products where both MRR and SR are to given equal amount of importance, optimized values for both MRR and SR are to be considered for predicting the output.

SCOPE OF FUTURE RESEARCH

This work can be even applied to several other fields or directions. Considering several other input parameters are considered this work can be enriched. Other responses like stress developed, modality etc. can also be obtained considering PAC machining method. There are various other solution platform like Simulated annealing, Synchronization or Hybridization of two parallel methods can also be done like Simulated annealing in Genetic Algorithm (SAGA) etc. for obtaining optimized results. Application of other optimization techniques too can be done and the results can be compared to obtain healthy conclusions.

ACKNOWLEDGMENT

The authors sincerely acknowledge the comments and suggestions of the reviewers that have been instrumental for improving and upgrading the paper in its final form.

REFERENCES

Bahram, A. (2009). Optimising the automated plasma cutting process by design of experiments. *International Journal of Rapid Manufacturing*, *1*(1), 19–40. doi:10.1504/IJRAPIDM.2009.028930

Bass, I. (2007). *Six sigma statistics with Excel and Minitab*. McGraw Hill.

Bhuvenesh, R., Norizaman, M. H., & Manan, M. S. A. (2012). Surface roughness and MRR effect on manual plasma arc cutting machining. *International Journal of Mechanical Engineering*, *6*, 131–134.

Bini, R., Colosimo, B. M., Kutlu, A. E., & Monno, M. (2007). Experimental study of the features of the kerf generated by a 200 A high tolerance plasma cutting system. *Journal of Materials Processing Technology*, *196*(1-3), 345–355. doi:10.1016/j.jmatprotec.2007.05.061

Callister, W. D. J., & Rethwich, D. G. (2010). Material Science and Engineering: An Introduction (8th ed.). Wiley.

Deng, J. (1989). Introduction to Grey System. *Journal of Grey System*, *1*, 1–24.

Fisher, R. A. (1951). *Design of experiments*. Oliver and Boyd.

Gane, N., Rogozinski, M. W., Polivka, F., Doolette, A. G., & Ramakrishnan, S. (1994). *Quality of cut in air plasma cutting*. Paper presented at the Washington Technology Industry Association 42nd annual National Welding Conference, Melbourne, Australia.

Gariboldi, E., & Previtali, B. (2004). High tolerance plasma arc cutting of commercially pure titanium. *Journal of Materials Processing Technology*, *160*(1), 77–89. doi:10.1016/j.jmatprotec.2004.04.366

Gullu, A., & Atici, U. (2006). Investigation of the effects of plasma arc parameters on the structure variation of AISI 304 and St 52 steels. *Materials & Design*, *27*(10), 1157–1162. doi:10.1016/j.matdes.2005.02.014

Ismail, M. I. S., & Taha, Z. (2011). Experimental design and performance analysis in plasma arc surface hardening, *World Academy of Science. Engineering and Technology*, *56*, 1052–1058.

Kadirgama, K., Noor, M. M., Harun, W. S. W., & Aboue-El-Hossein, K. A. (2010). Optimization of heat affected zone by partial swarm optimization in air plasma cutting operation. *Journal of Scientific and Industrial Research*, *69*, 439–443.

Karaboga, D. (2005). *An idea based on honey bee swarm for numerical optimization, Technical Report-TR06*. Erciyes University, Engineering Faculty, Computer Engineering Department.

Maity, K. P., & Bagal, D. K. (2015). Effect of process parameters on cut quality of stainless steel of plasma arc cutting using hybrid approach. *International Journal of Advanced Manufacturing Technology*, *78*(1-4), 161–175. doi:10.1007/s00170-014-6552-6

Minitab User Manual Release 16. (2010). State College, PA: MINITAB Inc.

Montgomery, D. C. (2001). *Design and analysis of experiments*. New York: John Wiley.

Ozek, C., Caydas, U., & Unal, E. (2012). A fuzzy model for predicting surface roughness in plasma arc cutting of AISI 4140 steel. *Materials and Manufacturing Processes*, *27*(1), 95–102. doi:10.1080/10426914.2011.551952

Phadke, M. S. (2012). *Quality Engineering using Robust Design*. New Delhi: Pearson.

Radovanovic, M., & Madic, M. (2011). Modeling the plasma arc cutting using ANN. *Nonconventional Technologies Review*, *4*, 43–48.

Ramakrishnan, S., Shrinet, V., Polivka, F. B., Kearney, T. N., & Koltun, P. (2000). Influence of gas composition on plasma arc cutting of steel. *Journal of Physics. D, Applied Physics*, *33*(18), 2288–2299. doi:10.1088/0022-3727/33/18/313

Ross, P. J. (1996). *Taguchi Techniques for Quality Engineering* (2nd ed.). New York: McGraw Hill.

Roy, R. K. (1990). *A primer on the Taguchi method Dearborn*. Michigan Society of Manufacturing Engineers.

Salonitis, K., & Vatousianos, S. (2012). Experimental investigation of the plasma arc cutting process. *Procedia CIRP*, *3*, 287–292. doi:10.1016/j.procir.2012.07.050

Sun, H., & Lee, S. (2005). Response surface approach to aerodynamic optimization design of helicopter rotor blade. *International Journal for Numerical Methods in Engineering*, *64*(1), 125–142. doi:10.1002/nme.1391

Taguchi, G. (1990). *Introduction to Quality Engineering*. Tokyo: Asian Productivity Organization.

Teulet, P., Girard, L., Razafinimanana, M., Gleizes, A., Bertrand, P., Camy-Peyret, F., & Richard, F. et al. (2006). Experimental study of an oxygen plasma cutting torch: II. Arc material interaction, energy transfer and anode attachment. *Journal of Physics. D, Applied Physics*, *39*(8), 1557–1573. doi:10.1088/0022-3727/39/8/015

Unitek Miyachi Group. (1999). Welding Material Control. *Technical Application Brief*, *2*, 1–5.

Vasant, P. M. (Ed.). (2014). *Handbook of Research on Artificial Intelligence Techniques and Algorithms*. IGI Global.

Vasant, P. M., Weber., G., & Diew, V. N. (Eds.). (2016). Handbook of Research on Modern Optimization Algorithms and Applications in Engineering and Economics. IGI Global.

Vasant, P. M., Barsoum, N., & Webb, J. (Eds.). (2011). *Innovation in Power, Control, and Optimization: Emerging Energy Technologies*. IGI Global.

Vasilev, K. V. (2003). Plasma-arc cutting – A promising method of thermal cutting. *Welding International*, *17*(2), 147–151. doi:10.1533/wint.2003.3096

Vo, D. N., & Schegner, P. (2012). An Improved Particle Swarm Optimization for Optimal Power Flow. In P. M. Vasant (Ed.), *Meta-Heuristics Optimization Algorithms in Engineering, Business, Economics, and Finance* (pp. 1–40). IGI Global.

Wang, J., Zhu, Z., He, C., & Yang, F. (2011). Effect of dual swirling plasma arc cutting parameters on kerf characteristics. *International Journal of Material*, *4*, 39–43.

Xu, W. J., Fang, J. C., & Lu, Y. S. (2002). Study on ceramic cutting by plasma arc. *Journal of Materials Processing Technology*, *129*(1-3), 152–156. doi:10.1016/S0924-0136(02)00600-3

Yang, L. J. (2001). Plasma surface hardening of ASSAB 760 steel specimens with Taguchi optimization of the processing parameters. *Journal of Materials Processing Technology*, *113*(1-3), 521–526. doi:10.1016/S0924-0136(01)00583-0

Yang, L. J. (2007). The effect of specimen thickness on the hardness of plasma surface hardened ASSAB 760 steel specimens. *Journal of Materials Processing Technology*, *185*(1-3), 113–119. doi:10.1016/j.jmatprotec.2006.03.117

Zajac, A., & Pfeifer, T. (2006). Restricting the heat-affected zone during the plasma cutting of high-alloy steels. *Welding International*, *20*(1), 5–9. doi:10.1533/wint.2006.3534

Chapter 6
Artificial Neural Network Training Algorithms in Modeling of Radial Overcut in EDM:
A Comparative Study

Raja Das
VIT University, India

Mohan Kumar Pradhan
Maulana Azad National Institute of Technology, India

ABSTRACT

This chapter describes with the comparison of the most used back propagations training algorithms neural networks, mainly Levenberg-Marquardt, conjugate gradient and Resilient back propagation are discussed. In the present study, using radial overcut prediction as illustrations, comparisons are made based on the effectiveness and efficiency of three training algorithms on the networks. Electrical Discharge Machining (EDM), the most traditional non-traditional manufacturing procedures, is growing attraction, due to its not requiring cutting tools and permits machining of hard, brittle, thin and complex geometry. Hence it is very popular in the field of modern manufacturing industries such as aerospace, surgical components, nuclear industries. But, these industries surface finish has the almost importance. Based on the study and test results, although the Levenberg-Marquardt has been found to be faster and having improved performance than other algorithms in training, the Resilient back propagation algorithm has the best accuracy in testing period.

DOI: 10.4018/978-1-5225-3035-0.ch006

INTRODUCTION

Due to the increasing trend of using lightweight, lean, and compact components in recent years, there has been growing interest in the advanced and tailor-made materials, with better properties such as high strength, high stiffness, good damping capability, low thermal expansion, higher fatigue characteristics. Besides, components made with these materials demands stringent design and close tolerances during manufacturing. The traditional manufacturing processes are unable to cope up the challenges exhibit by these advanced materials owing to the improved mechanical properties (Abbas, Solomon, & Bahari, 2007; Ho & Newman, 2003). They are hard and 'difficult to machine', strict high precision, higher surface quality standards lead to increase the scrap and rework that leads to increase the machining price. For the last seven decades, electrical discharge machining (EDM) has been extending inimitable capabilities to ma- chine "difficult to machine" materials with desired shape, size and required dimensional accuracy. It has been impressively applied for machining in the advance industries like automotive, medical, aerospace, consumer electronics and optoelectronic industries development. In the past, with the continuing advances of technology, there has been a significant enhancement in EDM technology also, to improve productivity, accuracy and the versatility of the process. The key interest in the active research was to choose the optimal setting of the process parameters in such a way that accuracy should increase and, concurrently, overcut or gap, tool wear and surface roughness should reduce (Pradhan, 2012; Anitha, Das, & Pradhan, 2012; Pradhan & Kumar, 2012). Moreover, a process can be identified better when a model replicates its behavior by its vital parameters. The factors that are significant for the system are to be recognized and different aspects of the process are to be correlated while constructing the model. It is expensive, unpractical or impossible to experiment directly with the process so a good model can be cost-effective to predict the actual process very closely (Das & Pradhan, 2014; Jena, Pradhan, Das, Acharjya, & Mishra, 2014; Pradhan & Das, 2015).

Experimental Setup and Procedure

To collect the data experiments were performed using a CNC Electrical discharge die sinking machine set up "Electronica Electraplus PS 50ZNC" presented in Figure 1. A pure copper electrode (99.9% Cu) of a diameter of 30 mm was used to machine the AISI D2 Tool steel, the photographic view of the specimen is depicted in Figure 2, and a commercial grade EDM oil (specific gravity = 0.763, freezing point= 94°C) was used as dielectric fluid, the power supply was linked with the tool electrode (Tool: positive polarity, work piece: negative polarity). Dielectric was pumped through the tube electrode laterally as shown in Figure 3, for effective flushing of machining debris from the working gap region with a pressure of 0.4 kgf/cm^2. Work piece material was initially circular bar of diameter 100 mm and was cut into specimens of thickness 10 mm. the top and bottom faces of the work piece were ground to make it flat and good quality surface finish before experimentation. the bottom of the cylindrical electrode was polished by a very fine grade emery sheet before each experimental run. every treatment of the experiment was run for 15 minutes and the time was measured with a stopwatch of accuracy 0.1s. The work piece as well because the tool was detached from the machine, clean and dried up, to form it free from the dirt, trash and dielectric. They were weighed, before and after machining, on a precision electronics balance (maximum capacity = 300 g, precision = 0.001 g). The diameter of the cavity machined on work piece was measured by a tool maker microscope (make: Carl Zeiss, Germany) with an accuracy of 1 μ m.

Artificial Neural Network

An ANN is a biologically inspired computational model that processes information. ANN has been shown to be highly flexible modeling tool with capability of learning the mathematical mapping between input and output. ANN is formed from several layers of neurons. The input layer of neurons is connected to the output layers of neurons through one or more hidden layers of neurons. Initially, ANN is trained and tested with experimental data to reach at an optimum topology and weights. A multilayer perceptron (MLP) is feed forward neural network with one or more hidden layers. During the training process ANN adjusts its weights to minimize the errors between the predicted result and actual output by using different back-propagation algorithms.

The Back-Propagation Neural Network (BPNN) with n input nodes, r output nodes and a single hidden layer of m nodes are shown in Figure 1. Each interconnection between the nodes has a weight associated with it. The Transfer function of the hidden and output nodes are tan-sigmoid S(•) and linear, respectively.

According to Figure 1 the net input to the jth hidden neuron is given by

$$y_j(x) = \sum_{i=1}^{n} w1_{ji} x_i + b1_j \tag{1}$$

where $w1_{ji}$ is the weight between the ith node of input layer and jth node of hidden layer and $b1_j$ is the bias at jth node of hidden layer. The output of the jth hidden node is defined by

$$z_j(x) = 2/\left(1 + \exp(-2x)\right) - 1 \tag{2}$$

Given an input vector x, the output, value $o_k(x)$ of the kth node of output layer is equal to the sum of the weighted outputs of the hidden nodes and the bias of the kth node output layer, and is given by

$$o_k = \sum_{j=1}^{m} w2_{kj} z_j + b2_k \tag{3}$$

where $w2_{kj}$ is the weight between the jth node of hidden layer and kth node of output layer (3), $b2_k$ is biasing term at the kth output node.

The output of ANN is determined by giving the inputs and computing the output from various nodes activation and interconnection weights. The output is compared to the experimental output and Mean Squared Error is calculated. The error value is then propagated backwards through the network and changes are made to the weights at each node in each layer by three different training algorithms.

ANN Training Algorithms

The present work describes three different artificial neural network (ANN) training algorithms Levenberg- Marquardt, conjugate gradient and resilient back propagation used in the study. This was done

Figure 1. Architecture of BPNN with single hidden layer

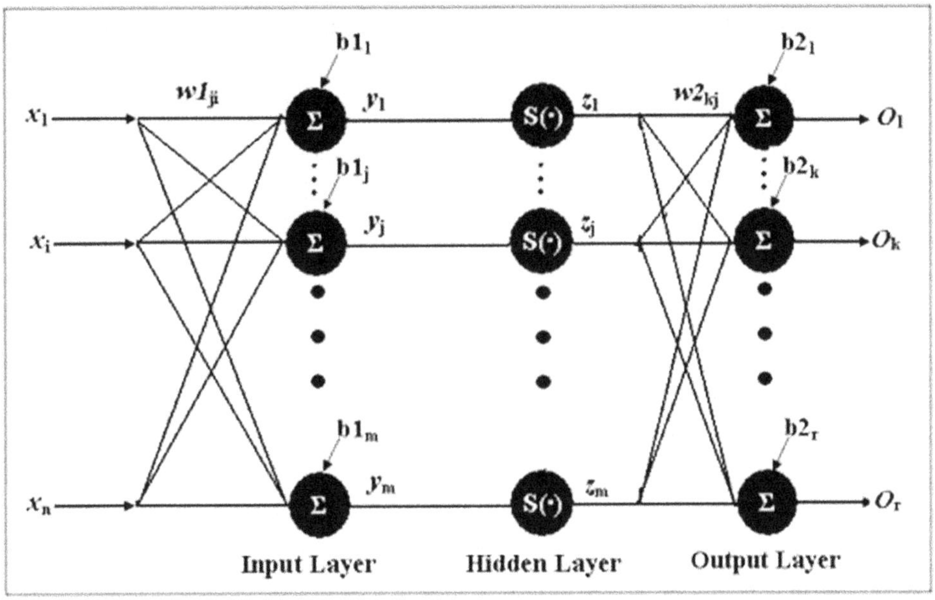

with a view to see which algorithm produces better results and has faster training for the application under consideration.

The objective of training is to reduce the global error E defined as

$$E = \frac{1}{P}\sum_{p=1}^{P} E_p$$

where P is the total number of training dataset; and E_α is the error for pth training data, E_p is calculated by the following formula

$$E_\alpha = \frac{1}{2}\sum_{q=1}^{r}\left(o_q - t_q\right)^2 \quad (4)$$

where r is the total number of output nodes, o_q is the network output at the qth output node, and t_q is the target output at the qth output node.

In every training algorithm, an attempt is made to reduce this global error (4) by adjusting the weights and biases.

Levenberg- Marquardt (LM) Algorithm

The approximated Hessian matrix $J^T J + \mu I$ is invertible; Liebenberg-Marquardt algorithm introduces another approximation to Hessian matrix.

$H = J^T J + \mu I$ (5)where J is the Jacobian matrix which contains first derivative of the network errors with respect to the weights and bias, μ is always positive, called combination coefficient and I is the identity matrix.

The Levenberg-Marquardt algorithm uses this approximation to the Hessian matrix in the following Newton-like update:

$$w_{k+1} = w_k - \left(J_k^T J_k + \mu I\right)^{-1} J_k e_k \tag{6}$$

where w is the weight vector and e is the error vector.

As the combination of the steepest descent algorithm and the Gauss–Newton algorithm, the Levenberg–Marquardt algorithm switches between the two algorithms during the training process. When the combination coefficient μ is very small (nearly zero), Gauss–Newton algorithm is used. When combination coefficient μ is very large, the steepest descent method is used. The Levenberg- Marquardt optimization technique is more powerful than the conventional gradient descent techniques.

Conjugate Gradient (CG) Algorithm

The standard conjugate gradient method is to minimize the differentiable function E by generating a sequence of approximation w_{k+1} iteratively according to

$$w_{k+1} = w_k + \alpha_k d_k$$

The scalar α_k is the step length, known in neural network notation as learning rate. The step length α_k can be determined by line search techniques in the way that $E\left(w_k + \alpha_k d_k\right)$ is minimize along the direction d_k, given w_k and d_k fixed.

The standard conjugate gradient algorithm begins the minimization process with initial estimate w_0 and an initial search direction

$$d_0 = -\nabla E\left(w_0\right) = -g_0$$

Each direction d_{k+1} is chosen to be linear combination of the steepest direction $-g_{k+1}$ and the previous direction d_k. We write

$$d_{k+1} = -g_{k+1} + \beta_k d_k$$

where the scalar β_k is to be determined by the requirement that d_k and d_{k+1} must fulfill the conjugate property. There are many formulae for the parameters β_k. One of them is Fletcher-Reeves formula and is given by

$$\beta_k = \frac{g_k^T g_k}{g_{k-1}^T g_{k-1}}$$

Conjugate gradient method has a second order convergence property without complex calculation of the Hessian matrix. A faster convergence established than first order steepest descent approach.

Resilient Back- Propagation (RP) Algorithm

The individual update value $\Delta_{ij}(k)$ for each weight $w_{ij}(k)$ can be expressed according to the learning rule for each case based on the observed behavior of the partial derivative during two successive weight-steps by the following formula:

$$\Delta_{ij}(k) = \begin{cases} \eta^{+}\Delta_{ij}(k-1), & if \frac{\partial E}{\partial w_{ij}}(k).\frac{\partial E}{\partial w_{ij}}(k-1) > 0 \\ \eta^{-}\Delta_{ij}(k-1), & if \frac{\partial E}{\partial w_{ij}}(k).\frac{\partial E}{\partial w_{ij}}(k-1) < 0 \\ \Delta_{ij}(k-1) & else \end{cases}$$

where

$$0 < \eta^{-} < 1 < \eta + .$$

It is evident that whenever that partial derivative of the equivalent weight w_{ij} varies its sign, which indicates that the last update was large in magnitude and the algorithm has skipped over a local minima, the update-value $\Delta_{ij}(k)$ is decreased by the factor η^{-}. If the derivative holds its sign, the update-value will to some extent increase in order to speed up the convergence.

When the update-value for each weight is settled in, the weight updates by a very simple rule.

$$w_{ij}(k+1) = w_{ij}(k) + \Delta w_{ij}(k)$$

where

$$\Delta w_{ij}(k) = \begin{cases} -\Delta_{ij}(k), & if \frac{\partial E}{\partial w_{ij}}(k) > 0 \\ \Delta_{ij}(k), & if \frac{\partial E}{\partial w_{ij}}(k) < 0 \\ 0, & else \end{cases}$$

If the partial derivative changes sign that is the previous step was too large and the minimum was missed, the previous weight-update is reverted:

$$\Delta w_{ij}\left(k\right) = -\Delta w_{ij}\left(k-1\right),\ if\ \frac{\partial E}{\partial w_{ij}}\left(k\right).\frac{\partial E}{\partial w_{ij}}\left(k-1\right) < 0$$

In order to avoid a double penalty of the update-value, there should be no adaptation of the update value in the succeeding step. In practice, this can be done by setting $\frac{\partial E}{\partial w_{ij}}\left(k-1\right) = 0$ in the Δ_{ij} update rule above.

Hence, the partial derivative of the errors must be accumulated for all training data. This indicates that the weights are updated only after the presentation of all of the training data. It is noticed that resilient back-propagation is much faster than the standard steepest descent algorithm.

Result and Discussion

In the present BPNN model, the inputs of the model are Ton, τ, Ip, V. The output of the model is Radial Overcut. In general, the architecture of multi-layer BPNN can have many layers where a layer represents a set of parallel processing units (or nodes). The three layers BPNN used in this study contain only one intermediate (hidden) layer. Multi-layer BPNN can have more than one hidden layer; however theoretical works have shown that a single hidden layer is sufficient for BPNN to approximate any complex non-linear functions. Indeed, many experimental results seem to confirm that one hidden layer may be enough for most forecasting problems. Therefore, in this study, one hidden layered ANNs with 6, 14 and 15 neurons for LM, CG and RP training algorithms respectively were used as shown in Figure 2. 44 set of data under EDM process was used for training, validation and testing for three ANNs. Out of 44 experimental data, 26 training, 9 validation, and 9 testing data sets are considered for the three ANNs to compare the performance. Three MATLAB language codes were written for the ANN algorithms.

Figure 2. Mean square error for ANNs with different training algorithms

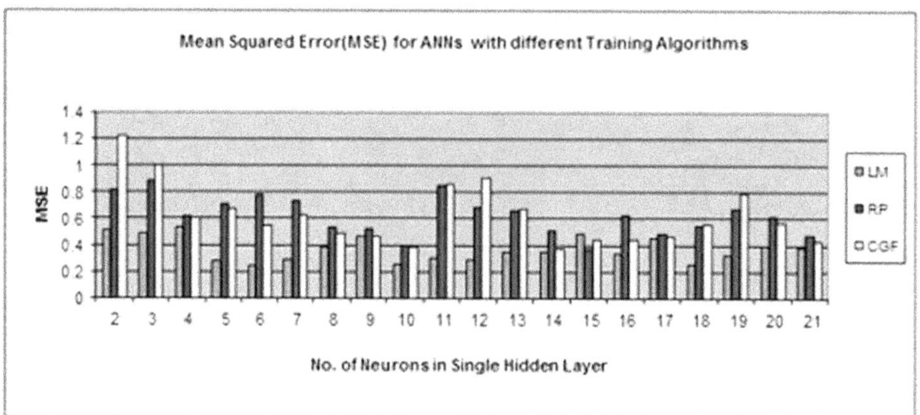

The ANN algorithms were compared according to mean squared errors (MSE) and mean percentile error (MPE) criteria. These criteria are defined as

$$Mean\ Squared\ Error = \frac{1}{N}\sum_{i=1}^{N}\left(Ra_{observed} - Ra_{predicted}\right)^2$$

$$Mean\ Percentile\ Error = \frac{1}{N}\sum_{i=1}^{N}\frac{\left|Ra_{observed} - Ra_{predicted}\right|}{Ra_{observed}} \times 100$$

For each combination ANN was trained using three different algorithms, that is, LM, CG, and RP. After training was over, the weights were saved and used to test network performance. The ANN results were transformed back to the original domain and MSE was computed. The iterations were stopped when the difference between two epochs was too small. The experimental results and predicted results of 'Ra' by the LM, CG and RP were plotted on the same scale, as shown in Figure 4. The performance of three neural network models is studied with the special attention to their regression plot and it is presented in Figure 3 (a)-3(b). The ANN with training algorithm LM has the best R value.

ANNs are compared separately with results obtained by experiments and the average absolute error obtained for all the three networks. ANNs with CG and RP models are poorer in predicting Radial Overcut. The test result accuracy measured in terms of MAE and MPE for nine test data are given in Table 1.

Figure 3. Regression analysis for ANN with CG

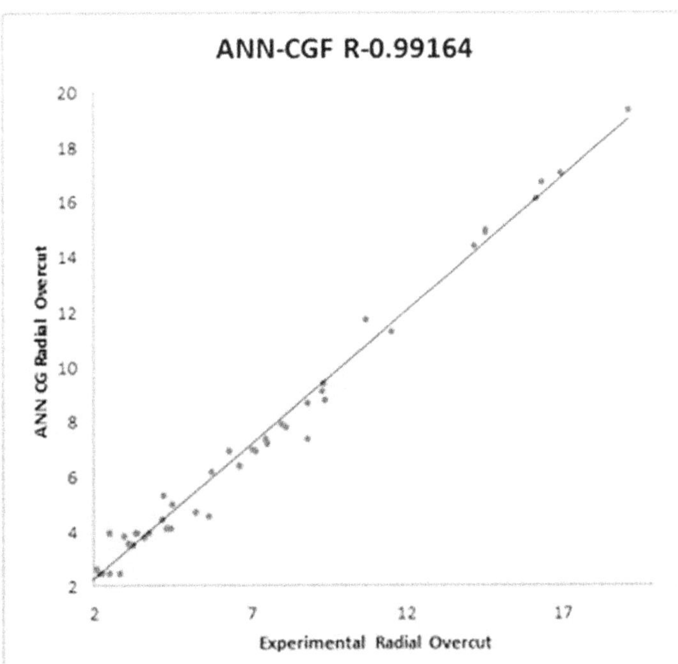

Figure 4. Regression analysis for ANN with LM

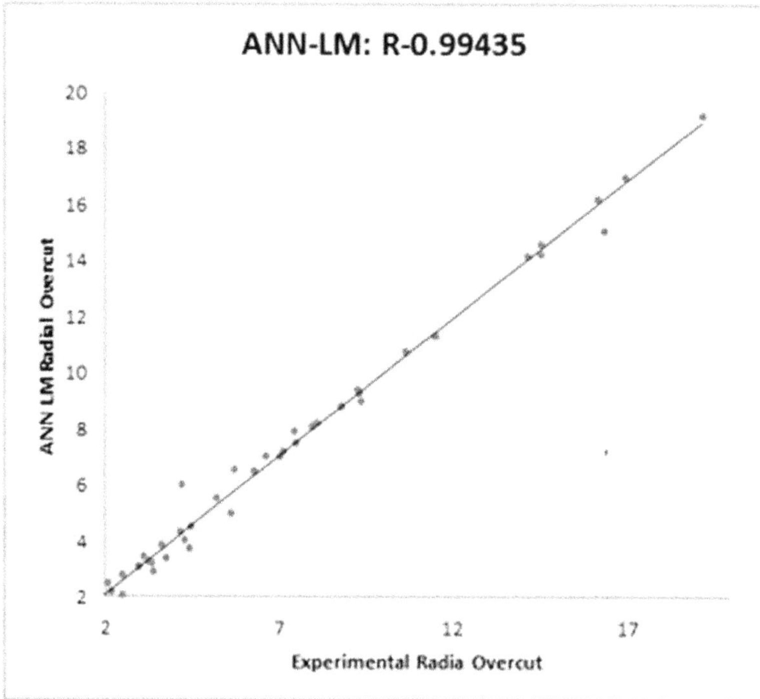

Figure 5. Regression analysis for ANN with RP

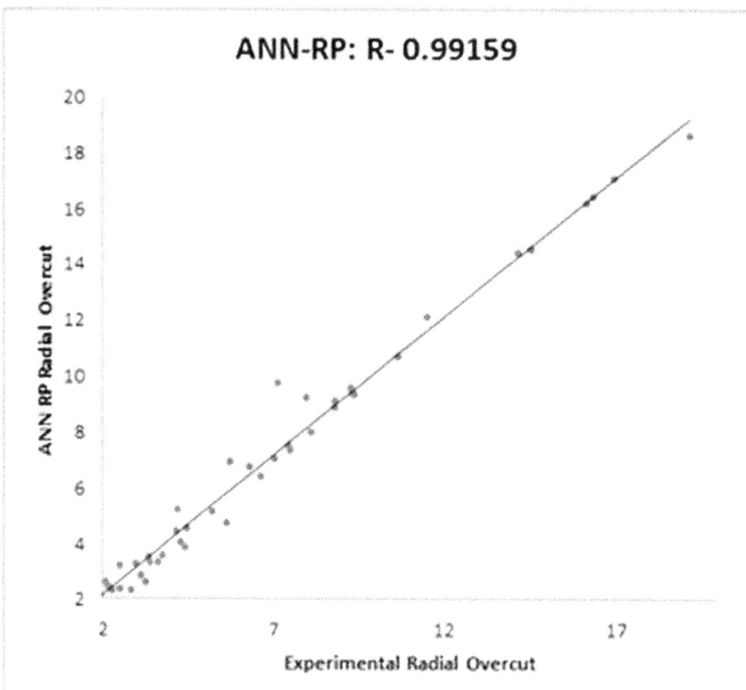

Figure 6. Comparison of ANNs with experimental values

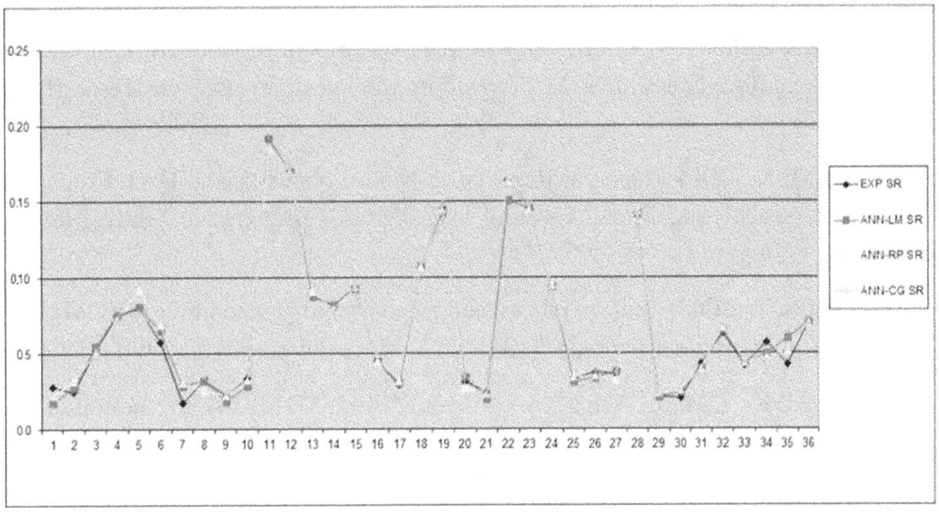

Table 1. MAE and MPE of testing data for training algorithms LM, CG and RP

	Number of Nodes in Hidden Single Layer	MAE	MPE
LM	6	0.243	9.674
CG	14	0.366	27.635
RP	15	0.370	11.634

CONCLUSION

In this chapter, three training algorithms LM, CG and RP were applied for the prediction of radial overcut of the Electrical discharge machined surface. This study indicated that the prediction of radial overcut is possible through the use of LM, CG and RP based neural network. The results obtained from widespread experiments conducted on ANSI D2 steel work piece materials with diverse machining parameters using copper electrode are compared and validate with the prediction. It was found to be close correlation with the experimental results. It was also observed that the LM model is quite analogous with CG and RP for surface roughness. The LM network demonstrated a slightly better performance compared to other models. And also, LM model predicted quite faster than the error goal reached in only 10 epochs while CG required 18 epochs and RP required 23 epochs. Conclusively speaking, the surface finish of EDMed surface cane be predicted by the above models with reasonable accuracy.

However, the prediction performance of this study may be improved further in three means. The first method is to include a few other variables that may affect the prediction performance. Second, optimal methods other than the GA may also be utilized to adjust the parameters of ANN model. We may even use models based on probabilistic neural networks for predicting the Material Removal Rate, Surface Roughness and Radial Overcut. Lastly, we could even propose an optimization based on the prediction outcomes of this study for future research, practical use and further validation.

REFERENCES

Anitha, J., Das, R., & Pradhan, M. K. (2012). Comparison of Neural Network Learning Algorithms for Prediction of Surface Roughness in EDM. *Journal of Mechatronics and Intelligent Manufacturing*, *3*(1/2), 3.

Das, R., & Pradhan, M. K. (2014). General Regression Neural Network and Back Propagation Neural Network Modeling for Predicting Radial Overcut in EDM: A Comparative Study. *Mechatronic and Manufacturing Engineering*, *8*(4), 799–805.

Ho, K. H., & Newman, S. T. (2003). State of the art electrical discharge machinings (EDM). *International Journal of Machine Tools & Manufacture*, *43*(13), 1287–1300. doi:10.1016/S0890-6955(03)00162-7

Jena, A. R., Pradhan, M. K., Das, R., Acharjya, D. P., & Mishra, M. (2014). Prediction of Radial Overcut in Electro-Discharge Machining through Artificial Neural Network. In *Advancements and Current Trends in Industrial Mechanical and Production Engineering*.

Mohd Abbas, N., Solomon, D. G., & Fuad Bahari, M. (2007, June). A review on current research trends in electrical discharge machining (EDM). *International Journal of Machine Tools & Manufacture*, *47*(7-8), 1214–1228. doi:10.1016/j.ijmachtools.2006.08.026

Pradhan, D., & Kumar, M. (2012). Multi-objective optimization of MRR, TWR and Radial Overcut of EDMed AISI D2 tool steel using response surface methodology, grey relational analysis and entropy measurement. *Journal for Manufacturing Science & Production*, *12*(1), 51–63. doi:10.1515/jmsp-2012-0004

Pradhan, M. K. (2012). Multi-objective optimization of MRR, TWR and radial overcut of EDMed AISI D2 tool steel using response surface methodology, grey relational analysis and entropy measurement. *J. Manuf. Science and Production*, *12*(1), 51–63.

Pradhan, M. K. (2010). *Experimental investigation and modelling of surface integrity, accuracy and productivity aspect in EDM of AISI D2 steel* [Ph.D. thesis]. National Institute of Technology, Rourkela, India.

Pradhan, M. K., & Das, R. (2015). Application of a general regression neural network for predicting radial overcut in electrical discharge machining of AISI D2 tool steel. *International Journal of Machining and Machinability of Materials*, *17*(3-4), 355–369. doi:10.1504/IJMMM.2015.071998

Pradhan, M. K., & Das, R. (2015). Application of ann modeling of radial overcut in electrical discharge machining. *Journal of Manufacturing Technology Research*, *7*(1/2), 39.

Chapter 7
Optimal Designs by Means of Genetic Algorithms

Lata Nautiyal
Graphic Era University, India

Preeti Shivach
Graphic Era University, India

Mangey Ram
Graphic Era University, India

ABSTRACT

With the advancement in contemporary computational and modeling skills, engineering design completely depends upon on variety of computer modeling and simulation tools to hasten the design cycles and decrease the overall budget. The most difficult design problem will include various design parameters along with the tables. Finding out the design space and ultimate solutions to those problems are still biggest challenges for the area of complex systems. This chapter is all about suggesting the use of Genetic Algorithms to enhance maximum engineering design problems. The chapter recommended that Genetic Algorithms are highly useful to increase the High-Performance Areas for Engineering Design. This chapter is established to use Genetic Algorithms to large number of design areas and delivered a comprehensive conversation on the use, scope and its applications in mechanical engineering.

INTRODUCTION

Designing process of a product includes a number of optimization problems. Such as, when we design an engine controller, the fuel economy and power performance should be optimized. And both of these factors are also affected by various other parameters such as temperature, pressure etc. Therefore, controlling fuel injection and air-fuel ratio with reference to these parameters is an optimizing problem and also a complex one. There are various issues when we deal with the engineering problems. First one is to improve design competency. Industries have to develop good products with in a time bound because of competition or requirements. Second issue is to optimize the design process. The third issue is to achieve robustness requirements.

DOI: 10.4018/978-1-5225-3035-0.ch007

Although design problems are typically demarcated more indefinitely and have a variety of precise answers, but the methodology may require backtracking as well as iteration. Designing process is a depending process and the solution depends on unpredicted difficulties and modifications as it matures.

The Wright brothers didn't come to know problems until they actually build and test their primary gliders. The fundamental five steps for resolving the problems associated with designing or design problems are as follows:

1. Describe the overall problem
2. Collect relevant information
3. Produce several solutions
4. Evaluate and choose a solution
5. Assess and execute the solution

The first phase is Describe the overall problem. Problem description step mainly consists of a list of requirements of the customer and also information like function and features of the product among other things.

In the second step, appropriate material is collected for the design and functional specifications of the product. For this purpose, similar products available in the market are reviewed. When the relevant information is collected successfully then the duty of the design teams, manufacturing teams, and marketing teams is to generate large number of options to accomplish the objectives.

Evaluate and choose a solution step is all about the comprehensive analysis of the solutions and results in finding out of the absolute design that finest fits the final needs of the customer. By applying this step, a sample is developed and tested against functionality to validate and possibly revise the design. This process is depicted in Figure 1.

Figure 1. Design Process

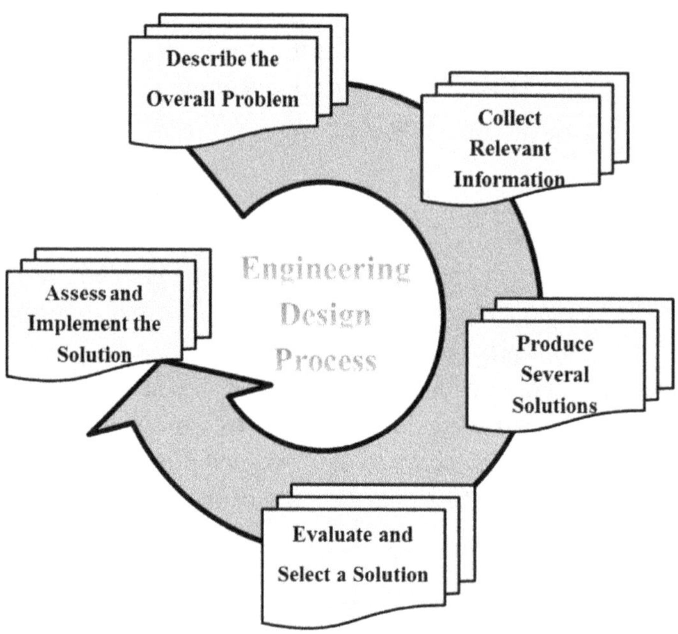

DESIGN PROBLEMS IN ENGINEERING

As contemporary computational and modeling technologies propagate all over the world, engineering design profoundly depends on computer modeling and simulation to quicken design cycles as well as to save the cost (Xiaopeng, 2007; Roosenburg & Eekeks, 1995). A complicated design problem will comprise of various design constraints. Discovering design space and finding best solutions are still major issues and the challenges for complex systems.

In a product design process, many complex multi-objective optimization problems occur. For example, in designing an engine controller, appropriate fuel injection times and air-fuel ratios have to be decided to improve engine fuel economy and power performance. But engine fuel economy and power are also affected by hundreds of other engine conditions, such as intake manifold pressure, intake manifold temperature, coolant temperature etc. How to control fuel injection time and air-fuel ratio with respect to these conditions to achieve the optimal fuel economy and power performance is an extremely complex problem. It is essential for engineers to progress the design by using simulation and optimization techniques. Although large number of challenging issues is there when complex engineering problems are solved. The first issue is all about finding the design efficiency. Current industries require a high quality product in a small time because of opposition or design cycle requirements. Traditional design processes can be much improved by using computational engineering tools. The second issue is how to optimize the complex design. The engineering optimization problems are normally high dimensional and with conflicting objectives. The optimization algorithms need to be introduced to help explore design space and find the optimal solution. The third issue is how to meet robustness requirements. Engineering design always has uncertainties due to manufacturing tolerance and perturbation in real operation. These three issues are the main focus of this dissertation.

Rapid prototyping helps to speed up the design process and explore research and development ideas. Engineers are able to build complex computational models to simulate many physical dynamics, such as combustion dynamics, fluid dynamics, and vibration dynamics. Model accuracy has been improving as we understand more about the system and computational power is enhanced. Industries are able to study prototyping before any manufacture production happens. However, even with the help of computational modeling, the design process is a long and tedious procedure and requires a lot of experiments and simulations to explore the design concept. How to improve design process efficiency is still one of major challenges in current industrial world.

GENETIC ALGORITHM

Genetic Algorithm (GA) starts with population of encoded solution and lead the population for optimal solution. Therefore, GA searches a space of possible solutions and finds the best solution. GA algorithms belong to zero-order optimization method. Three main tasks of these algorithms are; selection, mutation and crossover. Selection chooses individuals called parents that contribute to the population of next generation. Mutation operation mutates the distinct parents to generate children. Last operation joins two parents and generates children for future generation. Fittest is survived and unfit is died out (Tom, 2017). The success of a Genetic Algorithm depends on selection of parameter and genetic operators. They belong to class of evolutionary algorithms (Pisinger & Ropke, 2007). GA is different from conventional algorithms in the following manners (Golberg, 1989):

- Theses algorithms works on coding of design variables rather than variable itself.
- They use an objective function; no derivative is used.
- They start from a population not from a single point.
- They can be used with discrete, integer, continuous or a mix of the three.

For example, suppose the strings 10101 and 10010 are selected for crossover. Suppose we select an intertwining point of 2. Crossover is executed then the next iteration starts with selection. This practice carries on until a specific criterion is met.

1st Parent = 1 0 ¦ 1 0 1
2nd Parent = 1 0 ¦ 0 1 0
1st Child =1 0 0 1 0
2nd Child = 1 0 1 0 1

There are three fundamental operators originate in every genetic algorithm (Khan and Bajpai (2013)):

Reproduction

In this operation individuals are selected for generating next generation. The selection of an individual is based on the fitness value of that particular individual. Fitness value is calculated by using a fitness function. For every generation, the breeding operator selects individuals that are engaged into a mating pool. Random immigrants bring some entirely new randomly generated elements into the gene pool (Branke, 2000).

Crossover

In term of biology, crossover refers to mingling the chromosomes from the parents to generate new chromosomes for the descendants. Analogous to biological crossover, the GA also chooses two individuals randomly then checks whether crossover operation can be performed by using a factor called crossover probability. These two individuals are simply copied to next generation if the GA finds that the parameter is not up to and if crossover operation is performed then a point of joining is chosen randomly. New individuals are created and placed in next generation. A uniform crossover operator probability of 0.5 is recommended by various works such as Syswerda (1989) and Spears and De Jong (1991).

Mutation

In term of biology, mutation refers to changing the nucleotide arrangement of the genome. Unrepaired impairment to DNA or RNA genomes, faults in the course of replication is the causes of mutation. Mutation operation of GA is a genetic operation in which an operator is used to preserve the diversity from generations to generations. It basically changes one or more genomes in a chromosome from its original state. It may result in a totally different solution. Hence, a mutation operator has a great impact on the solution. It occurs for the duration of evolution according to a user-defined probability of mutation. Mutation probability should not be high because if it is then the search process will start moving towards a basic random search (Golberg, 2006; Yeh et al., 2004).

Upholding and introducing variety is the main goal of mutation operation. It should avoid local minima by stopping the population of individuals from becoming alike to each other. Therefore, it slows down the process of evolution. This logic also clarifies the point that many GA systems evade from only selecting the finest of the individuals in producing the next but rather a random selection with a weighting toward those that are fitter.

- **Examples:** Maximize the function $f(x) = x^2$, $x \in (0,31)$. x will be represented as five digit using integer. So, we will select randomly generated preset of solution as in (Goldberg (2006))

  Gene1 Gene2 Gene3 Gene 4
  01101 11000 0100 0 10011
  (13) (24) (8) (9)

Table 1. Reproduction

String No	Initial Population	Value of x	Fitness Function $f(x) = x^2$	Probability	Expected Count	Actual Count
1.	0 1 1 0 1	13	169	0.14	0.58	1
2.	1 1 0 0 0	24	576	0.49	1.97	2
3.	0 1 0 0 0	8	64	0.06	0.22	0
4.	1 0 0 1 1	19	361	0.31	1.23	1
Sum			1170	1.00	4.00	4
Average			293	0.25	1.00	1
Maximum			576	0.49	1.97	2

Table 2. Crossover

String No	Mating Pool	Cross Over Point	Offspring After Crossover	Value of x	Fitness Function $f(x) = x2$
1.	0 1 1 0 |1	4	0 1 1 0 0	12	144
2.	1 1 0 0 |0	4	1 1 0 0 1	25	625
3.	0 1 |0 0 0	2	1 1 0 1 1	27	729
4.	1 0 |0 1 1	2	1 0 0 0 0	16	256
Sum			**1754**		
Average			**439**		
Maximum			**729**		

VARIANTS OF GENETIC ALGORITHMS

Real Coded Genetic Algorithm

The term Real Coded Genetic Algorithm (RCGA) holds large amount of benefits as compared to binary coded equivalent when handling continuous search spaces with large dimensions and a great numerical precision is needed. In RCGA, each and every gene signifies a different set of the problem, and the size of the chromosome is reserved the same as the length of the outcome to the problem. Consequently, RCGA can handle large domains without sacrificing precision as the binary implementation. Additionally, RCGA retains the ability for the local tuning of the solutions; it similarly permits assimilating the domain knowledge so as to improve the performance of GA.

Binary Coded Genetic Algorithm

An algorithm which is based on a fundamental concept of probabilistic search algorithm that repetitively transmutes a group (called a population) of mathematical entities (typically fixed length binary character strings), each with an connected fitness value, into a new population of offspring entities using the theory of Charles Darwin's natural selection and using operations that are mottled after naturally occurring genetic operations, such as crossover (sexual recombination) and mutation is called as Binary Coded Genetic Algorithm. Following the model of evolution, they set up a population of individuals, where each distinct corresponds to a point in the search space. An objective function is applied to each individual to rate their fitness.

Differential Evolution

Differential Evolution (DE) methods are the replacement of traditional mutation and crossover operation by alternative differential operators. Machine intelligence and cybernetics are widely using these methods. It outperforms the GA in various cases. As in other evolutionary algorithms, two primary processes drive the evolution of a DE population: the variation process, which permits searching the diverse areas of the search space, and the selection process, which guarantees manipulation of the developed information about the fitness landscape.

Least Mean Square Algorithm

Least Mean Square (LMS) methods are stochastic gradient descent methods. Based on the current error rate, filters are designed to be adaptive. One can implement these methods without squaring, differentiating. These methods are used in adaptive filters to find the filter coefficients that relate to producing the least mean squares of the error signal (difference between the desired and the actual signal). It is a stochastic gradient descent method in which the filter is adaptive based on the error at the current time. The LMS algorithm can be implemented without squaring, averaging or differentiation and is simple and efficient process.

Sawtooth GA

There are various methods that are designed to enhance the robustness and effectiveness of computation. Standard GA is based on selection, crossover and mutation. Major parameter of GA is its population size which affects effectiveness of computation and robustness of the algorithm. If the population size is too small then the solution prematurely converges to a non-optimal solution, whereas if the population size is large then the computation efforts become considerable. Numerous approaches have been proposed that try to upsurge the range of the population and avoid premature convergence

WHY GENETIC ALGORITHM IN ENGINEERING?

As the growth of engineering industry, computer modeling and simulation are profoundly used to improve the designing process of engineering. That is also cost effective. Finding an optimal solution of an engineering problem is still a challenging task. Genetic algorithms are widely used solution for optimizing problems. Genetic Algorithms (GAs) are heuristic search algorithms that imitate the process of biological evolution (Goldberg, 1998). GA developed by Holland (1975), GAs are based on the principle given by Darwin. According to Darwin *Neither the most intelligent nor the most powerful but the species survive will be the most adaptive.* The search procedure of GAs finds the best and fittest design solutions. These algorithms are used in engineering because they are easy to use. Opposite to gradient methods, the GAs work with a set of solutions and are guided by probability. These are not deterministic.

Optimization techniques like goal programming, linear programming, branch and bound algorithm face the problem when the number of variables increases. Each optimization algorithm is intended to solve a particular type of problem. Some algorithms produce solution accurately but not cost efficient whereas some algorithms don't provide an optimal solution. Hence, while solving a problem we have to compromise between high accuracy and low accuracy. Genetic Algorithms (GAs) are heuristic search algorithms that imitate the process of biological evolution. Genetic algorithms are a very popular heuristic which have been successfully applied to many optimization problems (Bhoskar et al., 2015). These algorithms are a part of revolution in computer science field and downtrend the limits of other algorithms:

- GAs are most robust.
- While executing search in large state –space, GAs offer noteworthy profits over many other techniques.
- GA operates on coding of solution set, not the solutions themselves.
- GAs search from a population not a single solution.
- The rules are not deterministic rather probabilistic.

The positive side of GA is that it handles the constraints and objectives very easily. They can be easily applied to a wide range of jobs such as optimization, learning etc. Below are some advantages of using Gas:

- Each optimization problem that can be defined with the chromosome encoding can be solved by using GA.
- It solves problems with multiple solutions.
- Structural genetic algorithm gives us the possibility to solve the solution structure and solution parameter problems all at once.
- Genetic algorithm is a method which is very easy to understand and it practically does not demand the knowledge of mathematics.
- Genetic algorithms are easily moved to current simulations and models.

LIMITATIONS

- Certain optimization problems (they are called variant problems) cannot be solved by means of genetic algorithms. This happens because of inadequate fitness functions which produce bad chromosome blocks despite the fact that only noble chromosome blocks crossover.
- There is no absolute assurance that a genetic algorithm will find a global optimum. It happens very frequently when the populations have many subjects.
- Like other artificial intelligence techniques, the genetic algorithm cannot guarantee stable response time. This is a major reason of not applying Gas to real time applications.

APLLYING GENETIC ALGORITHM IN ENGINEERING DESIGN

One of the popular heuristic search algorithms is genetic algorithm. GA not only has all heuristic algorithms' characteristics, but also is a multi-directional search method. It originally is designed for single objective optimization problem since it uses a fitness to do evaluation. As GAs are applied to multi-objective optimization, the fitness concept has been extended to dominance rank, which is created for searching the Pareto Front. Since then, GAs begin to become popular in multi-objective optimization areas, especially in finding the Pareto Front.

GAs use dominance rank to push the population close to the Pareto Front and it has been proved to be an effective way to explore the Pareto Front. One of the difficulties in exploring the Pareto Front is the curse of dimension. As dimension of the problem increases, the Pareto Front becomes very complicated. GAs tends to be stuck in some local Pareto Front areas. To make the optimal solutions well covering the Pareto Front and quickly converting to the optimum is what most of research on GAs are focusing on. How to balance proximity and diversity in exploration is a multi - objective optimization problem itself.

There are many studies on diversity preservation, diversity estimation, and metric comparison to improve the population diversity. Many different GAs have been presented to improve diversity and convergence such as RAND, FFGA, NPGA, HLGA, VEGA, NSGA listed by Zitzler, Deb, and Thiele 2000. There are new research ideas such as co-evolutionary GAs and Clustered Oriented Gas (Packham & Parmee, 2000). No matter what GAs are, they are trying to make the search converge to the Pareto Front (or close to the Pareto Front) as quick as possible and make the population cover the Pareto Front (or close to the Pareto Front) as even as possible at the least computation time. Since these goals are conflicting themselves, we often find out that any GA is a tradeoff of these goals and it may perform well on some certain problems but bad on others.

Engineering Design Using GAs

Since GAs have shown excellent performance in optimization problems, especially in multiobjective optimization, engineering design optimization problems have been explored with GAs. Engineering design optimization problems normally are multi-objective problems with high dimensional design variables. They also are complicated in that system dynamics are always non-linear and with uncertainty. In addition, engineering design problems often have constraints on design variables. All these issues have been well addressed in different GAs.

As engineering design becomes more and more complex in modern industry, computer modeling is one of the essential methods to achieve reducing design cycle and improve design quality. Genetic algorithms have been used in a lot of complex design problems. There have been a number of activities from developing GA software for engineering design to improving GAs for engineering design.

Robustness in Engineering Design

Robustness is the key to designing products that work in a range of conditions. From this aspect, robustness is sometimes in higher priority than optimality. Engineering design has to deal with uncertain environment, manufacturing tolerance and un-modeled effects. Real industry problems have shown that uncertainty can result in failure in the field.

Previous researches focused on uncertain objective function problems, i.e., objective functions will return different values with the same design inputs. The techniques used for these problems are to estimate distribution of objective functions. These methods have been used to apply to mathematical problems to deal with uncertainty. However, this method is limited in that engineering design has to deal with uncertain design variables, especially curves. Few researches are oriented for this area at present.

GAs are useful to many scientific, engineering problems including:

- **Optimization:** GAs have been used in a wide variety of optimization tasks like circuit design, numerical optimization, video and sound quality optimization.
- **Automatic Programming:** GAs have been used to evolve computer programs for specific tasks, and to design many computational assemblies, such as cellular automata and sorting networks.
- **Machine and Robot Learning:** These have been used for various applications of machine learning such as classification and prediction etc. GAs have also been used to design learning classifier systems.
- **Ecological Models:** GAs have been used to model ecological phenomena such as biological arms races, host-parasite co-evolutions, symbiosis and resource flow in ecologies.
- **Population Genetics Models:** GAs have been used to study questions in population genetics, such as "under what conditions will a gene for recombination be evolutionarily viable?"
- **Models of Social Systems:** To analyze evolutionary facets of social systems, Gas are also used. Such as the evolution of interaction, and trail-following behavior in ants (Tom (2017))

Application in Mechanical Engineering

GA is a search based optimizing technique. It is used in Mechanical Engineering in the following ways:

- In optimizing processes like ECM, EDM, USM, etc.
- In operation sequencing & machining parameters selection.
- In designing airplane wings.
- Planning the cyclic preventive maintenance of the components.
- In physical distribution of a service. Organization uses GA to effective deliver their services to customer.
- In optimizing production planning and production scheduling activity of engineering.
- In designing Automobile suspension system.

CONCLUSION

With the advancement in contemporary computational and modeling skills, engineering design completely depends upon on variety of computer modeling and simulation tools to hasten the design cycles and decrease the overall budget. The most difficult design problem will include various design parameters along with the tables. Although to find out the design space and ultimate solutions to those problems are still biggest challenges for the area of complex systems. This chapter is all about suggesting the use of Genetic Algorithms to enhance maximum engineering design problems. The chapter recommended that Genetic Algorithms are highly useful to increase the High-Performance Areas for Engineering Design. This chapter established to use Genetic Algorithms to large number of design areas and delivered a comprehensive conversation on the use, scope and its applications in mechanical engineering.

REFERENCES

Bhoskar, T., Kukarni, O. K., Kulkarni, N. K., Patekar, S. L., Kakandikar, G. M., & Nandedkar, V. M. (2015). Genetic Algorithm and Its Applications to Mechanical Engineering: A Review. *4th International Conference on Materials Processing and Characterization. Materials Today: Proceedings, 2*, 2624–2630.

Branke, J. (2000). *Efficient evolutionary algorithms for searching robust solutions*. ACDM. doi:10.1007/978-1-4471-0519-0_22

Goldberg, D. E. (1998). The Design of Innovation: Lessons from Genetic Algorithms. Lessons for the Real World. University of Illinois at Urbana-Champaign.

Goldberg, D. E. (2006). Genetic algorithms in search, optimization & Machine learning. Pearson Education.

Holland, J. (1975). Adaptation in Natural and Artificial Systems. University of Michigan Press.

Khan, M. Z. R., & Bajpai, A. K. (2013). Genetic Algorithm and Its Application In Mechanical Engineering. *International Journal of Engineering Research & Technology, 2*(5), 677-683.

Packham, I. S. J., & Parmee, I. C. (2000). Data analysis and Visualization of Cluster-Orient Genetic Algorithm Output. *Proceedings of the International Conference on Information Visualization*, 173-178. doi:10.1109/IV.2000.859752

Pisinger, D., & Ropke, S. (2007). A General Heuristic for Vehicle Routing Problems. *Computers & Operations Research*, *34*(8), 2403–2435. doi:10.1016/j.cor.2005.09.012

Roosenburg, N., & Eekeks, J. (1995). *Product Design: Fundamentals and Methods*. John Wiley & Sons Inc.

Spears, W. M., & De Jong, K. A. (1991). On the Virtues of Parameterized Unıform Crossover. *Proceedings of the 4th International Conference on Genetic Algorithms*, 230-236.

Syswerda, G. (1989). Uniform Crossover in Genetic Algorithms. *Proceedings of the 3rd International Conference on Genetic Algorithms, 2-9.*

Tom, M. V. (2017). *Transportation System Engineering, Genetic Algorithm*. Indian Institute of Technology Bombay.

Xiaopeng, F. (2007). *Engineering design using genetic algorithms*. Retrospective Theses and Dissertations. Paper 15943.

Yeh, L. J., Chang, Y. C., & Chiu, M. C. (2004). Article. *Journal of Marine Science and Technology*, *12*(3), 189–199.

Zitzler, E., Deb, K., & Thiele, L. (2000). Comparison of multi-objective evolutionary algorithms: Empirical results. *Evolutionary Computation*, *8*(2), 173–195. doi:10.1162/106365600568202 PMID:10843520

Chapter 8
Analyzing Sustainable Food Supply Chain Management Challenges in India

Yogesh Kumar Sharma
Graphic Era University, India

Sachin Kumar Mangla
Graphic Era University, India

Pravin P. Patil
Graphic Era University, India

Surbhi Uniyal
Graphic Era University, India

ABSTRACT

The demand of food is increasing day by day, innovative agricultural practices and sustainable food supply chain management (SFSCM) has gained an emergent importance. Food industries across the globe mainly focus on the manufacturing of their own products to achieve sustainability. The importance of sustainable food supply chain management is to overcome the wastage in food manufacturing industries. In the present research, we identified eleven challenges in the SFSCM on the basis of literature review and expert opinion. The approach is an integration of fuzzy with DEMATEL which can be used for dividing the challenges into cause and effect group. Fuzzy DEMATEL method has continuously been used for the analysis of challenges and is the novel approach for decision making. Thus, this method can be implemented in many fields including automobiles, food industries, retail market etc. From the fuzzy DEMATEL results, it can be confirmed that the Safety and Security is one of the most influencing challenge and has the strongest association with other challenges.

DOI: 10.4018/978-1-5225-3035-0.ch008

INTRODUCTION

Since from the past decades, there is a development of green and sustainable supply chain management methods for reducing the environmental concerns and issues. Also the food production and consumption processes have been changed to overcome some of the major problems of the environment (Genovese and Acquaye, 2017). According to the report of India brand equity foundation (IBEF), 2017, the world food trade is rising day-by-day with the increase in the establishment of new food industries. The Indian food industries have evolved as a high income and growth area because of its enormous potential for value addition, especially in food processing industries. The current value of food industries is US$ 39.71 billion and it will be increase at a Compounded Annual Growth Rate (CAGR) of 11% to US$ 65.4 billion by 2018. The Government of India has constantly making efforts for the development of food processing industries which will account for approximately 32% of the country's total food market. The Ministry of Food Processing Industries (MOFPI) has already taken and is still taking many initiatives to increase investments in this business. Many proposals had been approved for foreign collaborations, joint ventures and export oriented units and in industrial licenses. Indian food processing industry is one of the largest industries accounting for 32% of the country's total food market and it secures fifth rank among other industries. The food industry accounts for around 14% of GDP (Gross Domestic Product), 13% of exports and 6% of total industrial investment. In context with organic food market, India is projected to increase its production by three times in next 3-4 years. The current value of Indian gourmet food market is US$ 1.3 billion which is continuously growing at a CAGR (Compound Annual Growth Rate) of 20%. Idea of fuzzy approach was used by (Bellman and Zadeh, 1970) in decision making theory. Fuzzy set theory is very useful for removing the uncertainty in the supply chain (Lee et at., 2004). According to the literature or researchers point of view it is clearly mentioned that fuzzy set theory is capable of managing the supply chain inventory (Petrovic et al., 1998, 2001). In the current problem Fuzzy DEMATEL is used to remove the biasness in the human judgments. Fuzzy approach is used for the better implementation of SFSCM in Indian food industry. To manage the increased demand of food throughout the world, SFSCM is very important. Supply Chain Management (SCM) is an upcoming and wide area and has been considered by scientists, researchers and academicians in the last years. One prominent research field is sustainability in SCM, namely Sustainable Supply Chain Management (SSCM). Both research and practical implementation have been growing steadily in the last decade in this specific area (Ahi and Searcy, 2013, 2014). The role of the food industry (retailers, manufacturers and food service) in helping consumers eat healthily and sustainably has been receiving considerable attention in recent years. Consumer perceptions thus show an increasing concern about food safety and about properties of the food they buy and eat. As the country experiences more pressure from globalization, the food industry sector is also subjected to the increased competition in the domestic market. The processors have to meet those challenges by responding very fast to avoid delays which can take them out of the business. If we combine both supply chain management and sustainability together one more interesting filed emerges i.e., sustainable food supply chain management (SFSCM), which is applied in recent years as a reaction to stakeholder pressures (Gold and Hahn, 2013). The consumers are very cautious about the food they eat and this is the duty of producers, retailers and manufacturers to provide them healthy and sustainable food. Also the consumers are well educated and they know what to eat and what not to, thus the food industry has to fulfill the consumer's demands in less time due to increase in globalization. For the analysis of SFSCM based challenges in Indian food industry, we used Fuzzy DEMATEL approach.

The paper is planned as follows: investigated the literature review related to SFSCM; solution methodology apply to solve the present problem in the paper; data collection for Fuzzy DEMATEL; results and discussion of identified challenges in SFSCM implementation; explained the conclusions, limitations and future opportunities of the present research in the food industry sector.

LITERATURE REVIEW

The literature review mainly focuses on SFSCM and analysis of challenges in the Indian food industry.

SFSCM

SFSCM means all the processes in the food industries such as production, procurement of materials and distribution, and also their inverse processes for collecting, returning of unused products to avoid wastage and a *socioeconomically* and *ecologically* sustainable recovery (Bloemhof, 2017). SSCM not only includes information about the food products, their management and capital flows but also maintains the cooperation between different companies. SSCM also results in taking goals of three dimensions of sustainable development viz; social, economic and environmental (Seuring and Mueller, 2008). In the modern food industry, processes have become developed, characterized by mass production. In addition, production, financing, and marketing have become globally incorporated to form global food supply chains (Manning et al., 2006; Trienekens et al., 2012). SSCM is an increasingly main topic in sustainability and (SCM), as companies respond to internal and external pressures from stakeholders, policymakers, and consumers, and from governments, and profit and non-profit organizations dedicated to ecological, public, and commercial responsibility (Ageron et al., 2012; Seuring and Müller, 2008). Managing food supply chains is a well-known problem due to the scale of this industry, the quantity of global food waste, and the relationship between food waste and global malnutrition (Parfitt et al., 2010). In the past years there were many food scandals and incidents which make consumers more attentive towards food products thus food products should be more sustainable to retain and regain consumer's trust. Consumer wants all the knowledge about the food products right from its origin to its delivery which includes safety and expiry date of the product. Sustainability is a broad area which comprises of social issues, environmental issues and expected returns. By this work we aim to discover the status of information technology which supports sustainability in the food supply chains and communication between stakeholders (Wognum et al., 2011).

Challenges in SFSCM

For the successful implementation of SFSCM in Indian food industries we identified eleven challenges based on experts opinion and literature review.

Table 1. Identified challenges to implement SFSCM in Indian food industry

S.No.	Challenges	Description	References
1	Food safety and security (CH1)	In the food production, food safety is an important issue and needs to be considered by stakeholders. From a safety point of view, food chains have numerous vulnerabilities.	(Beulens, 2005, Whipple, 2009 and Bloemhof, 2017)
2	Food quality (CH2)	Food quality means physical properties of food until the food is finally reached to the consumer. It not only includes microbial aspects but also flavors or texture.	(Grunert, 2005, Akkerman et al., 2010, Beta et al., 2017)
3	Transparency (CH3)	Transparency means that all the stakeholders have to share all the information regarding the product without delay and alteration.	(Hofstede, 2004, Trienekens, 2012 and Soysal, 2012)
4	Food wastage (CH4)	Food wastage is the major concern for our country now-a-days and is becoming a cause of global food security. This issue needs to be solved for protecting India's economy.	(Liu et al., 2013 Papargyropoulou et al., 2014, Girotto et.al., 2015)
5	Regulations and standards (CH5)	There should be strict rules and regulations for both food producers and stakeholders. Standards mean well-known norms or necessities for a product like material and packaging specifications. Standards should be maintained by industry associations, regulatory agencies and public organizations.	(Marucheck et al., 2011, Havinga and Verbruggen, 2017)
6	Traceability (CH6)	Keeping consumer's protection in view traceability is an essential and potential factor. Consumers can trace the food products and eliminate non-consumable food products thus can prevent from food safety issues.	(Badia et al., 2015, Souquet et al., 2017, Thomas et. al., 2017)
7	Transportation (CH7)	The food products should be transported well so that it reaches to consumers with all the nutritional qualities. The specific food should be transported in specific conditions to increase their shell life.	(Gustavsson et al., 2011, Egilmez and Park, 2014, Boukherroub et.al., 2017)
8	Supplier selection (CH8)	The relationship between buyer and supplier should be close enough so that buyer gets guaranteed by the food quality. Also supplier should not be fraud in any ways including financial matter and other issues regarding food.	(Liao and Rittscher, 2007, Van Weele and Van Tubergen, 2017)
9	Benchmarking (CH9)	The organizational products, services and processes should be evaluated in relation to its best practice and this is called benchmarking. Benchmarking is often used as a significant tool for the improvement of performance of an organization, competitive advantage and quality management.	(Yakovleva et.al., 2012, Swinburn et.al., 2013, 2015)
10	Information Technologies (CH10)	Efficiency, safety, globalization, refrigerants and efficiency are the most important constrains in the food delivery system. In recent years the market is based on information technology.	(Zailani et al., 2010, and Attaran, 2012)
11	Product life cycle management (CH11)	The management of the products lifecycle is crucial step for manufacturing any food and it is difficult to predict the lifecycle of the product. This includes the time from manufacturing to the final disposal. The data regarding products design, support and ultimate disposal of manufactured goods is hard to manage and it is important for improving safety and security of food product.	(Marucheck and Greis, 2011 and Stark 2015)

SOLUTION METHODOLOGY

Fuzzy Set Theory

Fuzzy set theory is mainly used where there is an inaccuracy in the information or knowledge present in the nature. According to Zadeh (1965), a fuzzy set theory was proposed in the decision making process where the information is insufficient in context with the problem (Bellman and Zadeh, 1970, Zimmermann, 2011). Fuzzy set theory evaluates the method for linguistic and vague terminology by capturing human injustice (Kumar et al., 2013). There are various examples in real world where decisions of decision makers are full of incorrectness and biasness. Linguistics variables are suitably expressed by fuzzy numbers and the fuzzy numbers like trapezoidal and triangular are suitable and commonly used as fuzzy numbers (Mangla et al., 2015).

Fuzzy Set

A fuzzy set is denoted as:

$$\tilde{E} = \left\{x, \mu_E\left(X\right)\right\}, \qquad x \in X \tag{1}$$

where $\mu_E\left(x\right): X \rightarrow \left[0,1\right]$ is the membership function of $\tilde{E}$ and $\mu_E\left(x\right) = 0$ then $\left(x\right)$ does not belong to the fuzzy set $\tilde{A}$. $\mu_E\left(x\right) = 1$ then $\left(x\right)$ completely belongs to the fuzzy set $\tilde{E}$.

Though, unlike the classical set theory, if $\mu\left(x\right)$ has a value between 0 and 1 then x partially belongs to the fuzzy set $\tilde{E}$ .That is, the pertinence of x is true with the degree of membership given by $\mu_E\left(x\right)$.

Fuzzy Numbers

A fuzzy number is a fuzzy set in which the membership function satisfies the conditions of normality

$$\sup \tilde{E}\left(X\right)_{x \in X = 1} \tag{2}$$

and of convexity

$C_1 C_2 C_j C_m$

$$\tilde{D} = \begin{matrix} E_1 & \tilde{X}_{11} & \tilde{X}_{12} & \tilde{X}_{13} & \tilde{X}_{14} \\ E_2 & \vdots & \vdots & \vdots & \vdots \\ E_3 & \tilde{X}_{n1} & \tilde{X}_{n2} & \tilde{X}_{nj} & \tilde{X}_{nm} \end{matrix} \tag{3}$$

For all $x_1, x_1 \in X$ and all $\lambda \in [0,1]$. The triangular fuzzy number is usually used in decision making because to its instinctive membership function,

$\widetilde{W} = \left[\widetilde{W_1} + \widetilde{W_2} + \widetilde{W_3} + \widetilde{W_4}\right]$, given by,

For practical applications a unique type of fuzzy number i.e., triangular fuzzy number (TFN) is generally chosen [Dubois and Prade, 1979]. In any (TFN) (p, q, r) its function is given by mathematically equation $\mu_E(x)$ as given in Eq. (4), where p ≤ q ≤ r. Also, (p, q, r,) represents the lower, mean and upper boundary of the TFN:

Mathematical Operations with Fuzzy Numbers

Given any real number K and two fuzzy triangular number $\tilde{E} = (p_1, q_1, r_1)$ and $\tilde{F} = (p_2, q_2, r_2)$, the main mathematical operations are given as follows: [Pedrycz and Gomide, 2007].

1. Addition of 2 TFNs

$$\tilde{E}(+)\tilde{F} = (p_1 + p_2, q_1 + q_2, r_1 + r_2) \quad p_1 \geq 0, p_2 \geq 0 \tag{5}$$

2. Subtraction of 2 TFNs

$$\tilde{E}(-)\tilde{F} = (p_1 - p_2, q_1 - q_2, r_1 - r_2) \quad p_1 \geq 0, p_2 \geq 0 \tag{6}$$

3. Multiplication of 2 TFNs

$$\tilde{E}(*)\tilde{F} = (p_1 * p_2, q_1 * q_2, r_1 * r_2) \quad p_1 \geq 0, p_2 \geq 0 \tag{7}$$

4. Division of 2 TFNs

$$\tilde{E}(\div)\tilde{F} = (p_1 \div p_2, q_1 \div q_2, r_1 \div r_2) \quad p_1 \geq 0, p_2 \geq 0 \tag{8}$$

5. Inverse of a TFN

$$\widetilde{E^{-1}} = \left(\frac{1}{r_1}, \frac{1}{q_1}, \frac{1}{p_1}\right) \geq 0 \tag{9}$$

6. Multiplication of a TFN by a constant

$$k * \tilde{A} = (k * p_1, k * q_1, k * r_1) \quad (10)$$

7. Division of a TFN by a constant

$$\frac{\widetilde{E}}{k} = \left(\frac{p_1}{k}, \frac{q_1}{k}, \frac{r_1}{k}\right) \quad (11)$$

Fuzzy DEMATEL

The Geneva Research Centre of the Battelle Memorial Institute who use DEMATEL method for the first time to solve MCDM problem (Gabus, and Fontela, 1972). DEMATEL approach is mainly used to detect the interrelationship among the factors and it is capable to solve complex problems (Tsai and Chou, 2009, Xia et al., 2015, Govindan et al., 2014, Jia et al., 2015). It is not able to handle the human bias and ambiguity in the data (Patil and Kant, 2013, Hsu et al., 2013). Fuzzy DEMATEL is a powerful tool to handle such situations. Presently many researchers and professionals used DEMATEL to find out the solutions of complex problems successfully (Wu WW, 2011, Mangla et al., 2015). Fuzzy DEMATEL is used in the present work to analyze the identified challenges in the implementation of SFSCM in Indian food industry; in this approach we combine fuzzy logic theory and the benefits of DEMATEL.

Figure 1. Research flow chart

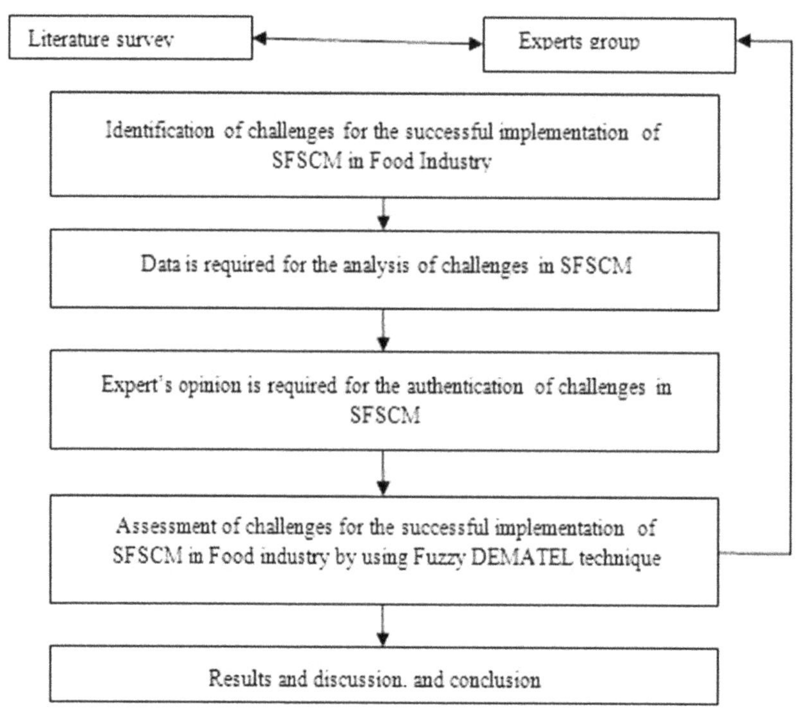

Fuzzy DEMATEL is very useful to find out the causal relations among the challenges and also helps the managers to construct policies for the successful implementation of SFSCM in food industries. In the given literature DEMATEL has been proved as a powerful tool to solve the complex MCDM problems. In many cases it is very difficult to decide the human decisions and its preferences into crisp values due to the fuzziness. For the present problem we applied fuzzy theory to the DEMATEL in order to measure the qualitative decisions on the interrelationships among challenges. Steps and equations used in Fuzzy DEMATEL are listed below ((Govindan et. al., 2013, 2015, Mangla et al., 2014 and Hernandez et al., 2014). Defining aim and assessment of factors: To collect the data and to attain the purpose of the problem, literature assessment and experts opinion is required. Possible associated factors that are important for the application of SFSCM are listed as assessment factors.

Step One: To define an experts opinion and evaluation criteria.

In the present step group of experts was created to give experts opinion on associated issues. The important challenges of SFSCM implementation in food sector were determined from literature and were finalized as the evaluation criteria.

Step Two: Formulate fuzzy direct evaluation matrix (F)

Once the assessment criteria are established, pair wise comparison is performed. For the comparison between challenges we need to design a five point linguistic scale in Table 2 which will helps the experts to identify the interrelation. For evaluating and transforming the linguistic information obtained from expert's judgments, the positive triangular fuzzy number (TFN) is used (see Table 2).

The TFN is denoted by quadruplet i.e., (pij, qij, rij), where p ≤ q ≤ r. Suppose $x^k_{ij=p^k_{ij},\ q^k_{ij},\ r^k_{ij}}$ where 1≤ k ≤ K, is the fuzzy evaluation that the kth expert gives about the degree to which *i* has an impact on factor *j*.

Step Three: To construct fuzzy initial direct relation matrix

Defuzzification is compulsory for the conversion of fuzzy numbers into crisp number. We can easily defuzzify the fuzzy evaluation matrix with the help of bisection area method. Equation 12 helps in the formation of fuzzy initial relation matrix (F). For fuzzy average matrix (A) we calculate the average of *k* fuzzy initial relation matrix (F) of n x n experts vies, where *k* is the number of experts.

Table 2. Linguistic scale (Mangla et al., 2015)

Linguistic Values (TFN)	Linguistic Terms
No effect	(0,0,0.25)
Very low effect	(0,0.25,0.5)
Low effect	(0.25,0.5,0.75)
High effect	(0.5,0.75,1.0)
Very high effect	(0.75,1.0,1.0)

Step Four: To attain the normalized initial direct relation matrix (D)

$$S = \min\left[\frac{1}{\max\sum_{j=1}^{n}\{m_{ij}\}}, \frac{1}{\max\sum_{i=1}^{n}\{m_{ij}\}}\right] \tag{12}$$

$$D = M \times S \tag{13}$$

Step Five: To obtain the total relation matrix

$$T = D\left(I-D\right)^{-1} \tag{14}$$

where, T= total relation matrix; I= Identity matrix

$$T = \left[t_{ij}\right]nxn$$

Step Six: To find the total of rows (R) and the total of columns (C)

$$R = \sum_{j=1}^{n} t_{ij} \quad {}_{nx1} \tag{15}$$

$$C = \sum_{j=1}^{n} t_{ij} \quad {}_{1xn} \tag{16}$$

Step Seven: To sketch a cause and effect graph with the help of *(r+c, r-c)* values.

If the value of *(r-c)* is positive, that challenge come in to the cause group, and if the value of *(r-c)* is negative that challenge come under effect group. A flowchart is given in the Figure 2. In which all steps are explained in Fuzzy DEMATEL approach.

Data Collection

In the current work, the data is collected in 3 steps:

1. Detection of significant challenges for the execution of SFSCM in Indian food industry.
2. Giving ranks to the challenges according to the calculated values and also numbering them accordingly.

Figure 2. Fuzzy DEMATEL flowchart for the successfully implementation of SFSCM in Indian food industry

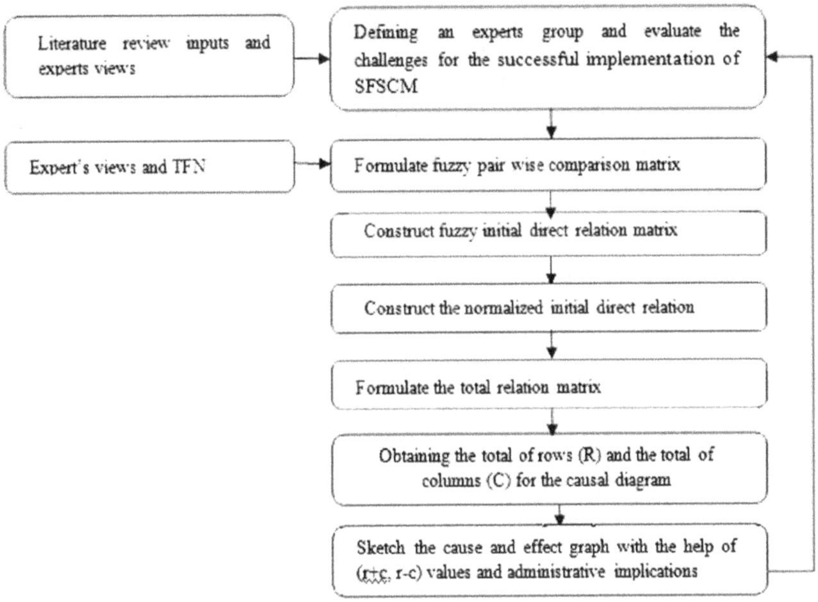

3. All the challenges were divided into cause and effect group according to their value of (*r+c* and *r-c*).

Eleven challenges (CH1, CH2, CH3, CH4, CH5, CH6, CH7, CH8, CH9, CH10, and CH11) were identified on the basis of literature survey and experts opinion in the beginning of the data collection. Experts from industries, academician in the particular field give their opinion for the identification of challenges. Among all, nine challenges were identified by literature survey and two from expert's opinion which were explained in Table 1.

RESULTS AND DISCUSSIONS

Step One: A group of experts (two senior members, one from the food industry and one from the academician) has created the aim of this present research, to find out the key challenges in the implementation of SFSCM in Indian food industries. Selected members were very much capable in their respective areas and good in decision making. Expert's selection was made on the basis of their knowledge and expertise in their particular area. Eleven challenges were identified from the given literature for the successful implementation of SFSCM in Indian food industry and validated by expert's views.

Step Two: Pair wise evaluation was made by experts between the key challenges for the implementation of SFSCM in Indian food industries using the linguistic scale given in the Table 2.

Step Three: To formulate the initial direct matrix, defuzzification process is used in which fuzzy number is converted into crisp number. Table 4 explained the initial direct relation matrix of key challenges for the implementation of SFSCM in Indian food industries.

Step Four: Explains about the fuzzy normalized initial direct relation matrix of identified important challenges for the implementation of SFSCM in Indian food industries which is illustrated in Table 5.

Step Five: Provides the total direct relation matrix of identified important challenges for the implementation of SFSCM in Indian food industries was obtained by equation 5 which is shown in Table 6.

Step Six: Total of rows (r) and total of columns (c) of the challenges for the implementation of SFSCM in Indian food industries was obtained by equation 15 and 16.

Step Seven: The value of *(r+c)* and *(r-c)* of the important challenges for the implementation of SFSCM in Indian food industries is explained in the Table 7.

The weight of eleven challenges in the perspective of SFSCM implementation in Indian food industry through the sum of *(r+c)* is given as CH1-CH4-CH6-CH7-CH9-CH5-CH8-CH11-CH10-CH2-CH3 given shown in the Table 7. (CH1) food safety and security, (CH4) food wastage, (CH5) regulations and standards, (CH6) traceability, (CH7) transportation, (CH9) benchmarking, (CH11) product life cycle management were grouped into cause group and (CH2) food quality, (CH3) transparency, (CH8) supplier selection, (CH10) information technology were grouped into effect group by using *(r-c)* values.

Cause Group Challenges

Challenges that come under cause group are having more importance and they are crucial so that focus should be made on them. In the cause group (CH6) traceability has higher (r-c) value (0.59), which directly affects the other challenges in the whole process. But the value of *(r+c)* of (CH6) is 3.13 which is very low as compared to other challenges. It is clearly seen from the picture that traceability affects the other challenges in the list but they receive very less influence in the return. Similarly other challenges in the cause group are ranked according to the *(r-c)* values affect the other challenges but not getting more influence in the return.

Effect Group Challenges

Effect group challenges gets easily influenced by the other challenges in the list, these challenges are not having that much of impact on other challenges but still they contribute significantly. In the effect group (CH2) food having the least *(r-c)* value of -0.57, which confirms that this challenge attain the maximum impact. (CH2) is a very important challenge in the food industry in terms of sustainability establishment, (CH2) directly affect the cause group. The other challenges in the effect group are sequentially arranged according to their priority. (CH8) transparency with *(r-c)* value of -0.46, (CH10) information technology with *(r-c)* value of -0.40 and transparency (CH3) with (r-c) value of -0.21. If industries and government give more emphasis on cause group challenges it automatically affect the effect group challenges.

CONCLUSION

In the present work, challenges were discussed for the successful implementation of SFSCM in Indian food industries. In SFSCM we combined three pillars of sustainability i.e. social, economic and environment. For achieving sustainability in the food sector it is mandatory for all the companies to consider

Table 3. Fuzzy direct evaluation matrix (F)

CH	CH1			CH2			CH3			CH4			CH5			CH6			CH7			CH8			CH9			Ch10			Ch11		
CH1	0.0	0.0	0.3	0.0	0.3	0.5	0.0	0.3	0.5	0.5	0.8	1.0	0.8	1.0	1.0	0.8	1.0	1.0	0.5	0.8	1.0	0.3	0.5	0.8	0.0	0.3	0.5	0.3	0.5	0.8	0.8	1.0	1.0
CH2	0.8	1.0	1.0	0.0	0.0	0.3	0.3	0.5	0.8	0.8	1.0	1.0	0.3	0.5	0.8	0.0	0.3	0.5	0.5	0.8	1.0	0.5	0.8	1.0	0.0	0.3	0.5	0.5	0.8	1.0	0.0	0.3	0.5
CH3	0.8	1.0	1.0	0.5	0.8	1.0	0.0	0.0	0.3	0.5	0.8	1.0	0.3	0.5	0.8	0.5	0.8	1.0	0.0	0.3	0.5	0.5	0.8	1.0	0.3	0.5	0.8	0.0	0.3	0.5	0.0	0.3	0.5
CH4	0.5	0.8	1.0	0.8	1.0	1.0	0.8	1.0	1.0	0.0	0.0	0.3	0.5	0.8	1.0	0.5	0.8	1.0	0.0	0.3	0.5	0.8	1.0	1.0	0.8	1.0	1.0	0.5	0.8	1.0	0.5	0.8	1.0
CH5	0.3	0.5	0.8	0.0	0.3	0.5	0.5	0.8	1.0	0.3	0.5	0.8	0.0	0.0	0.3	0.8	1.0	1.0	0.5	0.8	1.0	0.5	0.8	1.0	0.8	1.0	1.0	0.8	1.0	1.0	0.0	0.3	0.5
CH6	0.8	1.0	1.0	0.5	0.8	1.0	0.3	0.5	0.8	0.5	0.8	1.0	0.5	0.8	1.0	0.0	0.0	0.3	0.8	1.0	1.0	0.5	0.8	1.0	0.3	0.5	0.8	0.8	1.0	1.0	0.5	0.8	1.0
CH7	0.5	0.8	1.0	0.8	1.0	1.0	0.0	0.3	0.5	0.3	0.5	0.8	0.5	0.8	1.0	0.8	1.0	1.0	0.0	0.0	0.3	0.8	1.0	1.0	0.8	1.0	1.0	0.3	0.5	0.8	0.8	1.0	1.0
CH8	0.8	1.0	1.0	0.3	0.5	0.8	0.3	0.5	0.8	0.0	0.3	0.5	0.5	0.8	1.0	0.0	0.3	0.5	0.3	0.8	1.0	0.0	0.0	0.3	0.5	0.8	1.0	0.5	0.8	1.0	0.5	0.8	1.0
CH9	0.8	1.0	1.0	0.5	0.8	1.0	0.5	0.8	1.0	0.8	1.0	1.0	0.3	0.5	0.8	0.5	0.8	1.0	0.3	0.5	0.8	0.5	0.8	1.0	0.0	0.0	0.3	0.5	0.8	1.0	0.3	0.5	0.8
CH10	0.8	1.0	1.0	0.8	1.0	1.0	0.3	0.5	0.8	0.8	1.0	1.0	0.3	0.5	0.5	0.0	0.3	0.5	0.3	0.5	0.8	0.0	0.3	0.5	0.0	0.3	0.8	0.0	0.0	0.3	0.8	1.0	1.0
CH11	0.8	1.0	1.0	0.8	1.0	1.0	0.3	0.5	0.8	0.8	1.0	1.0	0.0	0.3	0.5	0.0	0.3	0.5	0.3	0.5	0.8	0.3	0.5	0.8	0.5	0.8	1.0	0.3	0.5	0.8	0.0	0.0	0.3

Table 4. Fuzzy initial direct relation matrix

Challenges	CH1	CH2	CH3	CH4	CH5	CH6	CH7	CH8	CH9	CH10	CH11
CH1	0.04	0.25	0.3	0.8	1.0	1.0	0.8	0.5	0.3	0.5	1.0
CH2	1.0	0.0	0.5	1.0	0.5	0.3	0.8	0.8	0.3	0.8	0.3
CH3	1.0	0.8	0.0	0.8	0.5	0.8	0.3	0.8	0.5	0.3	0.3
CH4	0.8	1.0	1.0	0.0	0.8	0.8	0.3	1.0	1.0	0.8	0.8
CH5	0.5	0.3	0.8	0.5	0.0	1.0	0.8	0.8	1.0	1.0	0.3
CH6	1.0	0.8	0.5	0.8	0.8	0.0	1.0	0.8	0.5	1.0	0.8
CH7	0.8	1.0	0.3	0.5	0.8	1.0	0.0	1.0	1.0	0.5	1.0
CH8	1.0	0.5	0.5	0.3	0.8	0.3	0.7	0.0	0.8	0.8	0.8
CH9	1.0	0.8	0.8	1.0	0.5	0.8	0.5	0.8	0.0	0.8	0.5
Ch10	1.0	1.0	0.5	1.0	0.5	0.3	0.5	0.3	0.3	0.0	1.0
Ch11	1.0	1.0	0.5	1.0	0.3	0.3	0.5	0.5	0.8	0.5	0.0

environmental factors, and also ensures that the product which they are making does not have any harmful effect on the environment. The companies are considering the whole product life cycle for the measurement of environment impact of the particular product. Also in future the production of the food depends on the consumers demand or priorities. There is a lot of pressure on food manufacturing companies by consumers, governments, retailers and suppliers for the development of sustainable products. Although the insufficient amount of natural resources, companies should consider the sustainability for making the product environmentally safe. The environmental safe products can be made by change in labeling, packaging types, size, weight and making initiatives to increase recycling and reuse. Indian food industry is still having many challenges in respect of sustainable implementation such as food safety and security, food wastage, food quality, information technology and regulations and standards etc. Researchers and

Table 5. Normalized initial direct relation matrix (D)

Challenges	CH1	CH2	CH3	CH4	CH5	CH6	CH7	CH8	CH9	CH10	CH11
CH1	0.005	0.029	0.029	0.086	0.109	0.110	0.086	0.057	0.029	0.057	0.110
CH2	0.109	0.005	0.057	0.109	0.057	0.029	0.086	0.086	0.029	0.086	0.029
CH3	0.109	0.086	0.005	0.086	0.057	0.086	0.029	0.086	0.057	0.029	0.029
CH4	0.086	0.109	0.109	0.005	0.086	0.086	0.029	0.109	0.109	0.086	0.086
CH5	0.057	0.029	0.086	0.057	0.005	0.109	0.086	0.086	0.109	0.109	0.029
CH6	0.109	0.086	0.057	0.086	0.086	0.005	0.109	0.086	0.057	0.109	0.086
CH7	0.086	0.109	0.029	0.057	0.086	0.109	0.005	0.109	0.109	0.057	0.109
CH8	0.109	0.057	0.057	0.029	0.086	0.029	0.081	0.005	0.086	0.086	0.086
CH9	0.109	0.086	0.086	0.109	0.057	0.086	0.057	0.086	0.005	0.086	0.057
CH10	0.109	0.109	0.057	0.109	0.052	0.029	0.057	0.029	0.033	0.005	0.109
CH11	0.110	0.110	0.057	0.110	0.029	0.029	0.057	0.057	0.086	0.057	0.005

Table 6. Total relation matrix

Challenges	CH1	CH2	CH3	CH4	CH5	CH6	CH7	CH8	CH9	CH10	CH11
CH1	0.328	0.183	0.328	0.238	0.207	0.181	0.328	0.207	0.328	0.203	0.256
CH2	0.236	0.196	0.206	0.300	0.216	0.185	0.244	0.270	0.198	0.263	0.188
CH3	0.211	0.262	0.152	0.273	0.211	0.230	0.191	0.264	0.216	0.209	0.178
CH4	0.200	0.349	0.300	0.268	0.286	0.328	0.244	0.346	0.317	0.319	0.281
CH5	0.200	0.250	0.253	0.283	0.189	0.281	0.269	0.295	0.293	0.311	0.212
CH6	0.211	0.326	0.247	0.336	0.284	0.203	0.313	0.321	0.270	0.335	0.283
CH7	0.230	0.344	0.222	0.311	0.283	0.297	0.219	0.341	0.315	0.292	0.213
CH8	0.230	0.251	0.209	0.238	0.243	0.190	0.247	0.198	0.253	0.268	0.241
CH9	0.210	0.311	0.263	0.343	0.248	0.266	0.253	0.307	0.205	0.301	0.243
CH10	0.218	0.298	0.209	0.309	0.211	0.186	0.221	0.222	0.204	0.190	0.259
CH11	0.213	0.301	0.212	0.312	0.194	0.189	0.224	0.251	0.252	0.244	0.166

academicians are still working to handle the major problem of food globally. Selected challenges in the current research work mainly focused to achieve sustainability in the food supply chain. It was suggested that SFSCM is an emerging tool for the sustainable, durable and healthy food supply for all. This works mainly focused on the challenges in Indian food industries for the implementation of SFSCM. In this research an effort has been made to put practical use of SFSCM in Indian food industries by identifying the challenges. Here, it can be suggested that Fuzzy DEMATEL, a decision model can be used to clarify the relationship among the challenges as well as causal interactions related to the implementation of SFSCM in Indian food industries. Fuzzy set theory is used for the human biasness or judgment and DEMATEL is used for the interrelation between challenges. Based on the proposed model Fuzzy DEMATEL we analyzed eleven challenges for the implementation of SFSCM in Indian food industry.

Table 7. Assessment of cause and effect challenges

Challenges	Sum= ri	Sum= cj	r+c	Rank	r-c	Cause/effect
CH1	2.787	2.487	5.27	1	**0.30**	Cause
CH2	2.503	3.071	2.50	10	**-0.57**	Effect
CH3	2.398	2.601	2.40	11	**-0.21**	Effect
CH4	3.238	3.212	3.24	2	**0.03**	Cause
CH5	2.836	2.571	2.84	6	**0.27**	Cause
CH6	3.128	2.536	3.13	3	**0.59**	Cause
CH7	3.067	2.754	3.07	4	**0.31**	Cause
CH8	2.565	3.023	2.57	7	**-0.46**	Effect
CH9	2.950	2.849	2.95	5	**0.10**	Cause
CH10	2.528	2.936	2.53	9	**-0.4**	Effect
CH11	2.559	2.521	2.56	8	**0.04**	Cause

Figure 3. Cause and effect diagram of challenges for the implementation of SFSCM in Indian food industry

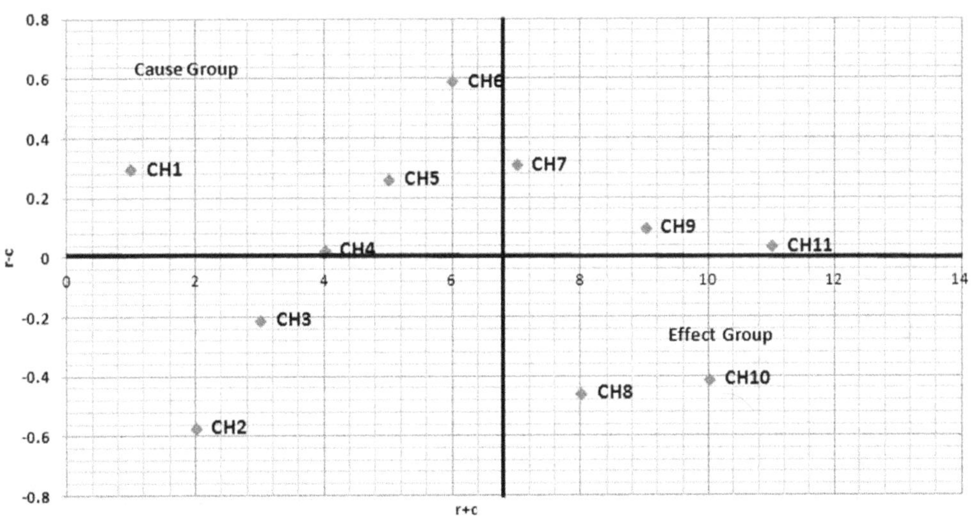

Challenges namely (CH1) food safety and security, (CH4) food wastage, (CH5) regulations and standards, (CH6) traceability, (CH7) transportation, (CH9) benchmarking, (CH11) product life cycle management are grouped into cause group and more attention is required by the higher management for better results. The remaining challenges are (CH2) food quality, (CH3) transparency, (CH8) supplier selection, (CH10) information technology and they come under effect group, if these challenges are improved it can automatically increase the success rate of SFSCM in Indian food industry.

LIMITATIONS AND FUTURE SCOPE

Present work has its own limitations as well. The eleven challenges were identified which were related to SFSCM implementation in Indian food industry. The other challenges were not identified. On the basis of present work challenges were identified as a future perspective. All identified challenges are made a pair wise comparison by Fuzzy DEAMATEL and experts views. Fuzzy DEMATEL is modified according to the situations in many foreign countries. For the future aspects these eleven challenges were also analyzed by Fuzzy analytic hierarchy process (AHP), analytical network process (ANP) approach to solve multi criteria decision making (MCDM) problem. Total interpretive structural modeling (TISM) is also used for the challenges for successful decision making in Indian food industry.

ACKNOWLEDGMENT

The authors acknowledge and express the gratitude for the support of the research facilities and funds provided by the Department of Mechanical and Automobile Engineering, Graphic Era University, Dehradun, India.

REFERENCES

Ageron, B., Gunasekaran, A., & Spalanzani, A. (2012). Sustainable supply management: An empirical study. *International Journal of Production Economics*, *140*(1), 168–182. doi:10.1016/j.ijpe.2011.04.007

Ahi, P., & Searcy, C. (2013). A comparative literature analysis of definitions for green and sustainable supply chain management. *Journal of Cleaner Production*, *52*, 329–341. doi:10.1016/j.jclepro.2013.02.018

Ahi, P., & Searcy, C. (2014). A stochastic approach for sustainability analysis under the green economics paradigm. *Stochastic Environmental Research and Risk Assessment*, *28*(7), 1743–1753. doi:10.1007/s00477-013-0836-5

Akkerman, R., Farahani, P., & Grunow, M. (2010). Quality, safety and sustainability in food distribution: A review of quantitative operations management approaches and challenges. *Spectrum (Lexington, Ky.)*, *32*(4), 863–904.

Attaran, M. (2012). Critical success factors and challenges of implementing RFID in supply chain management. *Journal of Supply Chain and Operations Management*, *10*(1), 144–167.

Badia-Melis, R., Mishra, P., & Ruiz-García, L. (2015). Food traceability: New trends and recent advances. A review. *Food Control*, *57*, 393–401. doi:10.1016/j.foodcont.2015.05.005

Bellman, R. E., & Zadeh, L. A. (1970). Decision-making in a fuzzy environment. *Management Science*, *17*(4), B-141–B-164. doi:10.1287/mnsc.17.4.B141

Beta, T., Nam, S., Dexter, J. E., & Sapirstein, H. D. (2017). Phenolic content and antioxidant activity of pearled wheat and roller-milled fractions. *LWT-Food Science and Technology*, *78*, 151–159.

Beulens, A. J., Broens, D. F., Folstar, P., & Hofstede, G. J. (2005). Food safety and transparency in food chains and networks Relationships and challenges. *Food Control*, *16*(6), 481–486. doi:10.1016/j.foodcont.2003.10.010

Bloemhof, J. M., & Soysal, M. (2017). Sustainable food supply chain design. In *Sustainable Supply Chains* (pp. 395–412). Springer International Publishing. doi:10.1007/978-3-319-29791-0_18

Boukherroub, T., Bouchery, Y., Corbett, C. J., Fransoo, J. C., & Tan, T. (2017). Carbon footprinting in supply chains. In *Sustainable Supply Chains* (pp. 43–64). Springer International Publishing. doi:10.1007/978-3-319-29791-0_3

Dubois, D., & Prade, H. (1979). Fuzzy real algebra: Some results. *Fuzzy Sets and Systems*, *2*(4), 327–348. doi:10.1016/0165-0114(79)90005-8

Egilmez, G., & Park, Y. S. (2014). Transportation related carbon, energy and water footprint analysis of US manufacturing: An eco-efficiency assessment. *Transportation Research Part D, Transport and Environment*, *32*, 143–159. doi:10.1016/j.trd.2014.07.001

Gabus, A., & Fontela, E. (1972). *World problems, an invitation to further thought within the framework of DEMATEL*. Geneva, Switzerland: Battelle Geneva Research Center.

Genovese, A., Acquaye, A. A., Figueroa, A., & Koh, S. L. (2017). Sustainable supply chain management and the transition towards a circular economy: Evidence and some applications. *Omega*, *66*, 344–357. doi:10.1016/j.omega.2015.05.015

Girotto, F., Alibardi, L., & Cossu, R. (2015). Food waste generation and industrial uses: A review. *Waste Management (New York, N.Y.)*, *45*, 32–41. doi:10.1016/j.wasman.2015.06.008 PMID:26130171

Gold, S., Hahn, R., & Seuring, S. (2013). Sustainable supply chain management in Base of the Pyramid food projects—A path to triple bottom line approaches for multinationals? *International Business Review*, *22*(5), 784–799. doi:10.1016/j.ibusrev.2012.12.006

Govindan, K., Kannan, D., & Shankar, K. M. (2014). Evaluating the drivers of corporate social responsibility in the mining industry with multi-criteria approach: A multi-stakeholder perspective. *Journal of Cleaner Production*, *84*, 214–232. doi:10.1016/j.jclepro.2013.12.065

Govindan, K., Kannan, D., & Shankar, M. (2015). Evaluation of green manufacturing practices using a hybrid MCDM model combining DANP with PROMETHEE. *International Journal of Production Research*, *53*(21), 6344–6371. doi:10.1080/00207543.2014.898865

Govindan, K., Khodaverdi, R., & Jafarian, A. (2013). A fuzzy multi criteria approach for measuring sustainability performance of a supplier based on triple bottom line approach. *Journal of Cleaner Production*, *47*, 345–354. doi:10.1016/j.jclepro.2012.04.014

Grunert, K. G. (2005). Food quality and safety: Consumer perception and demand. *European Review of Agriculture Economics*, *32*(3), 369–391. doi:10.1093/eurrag/jbi011

Gustavsson, J., Cederberg, C., Sonesson, U., Van Otterdijk, R., & Meybeck, A. (2011). Global food losses and food waste. *Food and Agriculture Organization of the United Nations, Rom.*

Havinga, T., & Verbruggen, P. (2017). Understanding complex governance relationships in food safety regulation: The RIT model as a theoretical lens. *The Annals of the American Academy of Political and Social Science*, *670*(1), 58–77. doi:10.1177/0002716216688872

Hernandez, R. R., Easter, S. B., Murphy-Mariscal, M. L., Maestre, F. T., Tavassoli, M., Allen, E. B., & Allen, M. F. et al. (2014). Environmental impacts of utility-scale solar energy. *Renewable & Sustainable Energy Reviews*, *29*, 766–779. doi:10.1016/j.rser.2013.08.041

Hofstede, G. J., Spaans, L., Schepers, H., Trienekens, J. H., & Beulens, A. J. M. (2004). *Hide or confide: the dilemma of transparency*. Academic Press.

Hsu, C. W., Kuo, T. C., Chen, S. H., & Hu, A. H. (2013). Using DEMATEL to develop a carbon management model of supplier selection in green supply chain management. *Journal of Cleaner Production*, *56*, 164–172. doi:10.1016/j.jclepro.2011.09.012

Jia, P., Govindan, K., & Kannan, D. (2015). Identification and evaluation of influential criteria for the selection of an environmental shipping carrier using DEMATEL: A case from India. *International Journal of Shipping and Transport Logistics*, *7*(6), 719–741. doi:10.1504/IJSTL.2015.072684

Kumar, S., Singh, B., Qadri, M. A., Kumar, Y. S., & Haleem, A. (2013). A framework for comparative evaluation of lean performance of firms using fuzzy TOPSIS. *International Journal of Productivity and Quality Management*, *11*(4), 371–392. doi:10.1504/IJPQM.2013.054267

Lee, H. L., Padmanabhan, V., & Whang, S. (2004). Information distortion in a supply chain: the bullwhip effect. *Management Science, 50*(12), 1875-1886.

Liao, Z., & Rittscher, J. (2007). A multi-objective supplier selection model under stochastic demand conditions. *International Journal of Production Economics*, *105*(1), 150–159. doi:10.1016/j.ijpe.2006.03.001

Liu, G., Liu, X., & Cheng, S. (2013). Food security: Curb Chinas rising food wastage. *Nature*, *498*(7453), 170–170. doi:10.1038/498170c PMID:23765482

Mangla, S., Kumar, P., & Barua, M. K. (2014). An evaluation of attribute for improving the green supply chain performance via DEMATEL method. *International Journal of Mechanical Engineering & Robotics Research*, *1*(1), 30–35.

Mangla, S. K., Kumar, P., & Barua, M. K. (2015). Flexible decision modeling for evaluating the risks in green supply chain using fuzzy AHP and IRP methodologies. *Global Journal of Flexible Systems Management*, *16*(1), 19–35. doi:10.1007/s40171-014-0081-x

Mangla, S. K., Kumar, P., & Barua, M. K. (2015). Prioritizing the responses to manage risks in green supply chain: An Indian plastic manufacturer perspective. *Sustainable Production and Consumption*, *1*, 67–86. doi:10.1016/j.spc.2015.05.002

Manning, L., Baines, R. N., & Chadd, S. A. (2006). Quality assurance models in the food supply chain. *British Food Journal*, *108*(2), 91–104. doi:10.1108/00070700610644915

Marucheck, A., Greis, N., Mena, C., & Cai, L. (2011). Product safety and security in the global supply chain: Issues, challenges and research opportunities. *Journal of Operations Management*, *29*(7), 707–720. doi:10.1016/j.jom.2011.06.007

Papargyropoulou, E., Lozano, R., Steinberger, J. K., & Wright, N., & bin Ujang, Z. (. (2014). The food waste hierarchy as a framework for the management of food surplus and food waste. *Journal of Cleaner Production*, *76*, 106–115. doi:10.1016/j.jclepro.2014.04.020

Parfitt, J., Barthel, M., & Macnaughton, S. (2010). Food waste within food supply chains: Quantification and potential for change to 2050. *Philosophical Transactions of the Royal Society of London. Series B, Biological Sciences*, *365*(1554), 3065–3081. doi:10.1098/rstb.2010.0126 PMID:20713403

Patil, S. K., & Kant, R. (2014). A hybrid approach based on fuzzy DEMATEL and FMCDM to predict success of knowledge management adoption in supply chain. *Applied Soft Computing*, *18*, 126–135. doi:10.1016/j.asoc.2014.01.027

Pedrycz, W., & Gomide, F. (2007). *Fuzzy systems engineering: toward human-centric computing*. John Wiley & Sons. doi:10.1002/9780470168967

Petrovic, D. (2001). Simulation of supply chain behaviour and performance in an uncertain environment. *International Journal of Production Economics*, *71*(1), 429–438. doi:10.1016/S0925-5273(00)00140-7

Petrovic, D., Roy, R., & Petrovic, R. (1998). Modelling and simulation of a supply chain in an uncertain environment. *European Journal of Operational Research*, *109*(2), 299–309. doi:10.1016/S0377-2217(98)00058-7

Seuring, S., & Müller, M. (2008). From a literature review to a conceptual framework for sustainable supply chain management. *Journal of Cleaner Production*, *16*(15), 1699–1710. doi:10.1016/j.jclepro.2008.04.020

Souquet, J., Liu, H., Liu, J., Fan, X., Liddell, J., Levison, P., ... Thomas, O. R. T. (n.d.). Tracking the movement of individual adsorbent particles in expanded beds. In *Book of Abstracts* (p. 76). Academic Press.

Soysal, M., Bloemhof, J., & van der Vorst, J. G. (2012). A Review of Quantitative Models for Sustainable Food Logistics Management: Challenges and Issues. *Proceedings in Food System Dynamics*, 448-462.

Stark, J. (2015). Product lifecycle management. In *Product Lifecycle Management* (pp. 1–29). Springer International Publishing. doi:10.1007/978-3-319-17440-2_1

Swinburn, B., Kraak, V., Rutter, H., Vandevijvere, S., Lobstein, T., Sacks, G., & Magnusson, R. et al. (2015). Strengthening of accountability systems to create healthy food environments and reduce global obesity. *Lancet*, *385*(9986), 2534–2545. doi:10.1016/S0140-6736(14)61747-5 PMID:25703108

Swinburn, B., Vandevijvere, S., Kraak, V., Sacks, G., Snowdon, W., Hawkes, C., ... & L'abbé, M. (2013). Monitoring and benchmarking government policies and actions to improve the healthiness of food environments: a proposed Government Healthy Food Environment Policy Index. *Obesity Reviews*, *14*(S1), 24-37.

Thomas, A. M., White, G. R., Plant, E., & Zhou, P. (2017). Challenges and practices in Halal meat preparation: A case study investigation of a UK slaughterhouse. *Total Quality Management & Business Excellence*, *28*(1-2), 12–31. doi:10.1080/14783363.2015.1044892

Trienekens, J. H., Wognum, P. M., Beulens, A. J., & van der Vorst, J. G. (2012). Transparency in complex dynamic food supply chains. *Advanced Engineering Informatics*, *26*(1), 55–65. doi:10.1016/j.aei.2011.07.007

Tsai, W. H., & Chou, W. C. (2009). Selecting management systems for sustainable development in SMEs: A novel hybrid model based on DEMATEL, ANP, and ZOGP. *Expert Systems with Applications*, *36*(2), 1444–1458. doi:10.1016/j.eswa.2007.11.058

van Weele, A., & van Tubergen, K. (2017). Responsible purchasing: moving from compliance to value creation in supplier relationships. In *Sustainable Supply Chains* (pp. 257–278). Springer International Publishing. doi:10.1007/978-3-319-29791-0_11

Whipple, J. M., Voss, M. D., & Closs, D. J. (2009). Supply chain security practices in the food industry: Do firms operating globally and domestically differ? *International Journal of Physical Distribution & Logistics Management*, *39*(7), 574–594. doi:10.1108/09600030910996260

Wognum, P. N., Bremmers, H., Trienekens, J. H., van der Vorst, J. G., & Bloemhof, J. M. (2011). Systems for sustainability and transparency of food supply chains–Current status and challenges. *Advanced Engineering Informatics*, *25*(1), 65–76. doi:10.1016/j.aei.2010.06.001

Wu, W. W. (2012). Segmenting critical factors for successful knowledge management implementation using the fuzzy DEMATEL method. *Applied Soft Computing*, *12*(1), 527–535. doi:10.1016/j.asoc.2011.08.008

Wu, W. W., Lan, L. W., & Lee, Y. T. (2011). Exploring decisive factors affecting an organizations SaaS adoption: A case study. *International Journal of Information Management*, *31*(6), 556–563. doi:10.1016/j.ijinfomgt.2011.02.007

Xia, X., Govindan, K., & Zhu, Q. (2015). Analyzing internal barriers for automotive parts remanufacturers in China using grey-DEMATEL approach. *Journal of Cleaner Production*, *87*, 811–825. doi:10.1016/j.jclepro.2014.09.044

Yakovleva, N., Sarkis, J., & Sloan, T. (2012). Sustainable benchmarking of supply chains: The case of the food industry. *International Journal of Production Research*, *50*(5), 1297–1317. doi:10.1080/00207543.2011.571926

Zadeh, L. A. (1965). Fuzzy sets. *Information and Control*, *8*(3), 338–353. doi:10.1016/S0019-9958(65)90241-X

Zailani, S., Arrifin, Z., Abd Wahid, N., Othman, R., & Fernando, Y. (2010). Halal traceability and halal tracking systems in strengthening halal food supply chain for food industry in Malaysia (a review). *Journal of Food Technology*, *8*(3), 74–81. doi:10.3923/jftech.2010.74.81

Chapter 9

A Hybrid Genetic Scatter Search Algorithm to Solve Flexible Job Shop Scheduling Problems:
A Hybrid Algorithm for Scheduling Problems

Mariappan Kadarkarainadar Marichelvam
Mepco Schlenk Engineering College, India

Geetha Mariappan
Kamaraj College of Engineering and Technology, India

ABSTRACT

Scheduling is one of the most important problems in production planning systems. It is a decision-making process that plays a crucial role in many Industries. Different scheduling environments were addressed in the literature. Among them Flexible job-shop problem (FJSP) is an important one and it is an extension of the classical JSP that allows one operation which can be processed on one machine out of a set of alternative machines. It is closer to the real manufacturing situation. Because of the additional needs to determine the assignment of operations on the machines, FJSP is more complex than JSP, and incorporates all the difficulties and complexities of JSP. This chapter addresses a hybrid genetic scatter search algorithm for solving multi-objective FJSP. Makespan and flow time are the objective functions considered in this chapter. The computational results prove the effectiveness of the proposed algorithm for solving flexible job-shop scheduling problem.

INTRODUCTION

Meticulous planning and scheduling is one of the most important issues tackled by many researchers in production management. Scheduling is defined as a type of decision-making process that plays a crucial role in our daily life. It refers to setting of operation that it will start dates so that jobs will be completed

DOI: 10.4018/978-1-5225-3035-0.ch009

with their due date. In manufacturing areas, the objective of scheduling is to satisfy the customers by minimizing the lead time and so on. Different types of scheduling environments were addressed by Pinedo (1995). Among them the flexible job shop environment plays a vital role as many industries resemble it. Flexible job-shop problem (FJSP) can be considered as an addition of the standard JSP that permits one operation which can be handled on one machine out of a set of different machines (Zhang et al., 2009). It is closer to the real manufacturing situation. Because of the additional needs to determine the assignment of operations on the machines, FJSP is more complex. The FJSP are non-deterministic polynomial time hard (NP-hard) type combinatorial optimization problems which means scarcely any algorithm exist can solve the problem in polynomial time. Hence, the exact algorithms cannot be used to solve the problems. Researchers have suggested many heuristics and meta-heuristics to resolve these problems. Tabu search (TS), ant colony optimization (ACO), artificial immune system (AIS), particle swarm optimization (PSO) and genetic algorithm (GA) have been projected to solve the FJSP. In this chapter, a hybrid genetic scatter search algorithm (HGSSA) is recommended to solve the FJSP. The objective considered in this chapter is to minimize the weighted sum of makespan and mean flow time. The layout of a flexible job shop environment is given in Figure 1.

BACKGROUND

The FJSP was first tackled by Brucker and Schile (1990). Brandimarte (1993) developed a Tabu search (TS) algorithm for solving the FJSP. Hurink et al. (1994) also addressed a TS algorithm to solve the job shop scheduling problem with multi-purpose machines. Mastrolilli and Gambardella (2000) projected local search algorithms and two neighborhood functions to solve the FJSP. Kacem et al. (2002) tackled the FJSP using the Pareto-optimality approach to minimize three objective functions namely makespan,

Figure 1. Layout of a Flexible Job Shop Environment

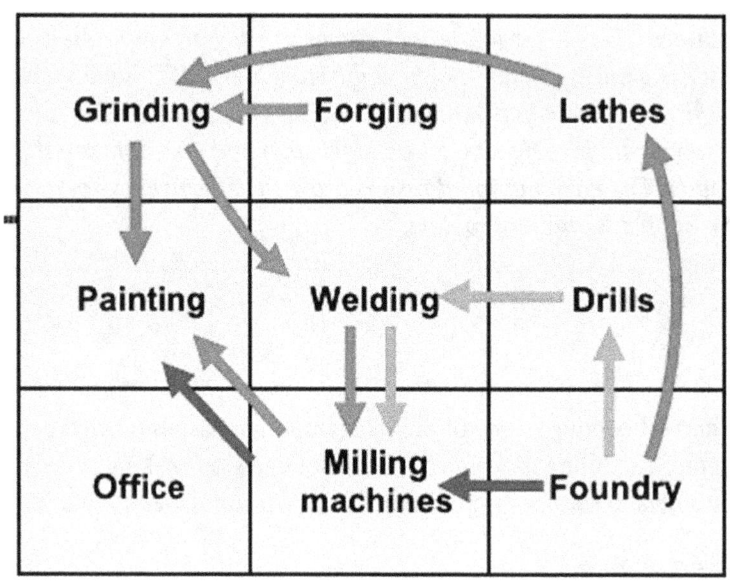

the total workload of machines and the workload of the most loaded machine. Zhang and Gen, 2005 proposed a multistage operation-based genetic algorithm to solve the FJSP to minimize the makespan. Saidi-Mehrabad and Fattahi (2007) presented a TS algorithm to solve the FJSP to minimize the makespan. To appraise the performance of the proposed algorithm randomly generated test problems were used by them. Gao et al. (2008) applied a hybrid genetic and variable neighborhood descent algorithm for solving the FJSP. Pezzella et al. (2008) presented a genetic algorithm (GA) to solve the FJSP with makespan objective. The proposed GA combines different tactics for producing the primary population, choosing the individuals for reproduction and reproducing novel individuals. Zhang et al. (2009) established a combined particle swarm optimization and TS algorithm to solve the multi-objective FJSP. They have verified the performance of the proposed algorithm using the benchmark problems available in the literature and demonstrated the effectiveness of the proposed algorithm. Bagheri et al. (2010) recommended an artificial immune algorithm (AIS) to explain the FJSP to minimize the makespan. Different mutation operators were introduced by them to generate new individuals. Experiments were conducted with benchmark problems to validate the performance of the proposed algorithm. Defersha and Chen (2010) addressed a parallel genetic algorithm (PGA) for FJSP with sequence dependent setup times to minimize the makespan. Xing et al. (2010) proposed an integrated ant colony optimization (ACO) model and knowledge model to solve the FJSP to minimize the makespan. The performance of the algorithm was validated with the benchmark problems. Yazdani et al. (2010) solved the FJSP by a parallel variable neighborhood search (PVNS) algorithm. The objective was to minimize makespan time. Bagheri and Zandieh (2011) tackled the FJSP with sequence-dependent setup times. The objective was to minimize the makespan and mean tardiness. A variable neighborhood search (VNS) algorithm was developed by them. A greedy randomized adaptive search procedure was presented by Rajkumar et al. (2011) to solve the FJSP. They considered the limited resource constraints and the objective was to minimize the makespan, maximum workload and total workload. Baykasoğlu et al. (2014) suggested a Teaching–learning based optimization (TLBO) algorithm to solve the FJSP. The objective is to minimize the makespan. An extensive experimental work was carried out to illustrate the effectiveness of the proposed algorithm. Recently, multi-objective FJSP was tackled by Karthikeyan et al. (2015) using a hybrid discrete firefly (HDF) algorithm. A local search method was combined with the discrete firefly algorithm. A scatter search (SS) was proposed by González et al. (2015) to tackle the FJSP to minimize the makespan. Xu at al. (2015) also developed an effective TLBO algorithm to solve the FJSP. They considered the fuzzy processing time of jobs in their research.

Academics introduced a lot of meta-heuristic algorithms to solve the combinatorial optimization problems. Scatter search algorithm (SSA) is one sort of meta-heuristics created by Glover (1977). He utilized the SSA as a heuristic for integer programming. Noorul Haq et al. (2006) applied the SSA to minimize makespan for flow shop scheduling problems. Rahimi-Vahed et al. (2008) developed a multi-objective scatter search algorithm (MOSSA) to minimize the weighted mean completion time and weighted mean tardiness to solve the no-wait flow shop scheduling problems. Saravanan et al., 2008 have assessed the performance of SSA to minimize the makespan for flow shop scheduling problems. Ranjbar et al. (2009) recommended a hybrid scatter search algorithm (HGSSA) to minimize the makespan for project scheduling problems. Saravanan and Noorul Haq, 2010 also evaluated the performance of SSA for the job shop scheduling problems to minimize the makespan. Baradaran et al. (2010) also established a HGSSA to minimize the makespan for a PERT type network for resource constrained project scheduling problem. They applied path relinking algorithm and two operators to solve the problem. The efficiency of the proposed algorithm was tested for real world problems and benchmark problems.

Tavakkoli-Moghaddam et al. (2010) suggested a SSA to minimize the makespan, intracellular movement, tardiness, and sequence-dependent setup costs, simultaneously for a group scheduling problem in a cellular manufacturing system. Marichelvam and Prabaharan (2014) appraised the performance of an improved hybrid genetic scatter search algorithm (IHGSSA) to solve the multi-stage hybrid flow shop scheduling problems with missing operations. They hybridized the constructive heuristics, genetic and scatter search algorithm. They tested the performance of the proposed algorithm using industrial data and random problem instances. Recently, several meta-heuristic algorithms were proposed by the researchers to tackle different optimization problems. Kumar et al. (2016) proposed grey wolf optimizer algorithm for system reliability optimization. Kumar et al. (2016) also addressed proposed cuckoo search algorithm for reliability optimization of complex system. Particle swarm optimization algorithm (Pant and Singh, 2011, Pant et al., 2015, Pant et al., 2017) and multi-objective particle swarm optimization algorithm (Pant et al., 2015, Kumar et al. 2017) were also addressed in the literature to solve the reliability optimization problems.

From the above literature review, one can easily conclude that the applications of the HGSSA to solve the FJSP require further extensive studies. Hence, in this paper, a hybrid genetic scatter search algorithm is proposed to minimize weighted sum of makespan and mean flow time in FJSP.

MAIN FOCUS OF THE CHAPTER

This chapter considers the FJSP with makespan and mean flow time objective functions. Makespan is the completion time of the last job in the production system and it is used to improve the efficiency of the production system (Tosun and Marichelvam, 2016). Mean flow time is the amount of average time job spends in a business process from beginning to end and it is used to minimize the inventory (Marichelvam and Geetha, 2014a). The objective of this chapter is to minimize the weighted sum of makespan and mean flow time. In this paper, the following assumptions are considered (Zhang et al., 2010)

- All machines are accessible at time zero.
- All jobs are ready at time zero.
- Each machine can do only one operation at a time.
- Each operation can be handled without interruption on one of a set of available machines.
- Recirculation takes place whilst a job ought to go to a machine more than once.
- The order of operations for each job is predefined and cannot be modified.

Genetic Algorithm

Genetic algorithm was developed by Holland (1975. GA was developed based on the process of natural evolution. In the GA, a sequence of genes represents a solution and is called as a chromosome. An intelligently selected set of solutions is called as a population. The population at a given time is called as a generation. The GA starts with an initial population of solutions whose potential is determined by the fitness function. The fitness function evaluates the chromosomes with respect to the objective function of the optimization problem considered. The population size remains same from generation to generation. The population size has played a vital role on the performance of the GA. The next generation of population is obtained by selection, crossover and mutation. The fitness of the new population is measured and the old population is replaced by the fittest chromosomes in the new population. The above steps will be repeated until the termination criterion is reached.

Scatter Search Algorithm

The initial solutions are randomly generated in most of the meta-heuristics. However, in the SSA the initial solutions are generated non-randomly. The five steps in scatter search algorithm are given by Glover et al. (2000).

1. **Diversification Generation Method:** This is used to generate a collection of diverse trial solutions.
2. **An Improvement Method:** This is used to transform a trial solution into one or more enhanced trial solutions.
3. **Reference Set Update Method:** This is used to build and maintain a reference set consisting of the b "best" solutions found.
4. **A Subset Generation Method:** This is used to operate on the reference set, to produce a subset of its solutions as a basis for creating combined solutions.
5. **A Solution Combination Method:** This is used to transform a given subset of solutions produced by the subset generation method into one or more combined solution vectors.

The Pseudo-code of SSA is given below (El-Sayed et al., 2008):

```
Begin
Repeat
Create Population;
Repeat
Generate Reference Set;
Repeat
Select Subset;
Combine Solutions;
Improve Solution;
Until (StoppingCriterion1);
Update Reference Set (RefSet);
Until (StoppingCriterion2);
Until (StoppingCriterion3);
End;
```

Hybrid Genetic Scatter Search Algorithm (HGSSA)

The proposed algorithm is based on the work done by Marichelvam and Prabaharan (2014). To improve the performance of individual meta-heuristic algorithm, academics attempted to combine two or more meta-heuristics. In this chapter, it is proposed to combine the GA with SSA. Moreover, the dispatching rules are also incorporated with the meta-heuristics. In the HGSSA, the parameters such as the initial population size (N), reference set size (b), size of the high quality solutions (b1) and diverse solutions (b2), crossover probability (Pc), mutation probability (Pm), fitness function, number of generations (G) are defined first. In the GA, the initial population of solutions is randomly generated. But, in the SSA the initial solutions are generated by the method of diversification. In the HGSSA proposed in this chapter, two solutions are generated using two dispatching rules namely shortest processing time (SPT) rule and longest processing time (LPT) rule and the remaining solutions are generated randomly. Then, the solutions generated in the previous step are evaluated. The objective function considered in this work is minimization of weighted sum of makespan and mean flow time. For the minimization problems, the fitness function is the reciprocal of the objective function. A reference set is built based on the results obtained above step. The reference set contains of both high-quality solutions and diverse solutions. The reference set size is b in which there are b1 high-quality solutions and b2 diverse solutions. The reference set size is calculated by using the following equation.

$$b = b1 + b2 \tag{1}$$

Selection is used to make a new population of solutions from the current population. Diverse categories of selection techniques are available in the literature. In this chapter, a random selection technique is used. After selection, cross over is carried out. Cross over is a genetic operator. It is used to produce new off springs from the parents. Different types of cross over techniques were addressed in the literature. The two-point crossover described by Holland (1975) is used in this chapter. Mutation is another used to reduce the chances for the search process to be getting trapped in local optimal solutions. The swap mutation described by Luo et al. (2009) is used in this chapter. In the swap mutation two genes are randomly selected and their positions are exchanged. After mutation, new solutions are generated. These solutions are evaluated according to the fitness function (this is the reciprocal of weighted sum of makespan and mean flow time). Based on the fitness function values, the reference set is updated. If there is no improvements, the termination criterion is checked. The number of generations is used as the termination criterion and the optimal sequence is printed.

SOLUTIONS AND RECOMMENDATIONS

To demonstrate the performance of the HGSSA, an automobile spare parts manufacturing unit located in south India is considered in this chapter. The collaborative company is one of the leading manufacturers in India. The company produces carious components such as Oil seals, Oil rings, Reed valve assemblies, Moulded rubber product, Gaskets and Packing and sealing devices. Many of the components are exported to several countries including Indonesia, Thailand, Japan, Korea, China and United States of America. Among the different products data related with oil seals and moulded rubber product are used in this study. The processing time for the oil seal is given below in Table 1.

The Gantt chart schedule obtained by the HGSSA for the oil seal and moulded rubber part production are depicted in Figures 2 and 3 respectively.

To validate the performance of the proposed algorithm random problem instances are also developed. The parameters of the problem instances and the algorithm are presented in Table 3.

Table 1. Processing time for the oil seal

JOB	O_{ij}	M1	M2	M3	M4	M5	M6	M7	M8
J1	$O_{1,1}$	4	7	6	2	5	3	8	6
	$O_{1,2}$	11	3	4	8	6	5	5	9
	$O_{1,3}$	5	7	5	11	6	8	4	5
J2	$O_{2,1}$	3	5	7	4	11	10	6	7
	$O_{2,2}$	12	8	9	7	5	11	9	8
	$O_{2,3}$	14	5	8	9	7	4	11	10
J3	$O_{3,1}$	3	7	2	4	9	8	3	5
	$O_{3,2}$	6	1	4	5	8	11	2	3
	$O_{3,3}$	2	8	5	3	7	4	11	6
J4	$O_{4,1}$	6	1	8	3	4	7	2	4
	$O_{4,2}$	5	4	2	7	9	8	3	11
	$O_{4,3}$	3	8	5	11	8	9	6	4
J5	$O_{5,1}$	6	5	8	4	7	3	4	9
	$O_{5,2}$	5	9	7	3	5	11	7	4
	$O_{5,3}$	7	3	5	4	8	2	6	3
J6	$O_{6,1}$	2	8	1	7	6	3	4	5
	$O_{6,2}$	7	1	4	2	4	8	2	9
	$O_{6,3}$	9	3	8	1	2	5	7	6
J7	$O_{7,1}$	10	4	5	8	7	5	8	3
	$O_{7,2}$	4	5	2	7	9	11	7	6
	$O_{7,3}$	9	7	9	11	8	10	8	6
J8	$O_{8,1}$	8	7	10	9	11	17	6	8
	$O_{8,2}$	2	5	4	7	3	5	6	9
	$O_{8,3}$	7	6	9	8	9	8	11	10

Number of machines – 8
Number of jobs – 8
Number of operations – 24

The performance of the proposed algorithm is compared with the ant colony optimization (ACO) algorithm, Genetic algorithm (GA), Particle swarm optimization (PSO) algorithm, Scatter Search (SS) algorithm, Tabu Search (TS) algorithm and Variable Neighbouhood Search (VNS) algorithm. The proposed algorithm is tested for all possible combination of parameters. Hence 3*3*3*1*2*1*1*2*2 = 216 different setups are considered for the calibration of the proposed algorithm. Each problem instance is replicated 20 times with different initial random solutions. Relative deviation index (RDI) is one of the most important performance measures used in the scheduling literature. Hence, in this paper also RDI is measured to compare the performance of different algorithms. The RDI is calculated using the equation (2).

Table 2. Processing time for the moulded rubber part

JOB	O_{ij}	M1	M2	M3	M4	M5
J1	$O_{1,1}$	5	2	2	1	4
	$O_{1,2}$	5	4	8	6	9
	$O_{1,3}$	14	5	9	4	8
J2	$O_{2,1}$	2	8	4	3	7
	$O_{2,2}$	7	9	6	11	5
	$O_{2,3}$	9	5	4	8	7
J3	$O_{3,1}$	7	8	6	9	8
	$O_{3,2}$	8	1	4	6	7
	$O_{3,3}$	2	8	4	5	3
J4	$O_{4,1}$	1	5	8	4	2
	$O_{4,2}$	4	2	2	1	3
	$O_{4,3}$	2	3	8	5	7
J5	$O_{5,1}$	1	3	2	3	5
	$O_{5,2}$	4	5	4	6	1
	$O_{5,3}$	2	9	2	1	5

Number of machines – 5
Number of jobs – 5
Number of operations – 15

Figure 2. Gantt Chart for Oil Seal Production

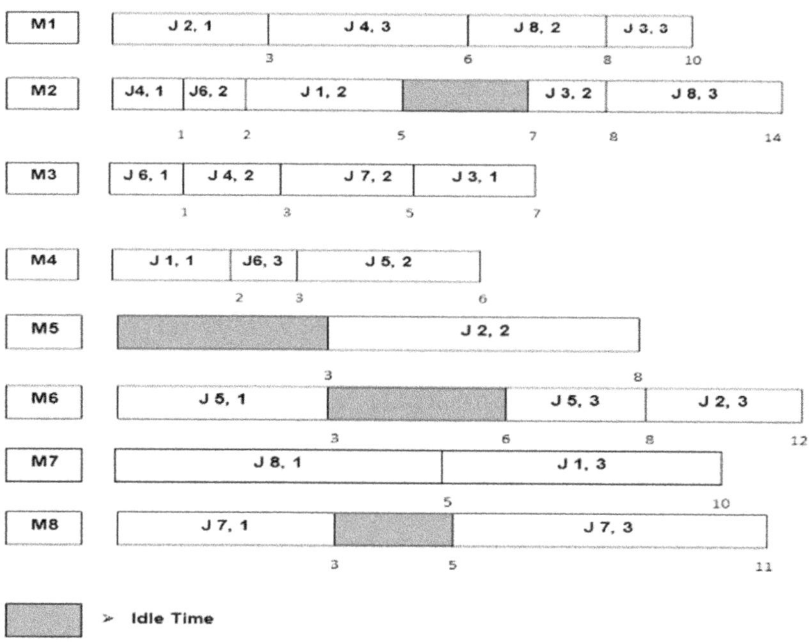

Figure 3. Gantt Chart for Moulded Rubber Part

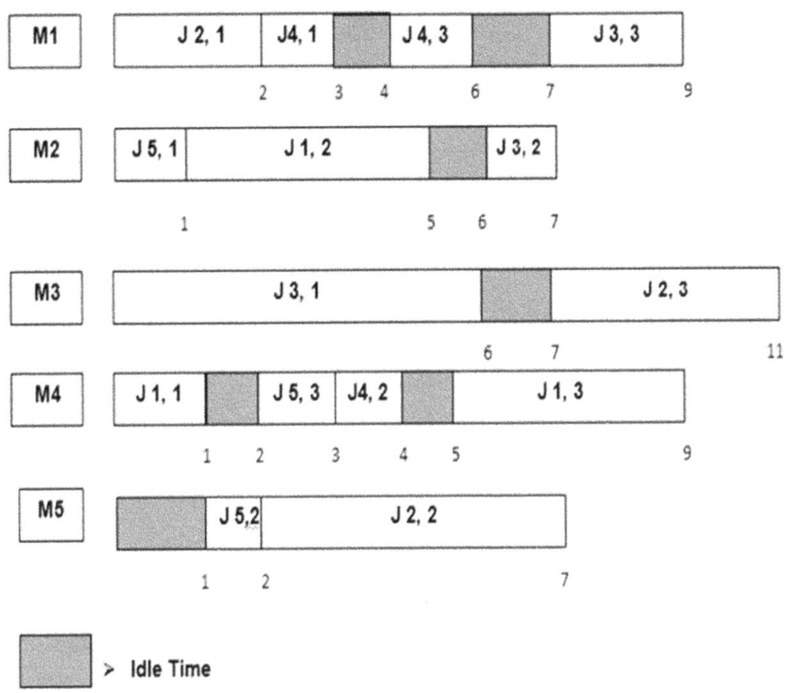

Table 3. Parameters of the random problem instances and algorithm

Sl. No.	Parameters	Values
1	Number of Jobs	5, 10 & 20
2	Number of Machines	2, 3 & 5
3	Number of operations	5, 10 & 20
4	Processing time distribution	Uniform (1-50)
5	Size of the population	50, 100
6	Number of good quality solutions	10
7	Number of diverse solutions	10
8	Probability of mutation	0.03 and 0.05
9	Cross over probability	0.5 and 0.6

$$\mathrm{RDI} = \sum_{l=1}^{R} \frac{(Z^* - Z_{meta})}{Z^*} \times 100 / R \qquad (2)$$

where,

Z^* = best objective function value
Z_{meta} = objective function value obtained by the different algorithms

R = number of runs (20)

From the calculated RDI vales, mean relative deviation index (MRDI) values are calculated for different algorithms by the following equation (3).

$$MRDI = \Sigma \text{RDI} / 216 \quad (3)$$

The MRDI comparison of different algorithm is presented in Figure 4.

Lower MRDI will be the indication of better performance of the proposed algorithm. From the figure, it is concluded easily that the performance of the proposed hybrid genetic scatter search algorithm is better than many other algorithms addressed in the literature for the random problem instances. This indicates the effectiveness of the proposed algorithm.

FUTURE RESEARCH DIRECTIONS

In this chapter, flexible job shop scheduling problems are considered with many assumptions. For instance, the setup time is not addressed in this chapter. Consideration of setup time and transportation time is a future research scope of this work. Consideration of due date related criteria such as earliness and tardiness would be another interesting scope of this research. The genetic scatter search algorithm may also be applied for other optimization problems. The proposed algorithm might be hybridized with other algorithms. Moreover, in this paper one type of crossover and mutation operators are used. It would be interesting to see the performance of the proposed algorithm with various crossover and mutation operators. Also, the incorporation of dispatching rules and constructive heuristics would be another important future scope of this research.

CONCLUSION

This chapter addresses the flexible job shop scheduling problems which have been proved to be NP-hard. An efficient HGSSA is proposed to tackle the FJSP to minimize the makespan and mean flow time. The

Figure 4. MRDI Comparison of Different Algorithms

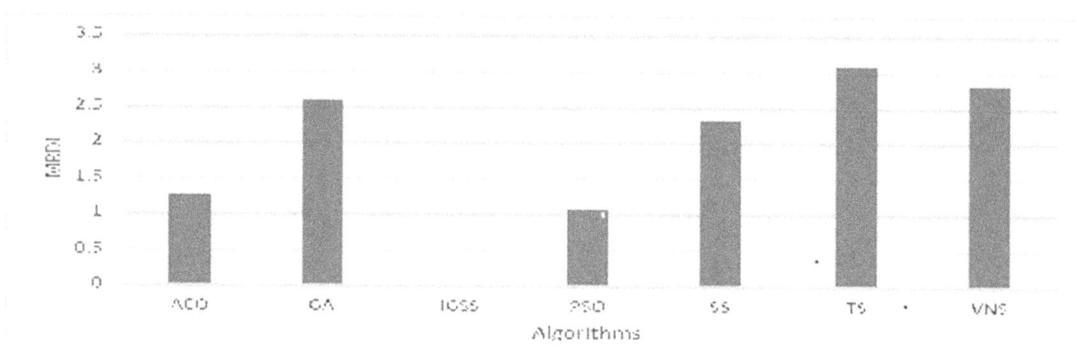

operators used in genetic algorithm such as cross over and mutation are entrenched in SSA to improve the solution diversity. The performance of the HGSSA is tested with an industrial data set from a leading automobile manufacturing unit. Gantt charts are drawn for the schedule obtained by the hybrid genetic scatter search algorithm. Moreover, extensive computational experiments are carried out by developing arbitrary problems with different setups. The outcomes of the HGSSA are equated using numerous former algorithms used in the literature. Computational results reveal the effectiveness of the HGSSA.

REFERENCES

Bagheri, A., & Zandieh, M. (2011). Bi-criteria flexible job-shop scheduling with sequence-dependent setup times—variable neighborhood search approach. *Journal of Manufacturing Systems*, *30*(1), 8–15. doi:10.1016/j.jmsy.2011.02.004

Bagheri, A., Zandieh, M., Mahdavi, I., & Yazdani, M. (2010). An artificial immune algorithm for the flexible job-shop scheduling problem. *Future Generation Computer Systems*, *26*(4), 533–541. doi:10.1016/j.future.2009.10.004

Baradaran, S., Fatemi Ghomi, S. M. T., Mobini, M., & Hashemin, S. S. (2010). A hybrid scatter search approach for resource-constrained project scheduling problem in PERT-type networks. *Advances in Engineering Software*, *41*(7-8), 966–975. doi:10.1016/j.advengsoft.2010.05.010

Baykasoğlu, A., Hamzadayi, A., & Köse, S. Y. (2014). Testing the performance of teaching–learning based optimization (TLBO) algorithm on combinatorial problems: Flow shop and job shop scheduling cases. *Information Sciences*, *276*, 204–218. doi:10.1016/j.ins.2014.02.056

Brandimarte, P. (1993). Routing and scheduling in a flexible job shop by tabu search. *Annals of Operations Research*, *41*(3), 157–183. doi:10.1007/BF02023073

Brucker, P., & Schlie, R. (1990). Job-shop scheduling with multi-purpose machines. *Computing*, *45*(4), 369–375. doi:10.1007/BF02238804

Defersha, F. M., & Chen, M. (2010). A parallel genetic algorithm for a flexible job-shop scheduling problem with sequence dependent setups. *International Journal of Advanced Manufacturing Technology*, *49*(1-4), 263–279. doi:10.1007/s00170-009-2388-x

El-Sayed, S. M., EL-Wahed, W. F. A., & Ismail, N. A. (2008). A hybrid genetic scatter search algorithm for solving optimization problems. *Constraints*, *500*, 1.

Gao, J., Sun, L., & Gen, M. (2008). A hybrid genetic and variable neighborhood descent algorithm for flexible job shop scheduling problems. *Computers & Operations Research*, *35*(9), 2892–2907. doi:10.1016/j.cor.2007.01.001

Glover, F. (1977). Heuristics for integer programming using surrogate constraints. *Decision Sciences*, *8*(1), 156–166. doi:10.1111/j.1540-5915.1977.tb01074.x

Glover, F., Laguna, M., & Marti, R. (2000). Fundamentals of Scatter Search and Path Relinking. *Controls and Cybernetics*, *39*(3), 653–684.

González, M. A., Vela, C. R., & Varela, R. (2015). Scatter search with path relinking for the flexible job shop scheduling problem. *European Journal of Operational Research*, *245*(1), 35-45. doi:10.1016/j.ejor.2015.02.052

Holland, J. H. (1975). *Adaptation in natural and artificial systems*. Ann Arbor: University of Michigan Press.

Hurink, J., Jurisch, B., & Thole, M. (1994). Tabu search for the job-shop scheduling problem with multi-purpose machines. *Operations-Research-Spektrum*, *15*(4), 205–215. doi:10.1007/BF01719451

Kacem, I., Hammadi, S., & Borne, P. (2002). Pareto-optimality approach for flexible job-shop scheduling problems: Hybridization of evolutionary algorithms and fuzzy logic. *Mathematics and Computers in Simulation*, *60*(3), 245–276. doi:10.1016/S0378-4754(02)00019-8

Karthikeyan, S., Asokan, P., Nickolas, S., & Page, T. (2015). A hybrid discrete firefly algorithm for solving multi-objective flexible job shop scheduling problems. *International Journal of Bio-inspired Computation*, *7*(6), 386–401. doi:10.1504/IJBIC.2015.073165

Kumar, A., Pant, S., & Ram, M. (2016). System reliability optimization using grey wolf optimizer algorithm. *Quality and Reliability Engineering International*. doi:10.1002/qre.2107

Kumar, A., Pant, S., Ram, M., & Singh, S. B. (2017) On Solving Complex Reliability Optimization Problem Using Multi-Objective Particle Swarm Optimization. In *Mathematics Applied to Engineering* (pp. 115-131).

Kumar, A., Pant, S., & Singh, S. B. (2017). *Reliability Optimization of Complex System by Using Cuckoos Search algorithm*. Mathematical Concepts and Applications in Mechanical Engineering and Mechatronics; doi:10.4018/978-1-5225-1639-2.ch005

Luo, H., Huang, G. Q., Zhang, Y., Dai, Q., & Chen, X. (2009). Two-stage hybrid batching flowshop scheduling with blocking and machine availability constraints using genetic algorithm. *Robotics and Computer-integrated Manufacturing*, *25*(6), 962–971. doi:10.1016/j.rcim.2009.06.001

Marichelvam, M. K., & Geetha, M. (2014a). Solving tri-objective multistage hybrid flow shop scheduling problems using a discrete firefly algorithm. *International Journal of Intelligent Engineering Informatics*, *2*(4), 284–303. doi:10.1504/IJIEI.2014.067190

Marichelvam, M. K., & Prabaharan, T. (2014). Performance evaluation of an improved hybrid genetic scatter search (IHGSS) algorithm for multistage hybrid flow shop scheduling problems with missing operations. *International Journal of Industrial and Systems Engineering*, *16*(1), 120–141. doi:10.1504/IJISE.2014.057946

Mastrolilli, M., & Gambardella, L. M. (2000). Effective neighborhood functions for the flexible job shop problem. *Journal of Scheduling*, *3*(1), 3–20. doi:10.1002/(SICI)1099-1425(200001/02)3:1<3::AID-JOS32>3.0.CO;2-Y

Noorul Haq, A., Saravanan, M., Vivekraj, A. R., & Prasad, T. (2006). A scatter search approach for general flowshop scheduling problem. *International Journal of Advanced Manufacturing Technology*, *31*(7-8), 731–736. doi:10.1007/s00170-005-0244-1

Pant, S., Anand, D., Kishor, A., & Singh, S. B. (2015). A Particle Swarm Algorithm for Optimization of Complex System Reliability. *International Journal of Performability Engineering.*, *11*(1), 33–42.

Pant, S., Kumar, A., Kishor, A., Anand, D., & Singh, S. B. (2015). Application of a Multi-Objective Particle Swarm optimization Technique to Solve Reliability Optimization Problem. In *Proceeding of IEEE Int. Conf. on Next generation Computing Technologies*, September 4-5 (pp. 1004-1007). doi:10.1109/NGCT.2015.7375272

Pant, S., Kumar, A., & Ram, M. (2017). Reliability Optimization: A Particle Swarm Approach. In Advances in Reliability and System Engineering (pp. 163-187). Springer International Publishing.

Pant, S., & Singh, S. B. (2011). Particle Swarm Optimization to Reliability Optimization in Complex System. In *Proceedings of the IEEE Int. Conf. on Quality and Reliability*, Bangkok, Thailand, Sept 14-17 (pp. 211-215). doi:10.1109/ICQR.2011.6031711

Pezzella, F., Morganti, G., & Ciaschetti, G. (2008). A genetic algorithm for the flexible job-shop scheduling problem. *Computers & Operations Research*, *35*(10), 3202–3212. doi:10.1016/j.cor.2007.02.014

Pinedo, M. (1995). *Scheduling: theory, algorithms, and systems*. Englewood Cliffs: Prentice-Hall.

Rahimi-Vahed, A. R., Javadi, B., Rabbani, M., & Tavakkoli-Moghaddam, R. (2008). A multi-objective scatter search for a bi-criteria no-wait flow shop scheduling problem. *Engineering Optimization*, *40*(4), 331–346. doi:10.1080/03052150701732509

Rajkumar, M., Asokan, P., Anilkumar, N., & Page, T. (2011). A GRASP algorithm for flexible job-shop scheduling problem with limited resource constraints. *International Journal of Production Research*, *49*(8), 2409–2423. doi:10.1080/00207541003709544

Ranjbar, M., De Reyck, B., & Kianfar, F. (2009). A hybrid scatter search for the discrete time/resource trade-off problem in project scheduling. *European Journal of Operational Research*, *193*(1), 35–48. doi:10.1016/j.ejor.2007.10.042

Saidi-Mehrabad, M., & Fattahi, P. (2007). Flexible job shop scheduling with tabu search algorithms. *International Journal of Advanced Manufacturing Technology*, *32*(5-6), 563–570. doi:10.1007/s00170-005-0375-4

Saravanan, M., & Noorul Haq, A. (2010). A scatter search algorithm for scheduling optimization of job shop problems. *International Journal of Product Development*, *10*(1-3), 259–272. doi:10.1504/IJPD.2010.029996

Tavakkoli-Moghaddam, R., Javadian, N., Khorrami, A., & Gholipour-Kanani, Y. (2010). Design of a scatter search method for a novel multi-criteria group scheduling problem in a cellular manufacturing system. *Expert Systems with Applications*, *37*(3), 2661–2669. doi:10.1016/j.eswa.2009.08.012

Tosun, Ö., & Marichelvam, M. K. (2016). Hybrid bat algorithm for flow shop scheduling problems. *International Journal of Mathematics in Operational Research*, *9*(1), 125–138. doi:10.1504/IJMOR.2016.077560

Xing, L. N., Chen, Y. W., Wang, P., Zhao, Q. S., & Xiong, J. (2010). A knowledge-based ant colony optimization for flexible job shop scheduling problems. *Applied Soft Computing*, *10*(3), 888–896. doi:10.1016/j.asoc.2009.10.006

Xu, Y., Wang, L., Wang, S. Y., & Liu, M. (2015). An effective teaching–learning-based optimization algorithm for the flexible job-shop scheduling problem with fuzzy processing time. *Neurocomputing*, *148*, 260–268. doi:10.1016/j.neucom.2013.10.042

Yazdani, M., Amiri, M., & Zandieh, M. (2010). Flexible job-shop scheduling with parallel variable neighborhood search algorithm. *Expert Systems with Applications*, *37*(1), 678–687. doi:10.1016/j.eswa.2009.06.007

Zhang, G., Gao, L., & Shi, Y. (2010, June). A genetic algorithm and tabu search for multi objective flexible job shop scheduling problems. In *Proceedings of the 2010 International Conference on Computing, Control and Industrial Engineering (CCIE)* (Vol. 1, pp. 251-254). IEEE. doi:10.1109/CCIE.2010.71

Zhang, G., Shao, X., Li, P., & Gao, L. (2009). An effective hybrid particle swarm optimization algorithm for multi-objective flexible job-shop scheduling problem. *Computers & Industrial Engineering*, *56*(4), 1309–1318. doi:10.1016/j.cie.2008.07.021

Zhang, H. P., & Gen, M. (2005). Multistage-based genetic algorithm for flexible job-shop scheduling problem. *Journal of Complexity*, *11*, 223–232.

ADDITIONAL READING

Baker, K. R., & Trietsch, D. (2009). *Principles of Sequencing and Scheduling*. New Jersey: John Wiley & Sons, Inc. doi:10.1002/9780470451793

KEY TERMS AND DEFINITIONS

Genetic Algorithm: A meta-heuristic algorithm used to solve the optimization problems.

Genetic Scatter Search Algorithm: A hybridization of genetic and scatter search algorithms.

Makespan: Makespan is defined as the completion time of the last job to leave the system.

Mean Flow Time: Mean flow time is defined as the average time spent by the jobs in the production system.

NP-Hard Problems: Non – deterministic polynomial time hard problems.

Scatter Search Algorithm: A meta-heuristic algorithm used to solve the optimization problems.

Scheduling: Scheduling is defined as a process of allocating resources over time to perform a collection of tasks.

Chapter 10
Electrical Discharge Coating by Copper–Tungsten Composite Electrode Prepared by Powder Metallurgy Route

Anshuman Kumar Sahu
National Institute of Technology Rourkela, India

Siba Sankar Mahapatra
National Institute of Technology Rourkela, India

ABSTRACT

Electrical discharge machining (EDM), a thermo-mechanical machining process, is used in producing complicated intrinsic cavity in difficult-to- machine materials with excellent surface finish. One of the major disadvantage of EDM process is the tool wear, which can be used advantageously for coating purpose. Coating is a unique method of EDM process by the use of electrode prepared via powder metallurgy route. Copper and tungsten powders in weight percentage of 30 and 70 respectively are used for the preparation of the tool electrode by varying the PM process parameters like compaction pressure and sintering temperature. The substrate on which coating is made is chosen as AISI 1040 stainless steel with EDM oil as the dielectric fluid. During coating, influence of parameters like discharge current, duty cycle and pulse-on-time on material deposition rate, tool wear rate and radial under deposition are studied. To find out the best parametric combination Grey Relational Analysis method combined with Harmony Search algorithm has been employed.

INTRODUCTION

Electrical discharge machining (EDM), a non-conventional machining process, is extensively used in aerospace, biomedical, chemical, automobile and tool and mold making industries. EDM is normally used to produce parts with complex geometry made of difficult-to-machine materials having reasonable surface finish. In EDM process, both work piece and tool electrode are electrically conductive and

DOI: 10.4018/978-1-5225-3035-0.ch010

immersed in a dielectric medium. The material removal occurs by electro-thermal process. When the potential difference is applied to the electrical circuit, an electric field is established due to the potential difference in the tool electrode and work piece gap. Therefore, free electrons on the electrode were emitted and accelerated towards the work piece by the electrostatic force. The accelerated electrons collide with the work piece surface and a high temperature around 8000-10000°C is generated. The material removal occurred by the melting and evaporation of tiny particles from the work piece surface due to the generation of successive sparks between work piece and electrode in the dielectric medium. Extensively high temperature generated at the work piece surface can easily melt and vaporize materials from difficult to machine materials with high hardness, strength and complex structural properties. The dielectric fluid closes the electrical circuit as well as swept out the removed material and enhanced the material removal rate (Mishra, 2012; Sahu, Mohanty, & Sahoo, 2017). During the process of EDM, material removals as well as tool wear occur due to the generation of very high temperature. The tool electrode wear phenomenon cannot be eliminated completely. But the wearing material from the tool can be used for coating purpose. Electrical discharge coating (EDC) is an exceptional method of EDM process where surface modification or surface alloying of the work piece occurs. The removed tool material is deposited on the work piece surface and form hard coated layer on the work piece surface. Hence, electrode prepared via powder metallurgy route is used for EDC process. By preparing electrode via powder metallurgy route, one can vary the properties of the tool electrode such as density, electrical conductivity as per requirement by varying different powder metallurgy process parameters like compaction pressure and sintering temperature. Therefore, tool material can be eroded sufficiently and transferred to the work piece surface. With the use hydrocarbon oil as dielectric fluid in EDC process, the dielectric fluid dissociated during the spark and combined with the tool material forming a hard-coated layer of metal carbide on the work piece surface. A hard ceramic layer of WC, TiC, SiC, TiN, B_4C, Cu_3Sn, CuSn, ZrB_2 can form on the work piece surface after EDC according to the type of the tool electrode and the dielectric fluid used during the EDC process. The EDC process can be used to improve the corrosion resistance and hardness of the work piece that can be used in the industries like aerospace, automobile, biomedical and chemical where the work piece materials are used in a wide variation of temperature with extreme environmental conditions. The EDC process is less expensive as compare to other complicated and costly coating processes like chemical vapor deposition (CVD) and physical vapor deposition (PVD).

In the EDC process, multiple conflicting performance characteristics need to be optimized in order to maximize productivity of the process. For example, material deposition rate need to be maximized whereas tool wear rate and radial under deposition are minimized for improving the EDC process. Therefore, proper selection of the various process parameters is an important issue in EDC process. From recent literature, it is found that different optimizations techniques have been used for the optimization of the EDM and EDC process to enhance the performance of the processes. Various techniques used for the purpose include Grey Relational Analysis (GRA), Satisfaction Function and Distance Based Approach, Utility Concept and Quantum Behaved Particle Swarm Optimization (QPSO) (Sahu, Mohanty, & Sahoo, 2017; Patowari, Saha, & Mishra, 2010; Rahul, Datta, Biswal, & Mahapatra, 2017; Mohanty, Mahapatra, & Singh, 2017). In this work, grey relational analysis (GRA) based Harmony search (HS) algorithm has been used to get optimal parametric setting to get best output responses of EDC process. Here, GRA method used to convert the multi responses into single response i.e. grey relational grade (GRG) and harmony search algorithm has been used to find the optimum parametric setting.

In this work, tool electrode prepared via powder metallurgy route has been used for the EDC process by varying parameters like compaction pressure, sintering temperature, discharge current, duty cycle and pulse-on-time. The effect of these parameters on the output responses like material deposition rate, tool wear rate and radial under deposition have been studied. The best parametric setting of the EDC process has been found out by grey relational analysis combined with harmony search algorithm to get optimum EDC performance.

BACKGROUND

For the EDC process, the electrodes are prepared by powder metallurgy (PM) route so that the electrode materials removed from the tool can be deposited on the work piece and form a deposited layer. The green compact electrodes as well as sintered electrodes are used for the EDC process. For surface modification of aluminum work piece, TiC-Cu green compact tool electrode is used. The effect of EDC process parameters like peak current, pulse-on-time, composition and compaction pressure of tool electrodes are studied along with process performance like surface roughness, coated layer thickness and micro-hardness of the coated surface. Here, the amount of Ti present on the coated surface decreases towards the parent material aluminum. Lower values of current provide a better surface with a lower layer thickness of coated layer (Das, & Misra, 2012). Powder metallurgy (PM) route is adopted for the preparation of copper-tungsten tool electrode for electrical discharge coating of work piece materials like C-40 steel, AISI D2 steel and Inconel 718. The transfer of tool material occurs with deposition of the material on the work piece. With the presence of hydrocarbon dielectric fluid like kerosene, the tool materials combine with the carbon of the dielectric fluid and form metallic carbide like tungsten carbide (WC). Deposition of tungsten carbide occurs on the work piece with an increase in micro-hardness of the work piece. Cu-W electrode prepared via powder metallurgy method and solid copper electrode have been used for the electro-discharge machining of the AISI D2 steel with kerosene as the dielectric fluid. Material removal rate (MRR) and surface roughness (Ra) have been analyzed for different process parameters like current, duty cycle and flushing pressure. Copper electrode gives higher material removal rate (MRR) whereas Cu-W electrode given better surface finish (Beri, Maheshwari, Sharma, & Kumar, 2008). Different types of electrodes like Cu electrode, Cu25-W75 and Cu20-W80 PM composite electrodes have been used in machining of Inconel 718 using EDM with EDM oil as the dielectric fluid. Surface modification of Inconel 718 surface has been made by the by EDM electrodes prepared by PM route. The tool material has been transferred to the work surface and formation of hard composite layers of carbides like Fe_6W_6C, $Cr_2F_{14}C$ and N_2Mo_4C has been observed (Beri, Maheshwari, Sharma, & Kumar, 2014). Different PM process parameters and EDM process parameters has been varied to analyze the EDC process. Cu-W PM electrodes have been used for the surface modification. Reverse polarity is used for the surface modification process of EDM. Material transfer rate (MTR) and average layer thickness (LT) have been evaluated for different process parameters and artificial neural network (ANN) is used to evaluate the performance of the EDC process (Patowari, Saha, & Mishra, 2010). Similarly, W-Cu PM electrodes have been used for the surface modification of C-40 grade plain carbon steel. The energy dispersion X-ray spectrograph (EDX) and X-ray diffraction (XRD) of the EDM machined surface indicate presence of Cu and W particles in the form of carbides. With the formation of the carbides, the micro-hardness of the machined surface is increased (Patowari, Saha, & Mishra,

2011; Patowari, Saha, & Mishra, 2015). The Schematic diagram of the EDC process is given in Figure 1. The tool wear materials combine with the carbon dissociated from the hydrocarbon fluid to form the metallic carbides. Therefore, the hardness of the EDC surface increased.

TiC-Cu and TiC/W-Cu tool electrodes have been prepared via powder metallurgy route and used in electro-discharge machining (EDM) as tool electrode. A better surface finish of the machined surface is observed with the decrease in relative density of the electrodes. The best EDM performance in terms of higher material removal rate, less tool wear rate with a better surface finish of the machined surface is found in 15% TiC addition with Cu-W electrode (Li, Wong, Fuh, & Lu, 2001a, 2001b). Two types of tool electrodes like solid copper and powder compact copper tool electrodes have been employed to study the EDC process of Ti-6Al-4V alloy. The electrode prepared via powder metallurgy process produces an alloying effect rather than material removal because of less strength of the compact electrodes. Powder metallurgy electrode produces thicker coating layer in positive polarity (Ho, Aspinwall, Voice, Box, & De, 2007). Likewise, the electrolytic copper powder is used to prepare EDM electrode by varying different PM process parameters like compaction pressure and sintering temperature. The EDM performance of these electrodes has been studied and found that these electrodes are used for material deposition rather than material removal (Samuel, & Philip, 1997).

Bronze tool composed of 90% copper and 10% tin is prepared by powder compaction process. The bronze tool electrode is used for the surface modification of mild steel by the electro-discharge process. It is observed that current exhibits more influence on the deposited layer because thickness of the deposited layer increases with increase in current. But frequency does not have significant influence on deposited layer. The EDS and XRD analysis of the work piece surface reveal the transformation of tool materials i.e. copper and tin to the surface of the work piece after EDM with the formation of different phases like Cu_3Sn, CuSn, Cu_6Sn_5 on the work piece surface. The bronze deposited layers exhibit

Figure 1. Schematic diagram of the EDC process (Patowari, Saha, & Mishra, 2010)

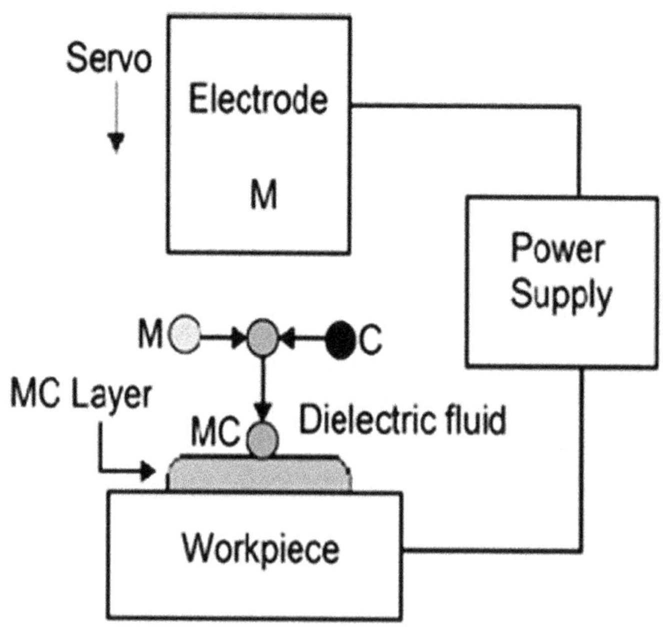

rougher surface properties (Gangadhar, Shunmugam, & Philip, 1991). Surface alloying of hardened AISI D2 Sendzimir rolls is performed by electro-discharge texturing using TiC-WC-Co and WC-Co electrode prepared via powder metallurgy route. The coated layers contain Ti and W which increase the micro-hardness of the roll surface. The compaction pressure and sintering temperature increase the physical, mechanical, electrical, thermal and microstructural properties of the electrode. This enhances coating performance by EDM (Simao, Aspinwall, El-Menshawy, & Meadows, 2002). TiC coating on the roll surface i.e. 2%Cr steel is made by electrical discharge coating process by the help of TiC sintered electrode. The hardness of the coated surface is found to be more than that of the parent materials. The wear resistance of coated roll surface by EDC is superior to the conventional chrome-plated rolls (Ueno, Fujita, Kimura, & Nakata, 2016). Ceramic coating of TiB_2 and TiC is applied on the surface of aluminum work piece by EDC process with the help of green compact Ti-B_4C electrode. Peak current is found to be the most significant parameters in the EDC process. With an increase in peak current, the tool wear rate is increased. The increase in tool wear rate causes to increase in deposition rate and layer thickness of the coated layer. The hardness of the coated layer is also improved after coating due to the formation of ceramic layers like TiB_2 and TiC (Ahmed, 2016).

In some cases, electrode prepared via powder metallurgy route is used for electrical discharge machining. For example, copper-titanium electrodes i.e. Cu-90%, Ti-10% and Cu-80%, Ti-20% are used for the machining of superni-800 super alloy. In order to obtain better material removal rate, it is recommended to use Cu-90% and Ti-10% electrode. However, conventional copper electrode must be used for minimizing tool wear rate (Bhanot, Beri, & Kumar, 2014). Copper-silicon carbide electrodes prepared through powder metallurgy method are used for the EDM of AISI D3 die steel as the work piece material and kerosene as the dielectric fluid. Here the copper-silicon carbide electrodes are prepared by varying the composition of the silicon carbide by 5%, 10% and 15% respectively. The microstructural analysis of the machined work piece surface reveal that the thickness of the recast layer increase with increase in gap current and with increase in the percentage of silicon carbide. Similarly, Cracks in the recast layers increase with increase in the gap current. As compare to other electrodes, Cu-SiC electrode having 5% SiC exhibit less tool wear. The XRD analysis of the machined surface reveal that maximum copper deposition occurs by using 5% SiC composite electrode (Choudhary, Kumar, & Singh, 2012). Copper-chromium composite electrode prepared via powder metallurgy route is used for the EDM of AISI 1045 medium carbon steel as work piece material and dielectric fluid as kerosene. The material removal rate and electrode wear rate are decrease with increase in chromium percentage in the composite tool electrode for both positive and negative polarity. The weight loss increases with increase in corrosion elapsed time for all the electrodes, but weight loss decreases with the increase in the percentage of chromium in the composite tool electrodes. Both material removal rate and electrode wear rate are increase with increase in sintering pressure and open circuit voltage. Similarly, material removal rate and electrode wear rate are more by using copper-chromium composite electrode as compare to a solid copper electrode. But better surface finishes produce on the work piece by using solid copper metal electrode. The micro-harness value decreases from the machined surface towards the base metal. The chromium elements are transferred to the machined surface during machining and enhanced the corrosion resistance of the machined work piece material. The thickness of the recast layer is more by using copper electrode as compare to the copper-chromium composite electrode (Tsai, Yan, & Huang, 2003). A metal matrix composite (ZrB_2-Cu) is used as the tool electrode in EDM for the machining of mild steel. By using ZrB_2-40% Cu as electrode, the material removal rate is more and tool wear rate is less as compare to the common copper electrode. But the diametric over cut and surface roughness are found to be more in ZrB_2-Cu electrode as compare to the copper electrode. Material removal rate, diametric over cut and tool wear rate are increased with increase in pulse on time. Similarly, roughness of the machined surface as well as tool surface after machining are increased with increase in pulse on time. Material removal rate and tool wear rate are also increase with increase in current (Khanra, Sarkar, Bhattachary, Pathak, & Godkhindi, 2007). Cu-ZrB_2 and Cu-Ti-Si composite electrodes are prepared by powder metallurgy route by varying different compaction pressure and sintering temperature to study the EDM performance. With increase in the percentage of ZrB_2 in electrode, the material removal rate decreases along with increase in tool wear rate and average surface roughness. But electrode prepared with high sintering temperature give better EDM performance as compare to the electrode prepared at less sintering temperature. In the case of Cu-Ti-Si electrode, the tool wear rate and average surface roughness increase with increase in percentage of the TiSi. But material removal first increase and then decrease with increase in the percentage of the TiSi. By analyzing the EDM performance of all the electrodes it is found that Cu-TiSi-Gr electrode cannot be used as EDM electrode due to high electrode rate, low surface finish of machined surface and lower material removal rate (Zaw, Fuh, Nee, & Lu, 1999). Ultrasonic electro-deposition technique is used to prepare Cu-SiC tool electrodes for EDM. The electrode prepared with low ultrasonic power contain more SiC particles as compare to the electrodes prepared at high ultrasonic power. The tool wear rate of the composite electrode is less as compare to copper electrode and tool wear rate of the Cu-SiC electrode is increase with increase in pulse current (Li, Niu, & Zheng, 2016).

In this work, to study the electrical discharge coating (EDC) process, copper-tungsten (Cu-W) electrode prepared via powder metallurgy (PM) route are used. Different powder metallurgy process parameters like compaction pressure (CP) and sintering temperature (ST) are varied during the preparation of tool electrodes. To study the EDC process along the powder metallurgy process parameters different EDC process parameters are also varied during the coating process like discharge current (I_p), duty cycle (τ) and pulse-on-time (T_{on}).

MAIN FOCUS OF THE CHAPTER

Electrical discharge coating process has been performed on AISI 1040 stainless steel with the help of copper-tungsten composite electrode prepared via powder metallurgy route and EDM oil as the dielectric fluid. To reduce the number of experiment, Taguchi's L_{18} orthogonal array has been used. The powder metallurgy process parameters like compaction pressure and sintering temperature have been varied along with electrical discharge coating process parameters like discharge current, duty cycle and pulse-on-time during the coating process. The effect of these process parameters on material deposition rate (MDR), tool wear rate (TWR) and radial under deposition (RUD) have been studied. To find out the best parametric combination that can simultaneously optimize three performance measures, Grey Relational Analysis (GRA) method combined with Harmony Search (HS) algorithm has been employed.

MATERIALS AND METHOD

Electrical discharge coating (EDC) process is an exceptional process of electrical discharge machining (EDM) process. For the EDC process in most of the cases powder metallurgy route has been used to prepare the tool electrode. The electrodes used in the EDC process are either green compact electrode or sintered electrode. In this work, copper-tungsten electrode was prepared via PM route and used as tool electrode in EDC of 1040 stainless steel of rectangular shape plate (100×30×6 mm) as work piece material. The chemical composition and properties of AISI 1040 stainless steel is presented in Table 1 and Table 2 respectively.

To prepare the electrodes for EDC process, copper and tungsten powders of mesh size 325 in weight percentage of 30 and 70 were used. Both the powders were mixed in a pulverizing mill for 10 hours. After complete mixing, the powder mixture was compacted by the help of a uniaxial compaction machine with different compaction pressure in a punch and die system. The electrode produced after uniaxial compaction is termed as the green electrode. This green electrode further sintered in a tubular furnace in an argon environment. The powder metallurgy process parameters like compaction pressure and sintering temperature were varied for the preparation of the electrodes as given in Table 3. The densities of the copper-tungsten tool electrodes were presented in Table 4.

The tool electrodes produced via PM route were of cylindrical shape with diameter 25mm. The electrical discharge coating process of AISI 1040 stainless steel by copper-tungsten tool electrodes prepared via PM route was carried out by a die sinking EDM (ELECTRA EMS 5535, India). Commercially available EDM oil (specific gravity: 0.763) was used as the dielectric fluid, which was a mixture of kerosene and water. In this work, straight polarity (i.e. work piece as anode and tool as cathode) has been used to conduct the experiment. The experiment of EDC process has been carried out by taking five process parameters with different levels. The PM process parameters like compaction pressure and sintering temperature have been taken along with EDC process parameters like discharge current (I_p), duty cycle (τ) and pulse-on-time (T_{on}). The different levels of the process parameters have been listed in Table 3. The different variable parameters used in this EDC process are discussed as follows.

Table 1. Chemical composition of AISI 1040 stainless steel

Element	Mn	C	S	P	Fe
Content (%)	0.75	0.4	0.04	0.03	Balance

Table 2. Properties of AISI 1040 stainless steel

Properties	Value
Density	7.845 g/cc
Melting point	1521^{0}C
Tensile strength	620 MPa
Yield strength	415 MPa
Elastic modulus	190-210 GPa
Poisson's ratio	0.27-0.30
Hardness	201 HB
Thermal conductivity	51.9 W/mK (at 0^{0}C)

1. **Compaction Pressure:** It is the pressure applied by the uniaxial compaction machine to the die and punch system to form green compact tool electrode. The compaction pressure directly affect the density of the tool electrode.
2. **Sintering Temperature:** It is the temperature at which the green compact tool electrodes are heated with constant maximum temperature for a period of time. At the sintering temperature the different materials of tool electrodes are combined with each other and form denser compound with higher electrical conductivity.
3. **Discharge Current:** It is also called as peak current. It is the most dominant process parameter that governs the spark energy. Higher value of discharge current increase the energy input in the EDC process.
4. **Duty Cycle:** It is the percentage of the pulse-on-time with respect to total cycle time.

$$\tau = \frac{T_{on}}{T_{on} + T_{off}} \tag{1}$$

5. **Pulse-on-Time:** It is also known as spark-on-time or pulse duration. It is the duration (per cycle) in which the current is allowed to flow through the electrode gap. The energy input is directly influence by the pulse-on-time.

The design of experiment (DOE) approach like Taguchi L_{18} orthogonal array has been used to plan the experiment. Here a mixed design of 1-factor-2-level and 4-factor-3-level has been used as in Table 3. The Taguchi L_{18} orthogonal array (OA) and the output responses for this experimental work were presented in Table 5. The electrical discharge coating process during coating and the work piece after coating were given in Figure 2.

Material Deposition Rate (MDR)

Material deposition rate (MDR) is defined as the rate at which material deposition occurred on the surface of the work piece. The MDR can be determined by the weight gain criteria of the work piece as shown in the Eqn. 2.

Table 3. Input parameters with different levels

Parameters	Unit	Level 1	Level 2	Level 3
A-Sintering temperature (ST)	(^{0}C)	700	900	-
B- Compaction pressure (CP)	(MPa)	100	150	200
C- Discharge current (I_p)	(A)	20	25	30
D- Duty cycle (τ)	(%)	42	50	58
E- Pulse-on-time (T_{on})	(μs)	100	200	300

Table 4. Densities of the tool electrodes at different PM process parameters

Sl. No.	Compaction Pressure (MPa)	Sintering Temperature (°C)	Density of Tool Electrodes (g/cm³)
1	100	700	7.8560
2	150	700	8.3140
3	200	700	8.5600
4	100	900	8.1142
5	150	900	8.1489
6	200	900	8.9682

Figure 2. (a) Electrical coating process, (b) Work piece after EDC

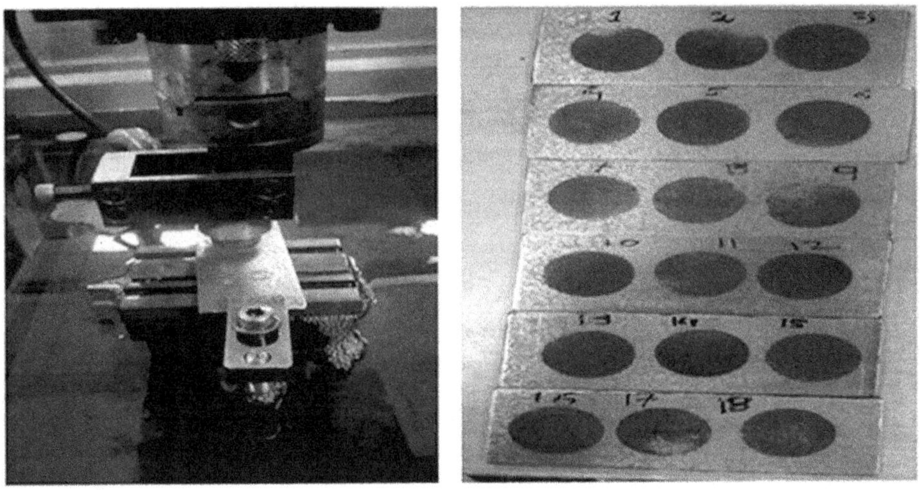

(a) (b)

$$MDR = \frac{\left(W_f - W_i\right)}{\left(t \times \rho_w\right)} \tag{2}$$

where

W_f = Final weight of the work piece after EDC,
W_i = Initial weight of the work piece before EDC,
t = Machining time,
ρ_t = Density of work piece 1040 stainless steel = 7.845 g/cm³.

Tool Wear Rate (TWR)

Tool wear rate (TWR) is defined as the rate at which material loss occurred from the tool electrode. The TWR calculated by the weight loss criteria of the tool electrode as given in the Eqn. 3.

$$TWR = \frac{\left(W_{ti} - W_{tf}\right)}{\left(t \times \rho_t\right)} \tag{3}$$

where

W_{ti} = Initial weight of the tool electrode before EDC,
W_{tf} = Final weight of the tool electrode after EDC,
t = Machining time,
ρ_t = Density of tool electrodes.

The densities of the tool electrodes were given in Table 4. The initial and final weight of the work piece and tool electrodes before coating and after coating were measured by a weight measurement machine of least count 0.05 gm (Figure 3).

Figure 3. Weight measurement machine

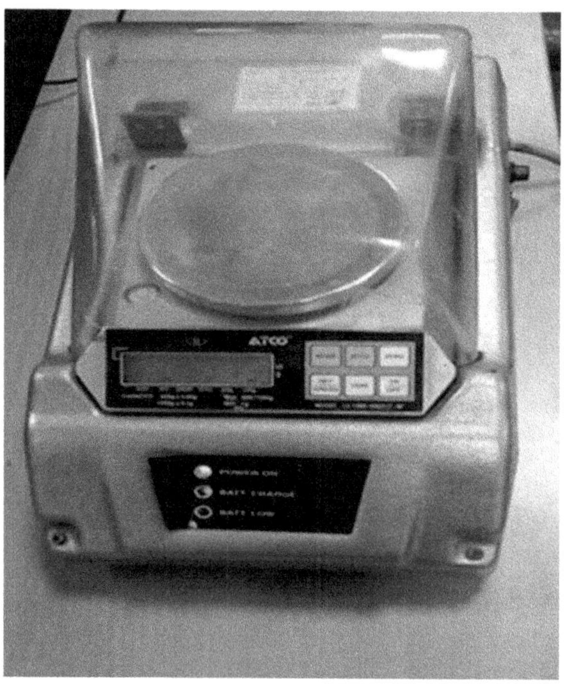

Radial Under Deposition (RUD)

Radial under deposition (RUD) is defined as the radial un-deposited space present of the work piece surface after EDC process. The RUD is calculated as shown in the Eqn. 4.

$$RUD = \frac{\left(D_o - D_i\right)}{2} \tag{4}$$

where

D_o = Diameter of tool electrode = 25mm
D_i = Diameter of the electrical discharge coated layer on the work piece surface

The diameters of the coated layer on the work piece were measured by taking the optical images of the specimen with 7× magnification by an optical microscope (SAMSUNG SDC-314B, Figure 4). The least count of the optical microscope is 0.001 mm. The diameters of the optical images of the specimen were measured by the image viewer application available in MATLAB R2014a. Four different diameters were measured and the average of these was taken as the diameter of the coated layers on the work piece (Figure 5).

Figure 4. Optical microscope

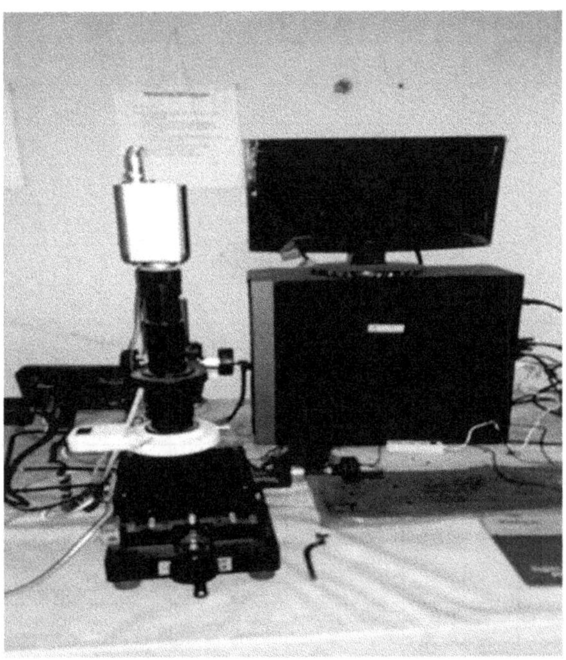

Figure 5. Measurement of diameter of coated layer by image viewer in MATLAB R2014a

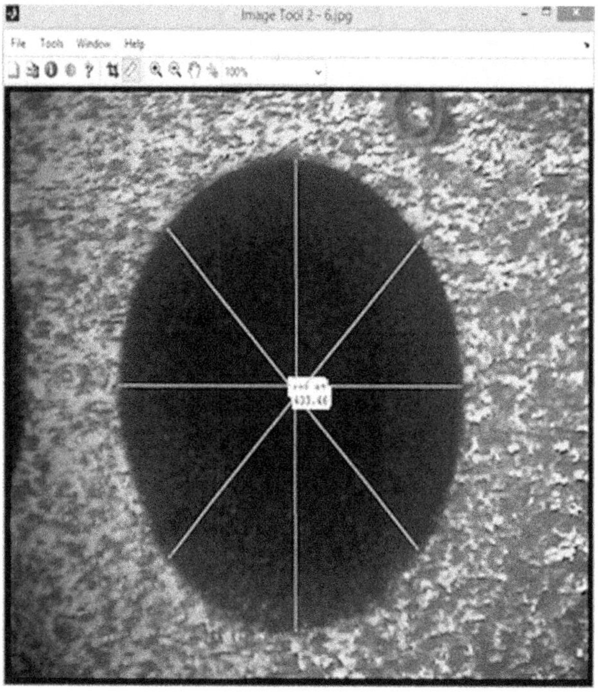

Table 5. Design of experiment (Taguchi L_{18} OA) and output responses

Expt. No.	A	B	C	D	E	MDR (mm³/min)	TWR (mm³/min)	RUD (mm)
	ST	CP	I_p	DC	T_{on}			
1	700	100	20	42	100	1.912	9.547	1.057
2	700	100	25	50	200	2.549	14.002	1.276
3	700	100	30	58	300	3.824	12.729	0.918
4	700	150	20	42	200	1.275	7.217	1.51
5	700	150	25	50	300	1.912	9.021	1.243
6	700	150	30	58	100	2.549	15.035	1.273
7	700	200	20	50	100	0.673	6.926	1.506
8	700	200	25	58	200	1.912	5.653	1.279
9	700	200	30	42	300	2.549	9.219	1.285
10	900	100	20	58	300	1.912	9.613	0.701
11	900	100	25	42	100	1.912	6.211	0.702
12	900	100	30	50	200	2.549	12.94	0.614
13	900	150	20	50	300	1.912	6.83	1.136
14	900	150	25	58	100	1.275	8.693	1.095
15	900	150	30	42	200	1.912	11.658	0.821
16	900	200	20	58	200	1.275	8.363	0.918
17	900	200	25	42	300	1.275	1.673	1.157
18	900	200	30	50	100	1.912	7.805	1.13

METHODOLOGY

Optimization is the process of minimization of the undesired parameters like time and cost with maximization of the desired benefits like quality and quantity of the product manufacture. So, by the optimization process, we can find out the optimized machining condition, so that by setting the optimal machining condition we can get maximum beneficial from the machining process that leads to the quality of the product and also decreased the cost of production by enhancing the machining performance. So, here we have used the optimization process to get the best parametric condition of the electrical discharge coating process. Different researchers have taken different optimization techniques to get the optimal setting of the machining of the processes. In this EDC process, Grey relational analysis (GRA) based Harmony search (HS) algorithm has been used to optimize the machining performance.

Grey Relational Analysis (GRA)

Julong Deng first time proposed the grey system theory (GST) in 1982. Here the researcher has proposed the grey relational analysis (GRA) to find out the relationship between the variables by using grey degrees (Julong, 1982; Julong, 1988). This grey relational analysis (GRA) method was used to optimize the process parameters in electrical discharge machining and wire electrical discharge machining process (Sahu, Mohanty, & Sahoo, 2017; Datta & Mahapatra, 2010). The detailed procedure of the grey relational analysis (GRA) method is explained as follows.

1. Calculate the scale value (Y_{ij}) of the observations.

For Lower is the better,

$$Y_{ij} = \frac{y_{ij}^{\max} - y_{ij}}{y_j^{\max} - y_j^{\min}} \tag{5}$$

For Higher is the better,

$$Y_{ij} = \frac{y_{ij} - y_{ij}^{\min}}{y_j^{\max} - y_j^{\min}} \tag{6}$$

where

y_{ij} = observed responses of the ith number of experiment in the jth response.
$y_j^{\max}$ = maximum value of the jth response.
$y_j^{\min}$ = minimum value of the jth response.

2. Calculate the Grey relational co-efficient (γ_{ij}).

$$\gamma_{ij} = \frac{\left(\Delta_j^{\min} + \xi \Delta_j^{\max}\right)}{\left(\Delta_{ij} + \Delta_j^{\max}\right)} \tag{7}$$

where,

$$\Delta_{ij} = \left|1 - Y_{ij}\right|$$

$$y_j^{\min} = \min\left(\Delta_{1j}, \Delta_{2j}, \ldots \Delta_{mj}\right),\ y_j^{\max} = \max\left(\Delta_{1j}, \Delta_{2j}, \ldots \Delta_{mj}\right)$$

$$\xi \in \left[0,1\right], \xi = 0.5$$

3. Grey relational grade (GRG_i)

$$GRG_i = \sum_{j=1}^{p} W_j \gamma_{ij} \tag{8}$$

where,

$$\sum_{j=1}^{p} W_j = 1$$

Harmony Search (HS) Algorithm

The Harmony Search (HS) algorithm is a music-based metaheuristic optimization technique, which is inspired by jazz music. Here the objective function is optimized by generating aesthetic music. Aesthetic quality of music is produced by adjusting band width, pitch, timbre and amplitude. The variables in the objective function are treated as the pitch of different musical instruments and the solution is termed as the harmony vector. When an aesthetic harmony was developed, it was stored as the best fitness of the objective function value in the memory. By continuously improvising the harmony matrix and replaced it by the best harmony, the best optimized harmony can be generated and the best parametric setting can be found out (Mahdavi, Fesanghary, & Damangir, 2007; Yang, 2009; Nayak, Mahapatra, Chatterjee, & Abhishek, 2015; Abhishek, Datta, & Mahapatra, 2016). The Harmony Search algorithm has been compared with others algorithm like Genetic Adaptive Search, Langrangian multiplier approach, Genetic Algorithm and it was found that Harmony Search algorithm has given a better result as compare to other methods (Mahdavi, Fesanghary, & Damangir, 2007). The procedure of HS algorithm is as follows.

1. Initialization of the problem and parameters of algorithm.
2. Initialization of the harmony memory.
3. Improvisation of the new Harmony.
4. Updating of the harmony memory.
5. Repeat the last two steps till the fulfillment of the termination criteria.

1. Initialization of the problem and parameters of algorithm

Interpretation of the optimization problem to be maximized as follows:
Maximize

$f(\mathrm{x})$

subjected to

$$x_i \in X_i = 1, 2, \ldots\ldots N \tag{9}$$

where

$f(\mathrm{x}) =$ Objective function,
$x_i =$ Set of decision variables,
$N =$ Number of decision variables,
$X_i =$ Set containing possible range of values for each decision variable.

The HS algorithm parameters like harmony memory size (HMS), harmony memory consideration rate (HMCR), pitch adjustment rate (PAR), arbitrary distance i.e. band width (bw) and maximum number of improvisation (stopping criteria) are initializes in this step.

2. Initialization of the harmony memory

The HM matrix is filled by the randomly generate vectors to form HMS as the solution vectors.

$$\mathrm{HM} = \begin{bmatrix} x_1^1 & x_2^1 & \cdots & x_N^1 \\ x_1^2 & x_2^2 & \cdots & x_N^2 \\ \vdots & \vdots & \vdots & \vdots \\ x_1^{HMS} & x_2^{HMS} & \cdots & x_N^{HMS} \end{bmatrix} \tag{10}$$

3. Improvisation of the new harmony

A new harmony vector, $x^{/} = \left(x_1^{/}, x_2^{/}, \ldots\ldots x_N^{/}\right)$, is generated based on the three rules (a) memory consideration, (b) pitch adjustment and (c) random selection. The generation of a new harmony is called as improvisation. In the memory consideration, the value of the first decision variables $\left(x^{/}\right)$ for the new vector is taken from the values of the specific harmony memory (HM) range, $x^{/} = \left(x - x_1^{/HMS}\right)$. Values of the other decision variables $x^{/} = \left(x_1^{/}, x_2^{/}, \ldots\ldots x_N^{/}\right)$ are taken in the same way. The value of HMCR varies between 0 and 1, which is the rate of choosing one value from the historical values stored in the HM, whereas (1-HMCR) is the rate of randomly selecting one value from the possible range of values.

$$x_j^{/} \rightarrow \begin{cases} x_j^{/} \in \left\{x_j^1, x_j^2, \ldots\ldots, x_j^{HMS}\right\} & \text{with probabilty PAR} \\ x_j^1 \in X_j & \text{with probablity (1-HMCR)} \end{cases} \tag{11}$$

Each component obtained by the memory consideration is examined to determine whether it should be pitch adjusted. This operation is using the PAR parameter that is the rate of the pitch adjustment as given below:

The pitch adjusting decision for $x^{/}$ is given as:

$$x^{/} \rightarrow \begin{cases} \text{Yes with probability PAR} \\ \text{No with probability (1-PAR)} \end{cases} \tag{12}$$

For the value of (1-PAR) sets no modification required in $x_i^{/}$. If the pitch adjustment decision is yes then modification is required as follows:

$$x^{/} = x^{/} \pm rand() \times bw \tag{13}$$

where,

$bw =$ arbitrary distance band width
$rand() =$ Random number varies between 0 and 1.

For the improvisation of the global search capacity of the HS algorithm, PAR and bw are dynamically adjusted with the generation number. The PAR is adjusted linearly as follows:

$$\text{PAR(gn)} = \text{PAR}_{\text{min}} + \frac{\text{PAR}_{\text{max}} - \text{PAR}_{\text{min}}}{\text{NI}} \times \text{gn} \tag{14}$$

where,

$\text{PAR(gn)} =$ the pitch adjustment rate for each generation,

NI = solution vector generation number (iteration performed in the algorithm),
gn= generation number,
PAR_{min} = minimum pitch generation number,
PAR_{max} = maximum pitch generation number.

The value of bw is decreased exponentially. Therefore, the higher value of bw increase the diversity of the solution and lower the value of bw tune the final solution.

$$bw(gn)=bw_{max}\times e^{(c\times gn)} \tag{15}$$

$$c=\frac{\ln\left(\frac{bw_{min}}{bw_{max}}\right)}{NI} \tag{16}$$

where,

bw(gn)= band width for each generation,
bw_{max} = maximum bandwidth,
bw_{min} = minimum bandwidth.

4. Updating of the harmony memory

If the new harmony vector, $x' = \left(x'_1, x'_2, \ldots\ldots\ldots x'_N\right)$ is better than the previous memory in the HM, which can be judge in term of objective function value (fitness function value), then the existing worst harmony is replaced by the new Harmony in existing HM.

5. Stopping criterion

Repetition of the steps 3 and 4 until the maximum number of improvisation (stopping criterion) is satisfied, then terminate the computation.

RESULT AND DISCUSSION

The experiments have been conducted and the process responses i.e. material removal rate, tool wear rate, radial under deposition were obtained at different parametric setting and presented in Table 5. The optimization process has selected the most suitable process variables. To increase the performance of electrical discharge coating process, MRR should correspond to Higher-is- Better (HB), TWR and RUD correspond to Lower-is-Better (LB). By following the procedure of the grey relational analysis (GRA) method as given in Eqn. 5-8, the scale value (Y_{ij}), grey relational co-efficient $\left(\gamma_{ij}\right)$ and grey relational grade (GRG_i) were calculated and tabulated in Table 6.

Table 6. Scale value (Y_{ij}), grey relational co-efficient (Υ_{ij}) and grey relational grade (GRGi)

Expt. No.	Y_{ij}	Y_{ij}	Y_{ij}	Δ_{ij}	Δ_{ij}	Δ_{ij}	Υ_{ij}	Υ_{ij}	Υ_{ij}	GRG_i
	MDR	TWR	RUD	MDR	TWR	RUD	MDR	TWR	RUD	
1	0.3932	0.4107	0.5056	0.6068	0.5893	0.4944	0.3112	0.3146	0.3346	0.3201
2	0.5954	0.0773	0.2612	0.4046	0.9227	0.7388	0.3560	0.2601	0.2875	0.3012
3	1	0.1726	0.6607	0	0.8274	0.3393	0.5000	0.2736	0.3733	0.3823
4	0.1911	0.5851	0	0.8089	0.4149	1	0.2764	0.3534	0.25	0.2933
5	0.3932	0.4501	0.2980	0.6068	0.5499	0.7020	0.3112	0.3226	0.2938	0.3092
6	0.5954	0	0.2645	0.4046	1	0.7355	0.3560	0.25	0.2881	0.2980
7	0	0.6069	0.0045	1	0.3931	0.9955	0.25	0.3589	0.2506	0.2865
8	0.3932	0.7021	0.2578	0.6068	0.2979	0.7422	0.3112	0.3852	0.2870	0.3278
9	0.5954	0.4353	0.2511	0.4046	0.5647	0.7489	0.3560	0.3195	0.2859	0.3205
10	0.3932	0.4058	0.9029	0.6068	0.5942	0.0971	0.3112	0.3136	0.4557	0.3602
11	0.3932	0.6604	0.9018	0.6068	0.3396	0.0982	0.3112	0.3732	0.4553	0.3799
12	0.5954	0.1568	1	0.4046	0.8432	0	0.3560	0.2713	0.5	0.3757
13	0.3932	0.6141	0.4174	0.6068	0.3859	0.5826	0.3112	0.3608	0.3159	0.3293
14	0.1911	0.4746	0.4632	0.8089	0.5254	0.5368	0.2764	0.3278	0.3253	0.3098
15	0.3932	0.2527	0.7690	0.6068	0.7473	0.2310	0.3112	0.2862	0.4062	0.3345
16	0.1911	0.4993	0.6607	0.8089	0.5007	0.3393	0.2764	0.3332	0.373	0.3276
17	0.1911	1	0.3940	0.8089	0	0.6060	0.2764	0.5	0.3113	0.3626
18	0.3932	0.5411	0.4241	0.6068	0.4589	0.5759	0.3111	0.3427	0.3173	0.3237

The analysis of variance (ANOVA) for the means of grey relational grade (GRG) was tabulated in Table 7 with R^2 = 94.8% and 95% confidence interval. The response table for the GRG was also presented in Table 8. The main effect plots for the GRG are plotted in Figure 6. The ANOVA and response table were generated by using MINITAB 16. From the ANOVA and response tables (Table 7-8), the compaction pressure was found to be the most significant parameter with percentage contribution of 34.05%, which influences the electrical discharge coating process followed by sintering temperature, pulse-on-time, interaction of compaction pressure and discharge current, pulse-on-time and duty cycle with the percentage contribution of 25.11%, 12.15%, 9.98%, 7.62% and 4.92% respectively. The interaction plot for the GRG values was plotted in Figure 7. Increase in compaction pressure and sintering temperature better electrical discharge coating process can be occurred i.e. the tool strength increased that leads to decrease in tool wear rate and uniform distribution of material deposition on the work piece, so that the radial under deposition decreased. With an increase in current and pulse-on-time the tool wear rate increased and more material deposition occurred on the work piece. So, the material deposition rate increased.

From the ANOVA table (Table 7) compaction pressure and sintering temperature were found to be most significance. With an increase in sintering temperature, the material deposition rate decreased as well as with an increase in compaction pressure material deposition rate decreased (Figure 8). It is due to the increased in the strength of the tool with the increase in compaction pressure and sintering temperature, so the tool wear rate also decreased with increase in compaction pressure and increase in sintering temperature (Figure 9). Similarly, with the increase in sintering temperature and compaction pressure

Table 7. ANOVA for grey relational grade (GRGi)

Source	DF	SS	MS	F	%Contribution
A	1	0.003889	0.003889	9.60	25.11
B	2	0.005273	0.002637	6.51	34.05
C	2	0.001180	0.000590	1.46	7.62
D	2	0.000762	0.000361	0.89	4.92
E	2	0.001881	0.000563	1.39	12.15
A*B	2	0.000147	0.000074	0.18	0.95
B*C	4	0.001545	0.000386	0.95	9.98
Residual Error	2	0.000810	0.000405		5.22
Total	17	0.015486			100

Table 8. Response table for grey relational grade (GRGi)

Level	A	B	C	D	E
1	0.3154	**0.3532**	0.3195	**0.3351**	0.3197
2	**0.3448**	0.3124	0.3318	0.3209	0.3267
3	-	0.3248	**0.3391**	0.3343	**0.3440**
Delta	0.0294	0.0409	0.0196	0.0142	0.0243
Rank	2	1	4	5	3

Figure 6. Main effects plot for GRG

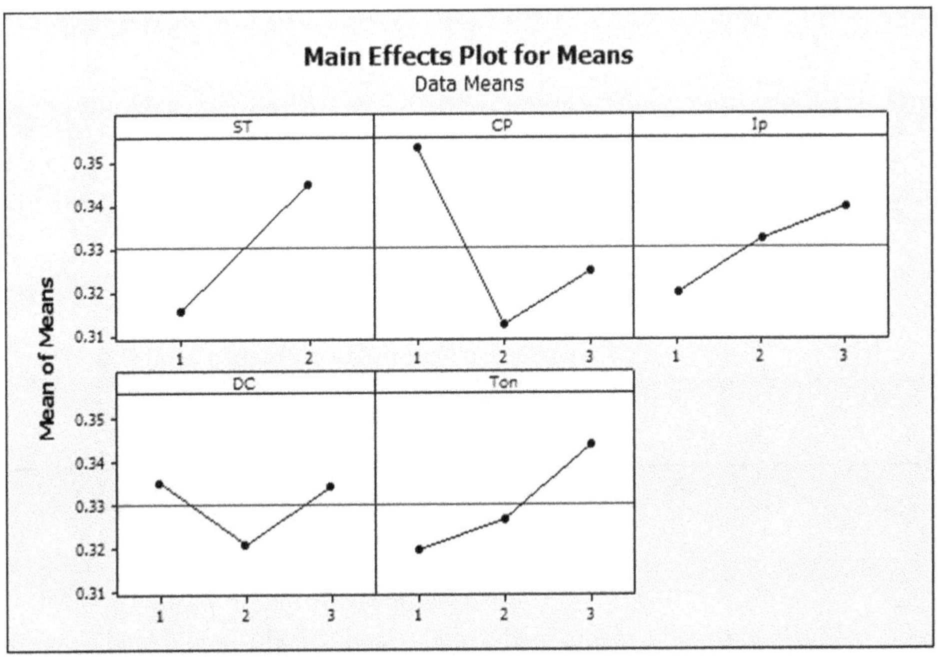

Figure 7. Interaction plot for GRG

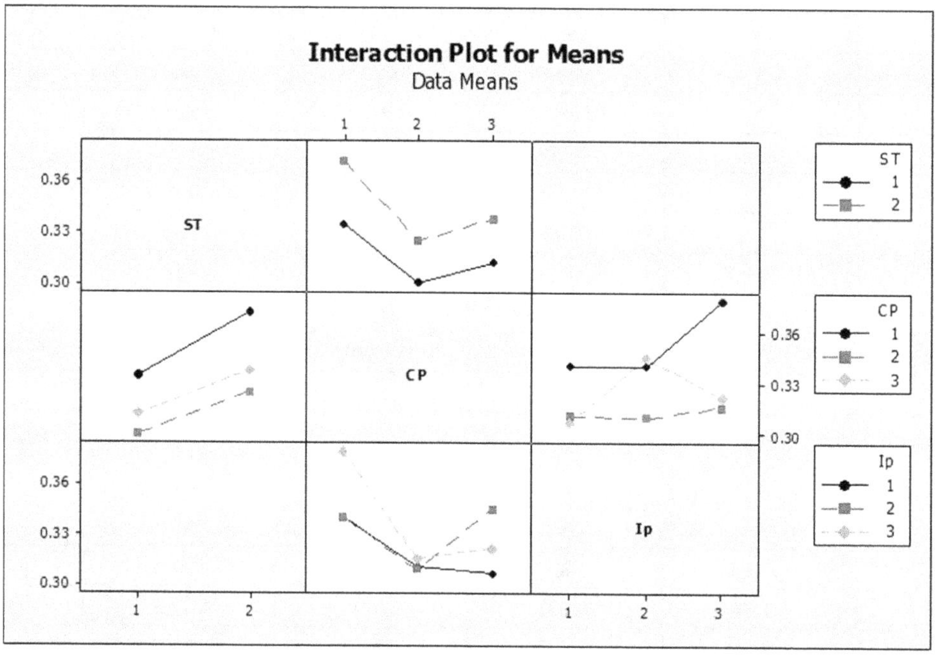

Figure 8. Effect of compaction pressure and sintering temperature on MDR

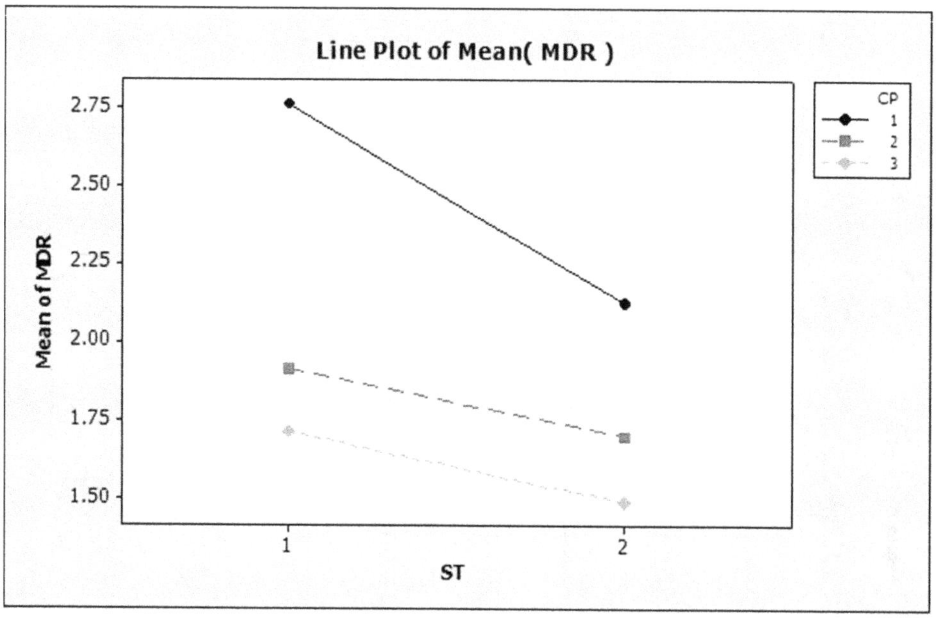

Figure 9. Effect of compaction pressure and sintering temperature on TWR

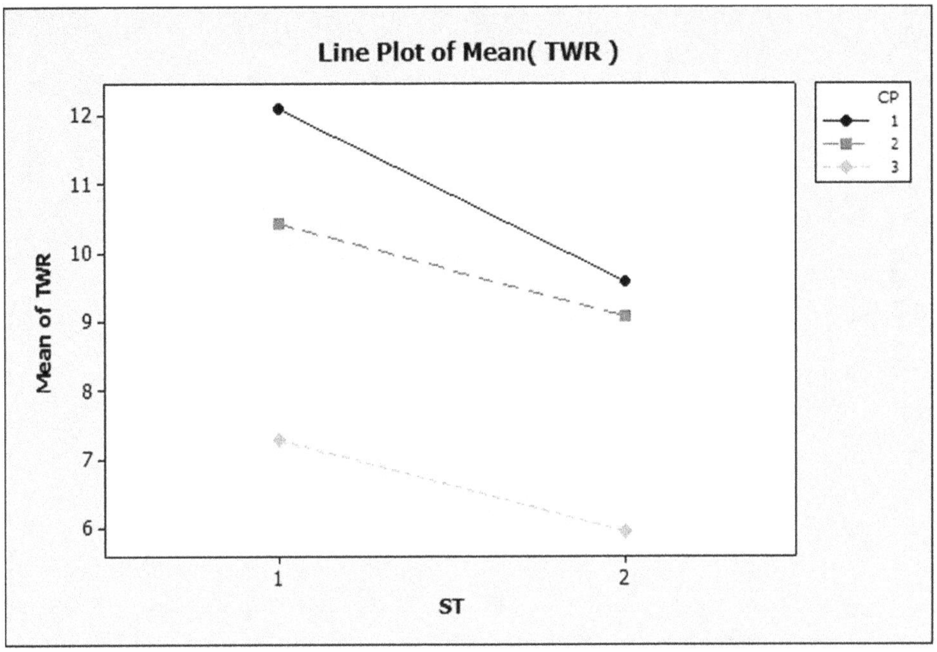

the radial under deposition decreased (Figure 10). This due to the increase in strength and density of the tool electrode at higher compaction pressure and sintering temperature, so uniform spark occurred that lead to uniform deposition with decreased in under deposition. With an increase in compaction pressure more uniform distribution of material deposition occurred and radial under deposition decreased (Figure 11, 12).Similarly, with the increase in sintering temperature, uniform distribution of the material deposition occurred with decreased in radial under deposition (Figure 13, 14).

A non-linear regression analysis has been done between the grey relational grade (GRG) and the EDC input process parameters were expressed by an objective function f(x) as represented in Eqn. (17) having R^2 value of 99.7%. The non-linear i.e. fitness function equation has been generated by using SYSTAT 13. This non-linear equation has been used as the objective function in the harmony search (HS) algorithm and the optimal parametric setting has been found out by executing the harmony search algorithm in MATLAB R2014a as discussed in Eqns. 9-16. The initial parameters setting for the HS algorithm are (a) maximum number of iteration as 5000, (b) harmony memory size as 6, (c) HMCR as 0.9, (d) PRA_{min} as 0.4 and (d) PAR_{max} as 0.9. The optimal parametric setting found out by the grey relational analysis based harmony search algorithm has been listed in Table 9. The graph of fitness value vs. number of iteration has been given in Figure 15.

$$f(x) = 0.311 \times A^{0.130} \times B^{-0.091} \times C^{0.057} \times D^{-0.010} \times E^{0.065} \tag{17}$$

Figure 10. Effect of compaction pressure and sintering temperature on RUD

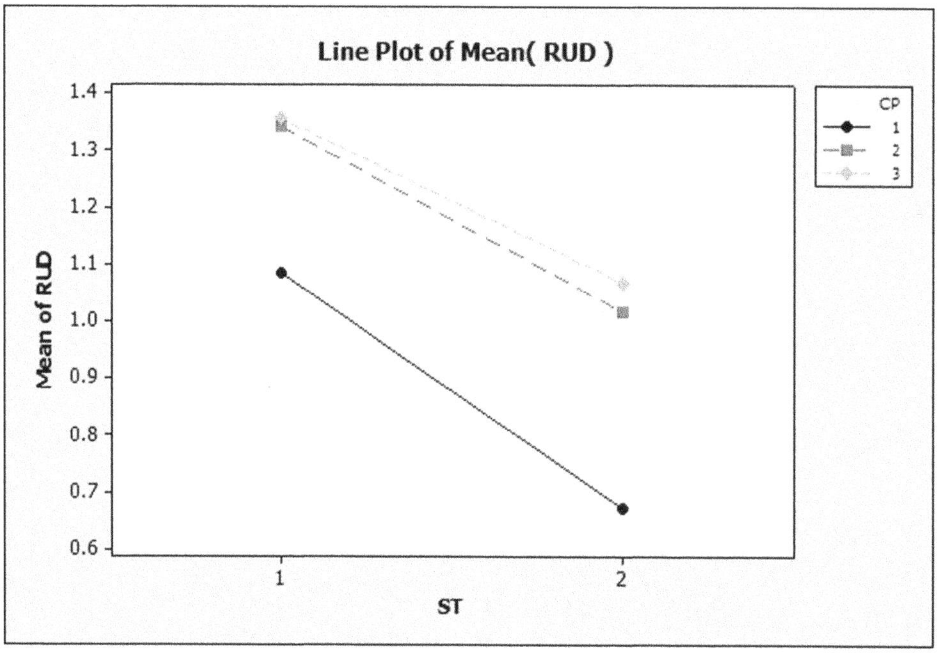

Figure 11. Optical microscope images of coated surface for (a) Expt. No.1 (CP=100MPa), (b) Expt. No.4 (CP=150MPa), (c) Expt. No.7 (CP=200MPa)

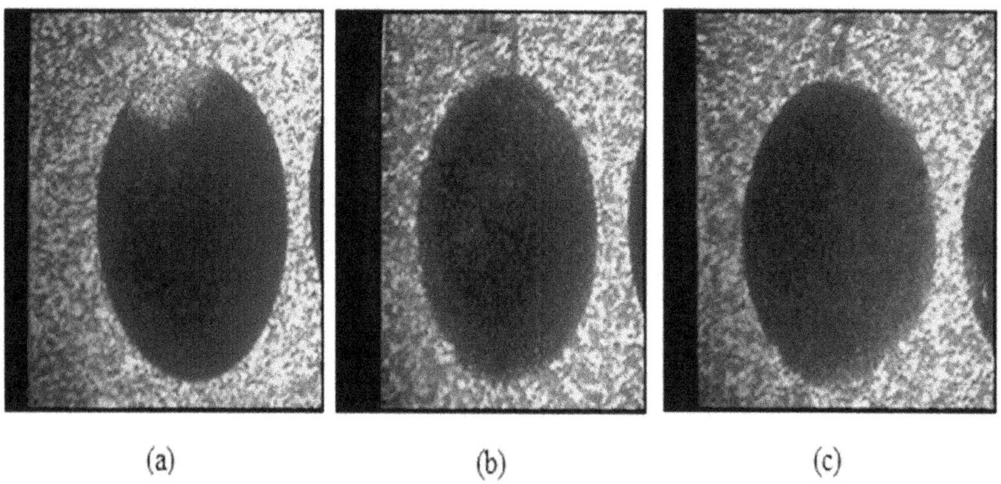

SEM and EDX Analysis

The scanning electron microscopy (SEM) images and energy dispersion X-ray spectrograph (EDX) of the coated specimen were done by using scanning electron microscope (JEOL JMS-6480LV). The SEM image of the coated surface was presented in Figure 16. From the SEM image, it was found that there were present of very small crack as well as bulk deposition and globules. The energy dispersion X-ray

Figure 12. Optical microscope images of coated surface for (a) Expt. No.2 (CP=100MPa), (b) Expt. No.5 (CP=150MPa) and (c) Expt. No.8 (CP=200MPa)

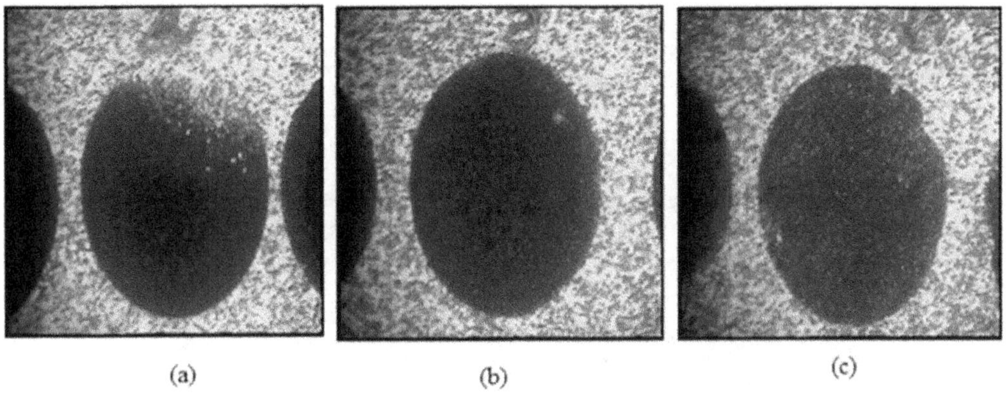

Figure 13. Optical microscope images of coated surface for (a) Expt. No.1 (ST=700^{0}C) and (b) Expt. No.10 (ST=900^{0}C)

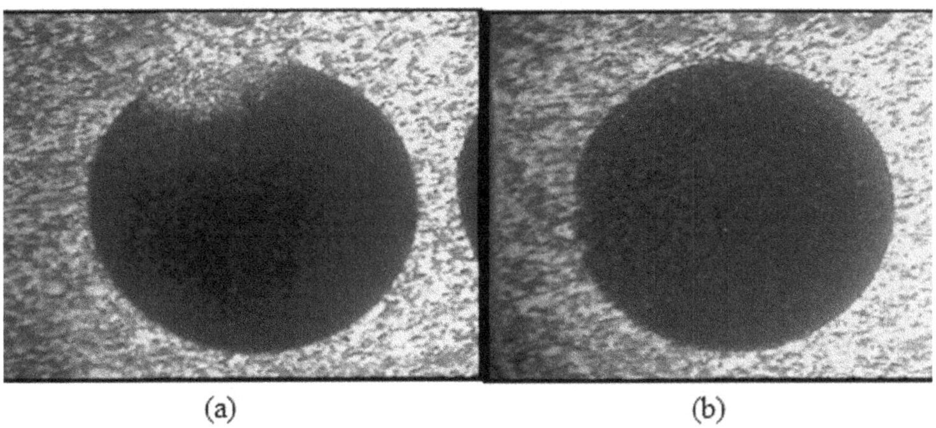

spectroscopy (EDX) of the electrical discharge coated specimen was also performed. From the EDX analysis, it was found that there is the presence of copper and tungsten element on the surface of the coated work piece (Figure 17, 18). So, the tool material was migrated from the tool electrode surface to the work piece and the coating of the specimen has occurred during the electrical discharge coating process. In this electrical discharge coating process EDM oil (a mixture of kerosene and water) was used as the dielectric fluid, so kerosene was dissociated during the electric spark and combine with tool electrode material and formed metal carbides, which deposited a hard-coated layer on the work piece surface. The hardness of the coated layer has been measured by the Vicker's micro-hardness tester (LECO LM248 AT). The micro-hardness of the coated layer has been varied between 350-450HV, which was more than the hardness of the work piece material. The maximum layer thickness of the coated layer was varied from 50-120μm. The coated layer thickness of the specimen of Expt. 6 has been presented in Figure 19.

Figure 14. Optical microscope images of coated surface for (a) Expt. No.2 (ST=700⁰C) and (b) Expt. No.11 (ST=900⁰C)

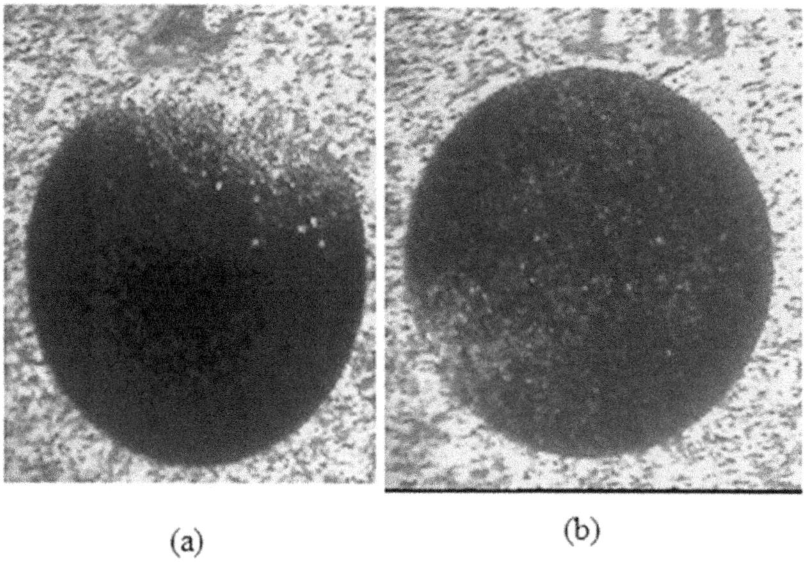

Table 9 Optimal parametric setting of the EDC process obtained by GRA based HS

Process parameters	Sintering Temperature	Compaction pressure	Discharge current	Duty cycle	Pulse-on-time	Fitness value
Optimal setting	900⁰C	100MPa	30A	42%	300μs	0.8390

FUTURE RESEARCH DIRECTIONS

The present work describes the electrical discharge coating by copper-tungsten composite electrode prepared via powder metallurgy route. The effect of the process parameters on material deposition rate (MDR), tool wear rate (TWR) and radial under deposition (RUD) were studied. The best parametric combinations have been found out by using Grey Relational Analysis (GRA) method combined with Harmony Search (HS) algorithm. However, there are enormous scope in the field of electrical discharge coating process by using different types of tool electrodes to produce hard metallic and ceramic coatings like TiC, SiC, TiN, B_4C, Cu_3Sn, CuSn, ZrB_2 etc. Different methods like spark plasma sintering and direct metal laser sintering can be used to prepare the tool electrode, which enhance the properties of tool electrode. The output responses like coated layer thickness, micro-hardness and material characteristics of the coated surface can be analyze. Different types of non-traditional multi-response optimization techniques can be used to get the optimal parametric setting. By finding best optimal setting of the EDC process, the best performance of the EDC process can be achieved, that leads to growth of the manufacturing industries.

Figure 15. Graph of fitness value vs. number of iterations

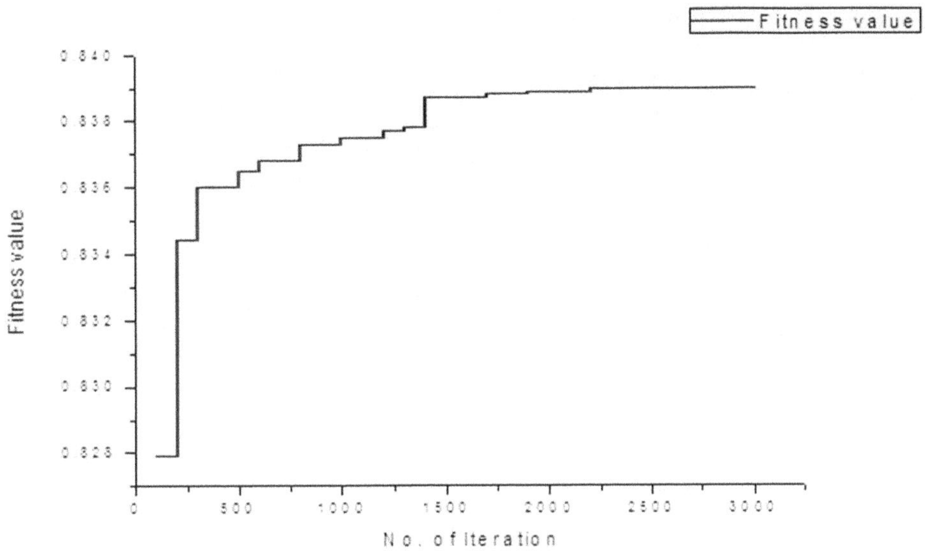

Figure 16.SEM of the coated specimen for Expt. No. 9

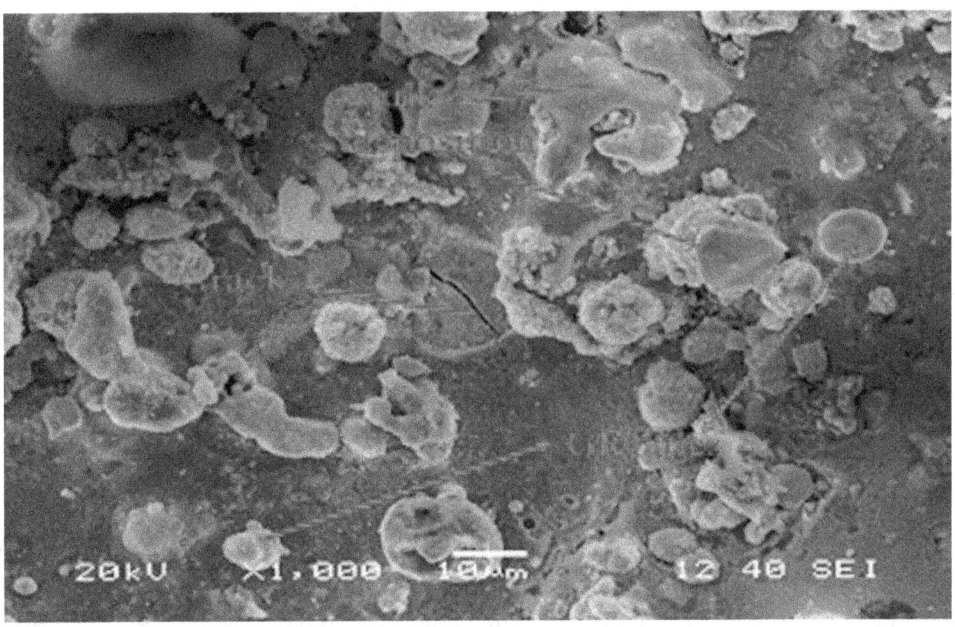

Figure 17.EDX analysis of the coated specimen for Expt. No. 9

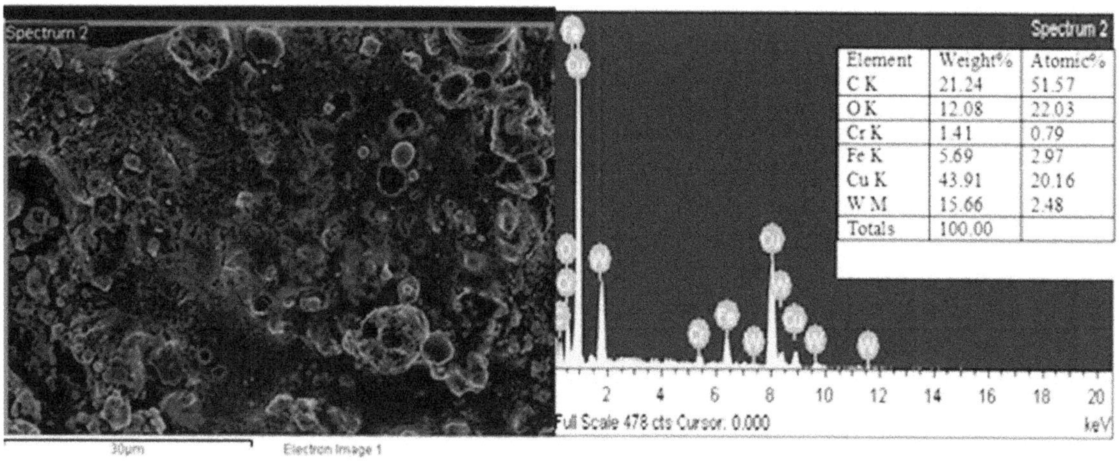

Figure 18.EDX analysis of the coated specimen for Expt. No. 14

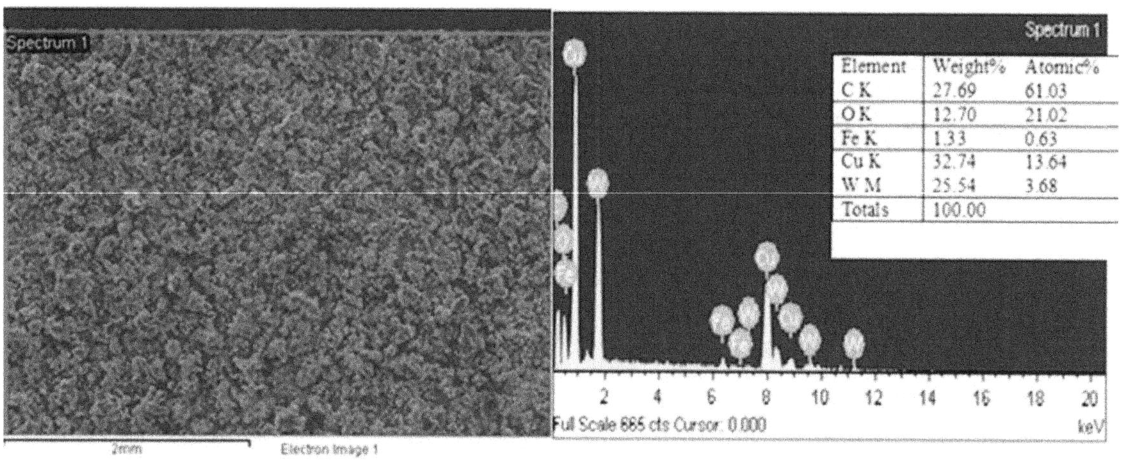

CONCLUSION

The electrical discharge coating (EDC) of AISI 1040 stainless steel was successfully performed by copper-tungsten composite electrodes prepared via the powder metallurgy (PM) route. The EDC process was optimized by grey relational analysis (GRA) based harmony search (HS) algorithm to obtain best parametric setting. By utilizing the above optimization technique, the optimum parametric setting for the satisfying of material deposition rate (MDR), tool wear rate (TWR) and radial under deposition (RUD) has been obtained as $A_2B_1C_3D_1E_3$ that are ST = 900^{0}C, CP = 100MPa, I_p = 30A, DC = 42%, T_{on} = 300μs. Compaction pressure and sintering temperature have the highest contribution toward the output responses with the percentage contribution of 34.05% and 25.11% respectively followed by pulse-on-time, interaction of compaction pressure and discharge current, pulse-on-time, duty cycle with the percentage contribution of 12.15%, 9.98%, 7.62% and 4.92% respectively. With an increase in sintering

Figure 19. Maximum coated layer thickness of specimen of Expt. 6

temperature, discharge current and pulse-on-time the EDC performance increased and with the increase in compaction pressure and duty cycle the EDC performance has been decreased. With the increase in compaction pressure and sintering temperature, material deposition rate and tool wear rate decreased. So, to get maximum material deposition rate these parameters must be decreased to increase tool wear rate. Here, tool wear rate has been directly influence the material deposition rate. Similarly, radial under deposition has been decreased with increase in sintering temperature and decrease in compaction pressure. Therefore, to get best performance of EDC process with respect to all three responses, the optimum parametric setting must be used, which leads to growth of the manufacturing industries. By utilizing the EDC process, the cost of coating process will decease as compare to other coating process like PVD, CVD etc. we are using now.

REFERENCES

Abhishek, K., Datta, S., & Mahapatra, S. S. (2016). Multi-objective optimization in drilling of CFRP (polyester) composites: Application of a fuzzy embedded harmony search (HS) algorithm. *Measurement*, *77*, 222–239. doi:10.1016/j.measurement.2015.09.015

Ahmed, A. (2016). Deposition and Analysis of Composite Coating on Aluminum Using Ti – B_4C Powder Metallurgy Tools in EDM. *Materials and Manufacturing Processes*, *31*(4), 467–474. doi:10.1080/10426914.2015.1025967

Beri, N., Maheshwari, S., Sharma, C., & Kumar, A. (2008). Performance Evaluation of Powder Metallurgy Electrode in Electrical Discharge Machining of AISI D2 Steel Using Taguchi Method, *International Journal of Mechanical, Aerospace, Industrial, Mechatronic and Manufacturing Engineering*, 2(2), 225–229.

Beri, N., Maheshwari, S., Sharma, C., & Kumar, A. (2014). Surface quality modification using powder metallurgy processed CuW electrode during electric discharge machining of Inconel 718. *Procedia Materials Science*, *5*, 2629–2634. doi:10.1016/j.mspro.2014.07.524

Bhanot, V. K., Beri, N., & Kumar, A. (2014). Machinability assessment of Superni-800 during EDM with powder metallurgy processed Cu-Ti electrode using the Taguchi method.

Choudhary, R., Kumar, P., & Singh, J. (2012). Analysis of Electro Discharged Machined Surfaces of AISI D3 Steel by Composite Tool Electrode. *International Journal of Surface Engineering & Materials Technology*, *2*(2), 3–8.

Das, A., & Mishra, J. P. (2017). Experimental investigation on surface modification of aluminum by electric discharge coating process using TiC/Cu green compact tool-electrode. *Mining Science and Technology*, *16*, 601–623.

Datta, S., & Mahapatra, S. S. (2010). Modeling, simulation and parametric optimization of wire EDM process using response surface methodology coupled with grey-Taguchi technique. *International Journal of Engineering Science and Technology*, *2*(5), 162–183. doi:10.4314/ijest.v2i5.60144

Datta, S., Biswal, B.B., & Mahapatra, S.S. (2017). A Novel Satisfaction Function and Distance-Based Approach for Machining Performance Optimization During Electro-Discharge Machining on Super Alloy Inconel 718. *Arabian Journal for Science and Engineering, 42*, 1999–2020. doi:10.1007/s13369-017-2422-5

Gangadhar, A., Shunmugam, M. S., & Philip, P. K. (1991). Surface modification in electrodischarge processing with a powder compact tool electrode. *Wear*, *143*(1), 45–55. doi:10.1016/0043-1648(91)90084-8

Ho, S. K., Aspinwall, D. K., Voice, W., Box, P. O., & De, D. (2007). Use of powder metallurgy (PM) compacted electrodes for electrical discharge surface alloying / modification of Ti – 6Al – 4V alloy. *Journal of Materials Processing Technology*, *191*(1-3), 123–126. doi:10.1016/j.jmatprotec.2007.03.003

Ju-long, D. (1982). Control problems of grey systems. *Systems & Control Letters*, *1*(5), 288–294. doi:10.1016/S0167-6911(82)80025-X

Julong, D. (1988). Introduction to Grey System Theory. *Journal of Grey System*, *1*, 1–24.

Khanra, A. K., Sarkar, B., Bhattachary, B., Pathak, L. C., & Godkhindi, M. M. (2007). Performance of ZrB2–Cu composite as an EDM electrode. *Journal of Materials Processing Technology*, *183*(1), 122–126. doi:10.1016/j.jmatprotec.2006.09.034

Li, L., Niu, Z. W., & Zheng, G. M. (2016). Ultrasonic Electrodeposition of Cu – SiC Electrodes for EDM. *Materials and Manufacturing Processes*, *31*(1), 37–41. doi:10.1080/10426914.2015.1025968

Li, L., Wong, Y. S., Fuh, J. Y. H., & Lu, L. (2001a). Effect of TiC in copper-tungsten electrodes on EDM performance. *Journal of Materials Processing Technology*, *113*(1-3), 563–567. doi:10.1016/S0924-0136(01)00622-7

Li, L., Wong, Y. S. U., Fuh, J. Y. H., & Lu, L. (2001b). EDM performance of TiC/copper-based sintered electrodes. *Materials & Design*, *22*(8), 669–678. doi:10.1016/S0261-3069(01)00010-3

Mahdavi, M., Fesanghary, M., & Damangir, E. (2007). An improved harmony search algorithm for solving optimization problems. *Applied Mathematics and Computation, 188*(2), 1567–1579. doi:10.1016/j.amc.2006.11.033

Mishra, P. K. (2012). *Nonconventional Machining*. New Delhi: Narosa Publishing House.

Mohanty, C. P., Mahapara, S. S., & Singh, M. R. (2017). An intelligent approach to optimize the EDM process parameters using utility concept and QPSO algorithm. *Engineering Science and Technology, an International Journal, 20*, 552–562.

Nayak, B. B., Mahapatra, S. S., Chatterjee, S., & Abhishek, K. (2015). Parametric Appraisal of WEDM using Harmony Search Algorithm. *Materials Today: Proceedings, 2*(4-5), 2562–2568. doi:10.1016/j.matpr.2015.07.207

Patowari, P. K., Saha, P., & Mishra, P. K. (2010). Artificial neural network model in surface modification by EDM using tungsten – copper powder metallurgy sintered electrodes. *International Journal of Advanced Manufacturing Technology, 51*(5-8), 627–638. doi:10.1007/s00170-010-2653-z

Patowari, P. K., Saha, P., & Mishra, P. K. (2011). Taguchi analysis of surface modification technique using W-Cu powder metallurgy sintered tools in EDM and characterization of the deposited layer. *International Journal of Advanced Manufacturing Technology, 54*(5-8), 593–604. doi:10.1007/s00170-010-2966-y

Patowari, P. K., Saha, P., & Mishra, P. K. (2015). An experimental investigation of surface modification of C-40 steel using W – Cu powder metallurgy sintered compact tools in EDM. *International Journal of Advanced Manufacturing Technology, 80*(1-4), 343–360. doi:10.1007/s00170-015-7004-7

Sahu, A. K., Mohanty, P. P., & Sahoo, S. K. (2017). Electro discharge machining of Ti-alloy (Ti6Al4V) and 316L Stainless Steel and Optimization of process parameters by Grey relational analysis(GRA) method. In D.I. Wimpenny et al. (Eds.), Advances in 3D Printing & Additive Manufacturing Technologies (pp. 65–78). Singapore: Springer Science+Business Media.

Samuel, M. P., & Philip, P. K. (2007). Power metallurgy tool electrodes for electrical discharge machining. *International Journal of Machine Tools & Manufacture, 37*(11), 1625–1633. doi:10.1016/S0890-6955(97)00006-0

Simao, J., Aspinwall, D., El-Menshawy, F., & Meadows, K. (2002). Surface alloying using PM composite electrode materials when electrical discharge texturing hardened AISI D2. *Journal of Materials Processing Technology, 127*(2), 211–216. doi:10.1016/S0924-0136(02)00144-9

Tsai, H. C., Yan, B. H., & Huang, F. Y. (2003). EDM performance of Cr / Cu-based composite electrodes. *International Journal of Machine Tools & Manufacture, 43*(3), 245–252. doi:10.1016/S0890-6955(02)00238-9

Ueno, M., Fujita, N., Kimura, Y., & Nakata, N. (2016). Journal of Materials Processing Technology Evaluation of coating and wear characteristics of roll surface coated with TiC by electrical discharge coating. *Journal of Materials Processing Technology, 236*, 9–15. doi:10.1016/j.jmatprotec.2016.04.025

Yang, X. (2009). Harmony Search as a Metaheuristic Algorithm. In Z.W. Geem (Eds.), Music-Inspired Harmony Search Algorithm: Theory and Applications, Studies in Computational Intelligence. Springer Berlin. doi:10.1007/978-3-642-00185-7_1

Zaw, H. M., Fuh, J. Y. H., Nee, A. Y. C., & Lu, L. (1999). Formation of a new EDM electrode material using sintering techniques. *Journal of Materials Processing Technology*, *89*(90), 182–186. doi:10.1016/S0924-0136(99)00054-0

KEY TERMS AND DEFINITIONS

Electrical Discharge Coating (EDC): Electrical discharge coating is the process of coating by applying the principle of electrical discharge machining.

Grey Relational Analysis (GRA): The Grey relational analysis is a multi-response optimization technique used for the conversion of multi-responses into single response and to get the best parametric setting to get best machining performance.

Harmony Search (HS) Algorithm: The Harmony Search (HS) algorithm is a music-based metaheuristic optimization technique, which is inspired by jazz music.

Material Deposition Rate (MDR): Material deposition rate is the rate at which material deposition occurred on the surface of the work piece.

Radial Under Deposition (RUD): Radial under deposition is the radial un-deposited space present of the work piece surface after EDC process.

Tool Wear Rate (TWR): Tool wear rate is the rate at which material loss occurred from the tool electrode.

Chapter 11
Novel Approaches to Prediction of a Future Number of Failures Based on Previous In-Service Inspections

Nicholas Nechval
University of Latvia, Latvia

Konstantin Nechval
Transport and Telecommunication Institute, Latvia

ABSTRACT

In this chapter, we present novel approaches to predictions of the number of failures that will be observed in a future inspection of a sample of units, based only on the results of the previous in-service inspections of the same sample. The failure-time of such units is modeled with a distribution from a two-parameter Weibull distribution. The different cases of parametric uncertainty are considered. The pivotal quantity averaging approach proposed here for constructing point prediction and simple prediction limits emphasizes pivotal quantities relevant for eliminating unknown parameters from the problems and represents a special case of the method of invariant embedding of sample statistics into a performance index applicable whenever the statistical problem is invariant under a group of transformations, which acts transitively on the parameter space. For illustration, a numerical example is given.

INTRODUCTION

This chapter extends the results of Nelson (2000) through the use of novel approaches to prediction of a future number of failures based on previous in-service inspections. Nelson's prediction limits were motivated by the following application.

Nuclear power plants contain large heat exchangers that transfer energy from the reactor to steam turbines. Such exchangers typically have 10,000 to 20,000 stainless steel tubes that conduct the flow of steam. Due to stress and corrosion, the tubes develop cracks over time. Cracks are detected during

DOI: 10.4018/978-1-5225-3035-0.ch011

planned inspections. The cracked tubes are subsequently plugged to remove them from service. To develop efficient inspection and plugging strategies, plant management can use a prediction of the added number of tubes that will need plugging by a specified future time.

Nelson (2000) presents the procedures to obtain predictions for the number of failures that will be observed in a future inspection of a sample of units, based only on the results of the first in-service inspection of the same sample. The failure-time of such units is modeled with a two-parameter Weibull distribution indexed by scale and shape parameters β and δ, respectively. The case is considered when the scale parameter β is unknown, but the shape parameter δ is known.

Prediction of a Future Number of Failures via Nelson's Procedures

Nelson (2000) presents point prediction and simple prediction limits for the number of failures that will be observed in a future inspection of a sample of units. The past data consist of the cumulative number of failures in a previous inspection of the same sample of units. Life X of such units is modeled by a Weibull distribution with the probability density function

$$f_\theta(x) = \frac{\delta}{\beta}\left(\frac{x}{\beta}\right)^{\delta-1} \exp\left[-\left(\frac{x}{\beta}\right)^{\delta}\right], \quad x > 0 \tag{1}$$

and cumulative distribution function

$$F_\theta(x) = 1 - \exp\left[-\left(\frac{x}{\beta}\right)^{\delta}\right], \quad x > 0 \tag{2}$$

where $\theta = (\beta,\delta)$, $\beta > 0$ and $\delta > 0$ are the scale and shape parameters, respectively.

To illustrate the procedures of constructing point prediction and simple prediction limits for the number of failures that will be observed in a future inspection of a sample of units, based only on the results of the first in-service inspection of the same sample, Nelson (2000) used the following example.

Steam Generator Example

For example, a steam generator with $n = 20000$ tubes, of which $k_\bullet = 8$ have failed by the inspection at age t_1=3.0 years, is considered. It is assumed that a Weibull shape parameter δ=3.3. It is necessary to find a point prediction as well as prediction interval for the future number of tubes that will need plugging by a future inspection at age t_2 = 10.0 years. Table 1 illustrates a situation of within-sample prediction based on the cumulative number of failures ($k_\bullet$ = 8).

It is assumed that the failure times are independent observations from a Weibull distribution with unknown scale parameter β and given (known) shape parameterδ (=3.3). Thus,

$$p = 1 - \exp\left[-\left(t_1 / \beta\right)^{\delta}\right] \tag{3}$$

Table 1. The situation of within-sample prediction of the number of future cracked heat exchanger tubes via the cumulative number of failures ($k_{\bullet} = 8$) and procedure of maximum likelihood

What Is Known?
Units at start: n=20000
First inspection was at the age $t_1 = 3.0$ years
Cumulative number of failures in the interval $[0, t_1]$ was $k_{\bullet} = 8$
Future inspection will be at the age $t_2 = 10.0$ years
What Should Be Found?
Point prediction for the number $l_{\bullet}$ of future failures in the interval $(t_1, t_2]$
Prediction interval $[\underset{\sim}{l}_{\bullet}, \tilde{l}_{\bullet}]$ for the number $l_{\bullet}$ of future failures in the interval $(t_1, t_2]$

is the probability that a tube will fail in the interval $[0, t_1]$,

$$q = \exp\left[-(t_1 / \beta)^{\delta}\right] - \exp\left[-(t_2 / \beta)^{\delta}\right] \tag{4}$$

is the probability that a tube will fail in the interval $(t_1, t_2]$,

$$r = 1 - p - q = \exp\left[-(t_2 / \beta)^{\delta}\right] \tag{5}$$

is the probability that a tube will fail in the interval (t_2, ∞).

Point Prediction

Based on the observed number $k_{\bullet}$ of failures, the maximum likelihood (ML) estimate of β is given by

$$\beta^{\mathrm{ML}} = \arg\max_{\beta}\left[\ln\left(\binom{n}{k} p^{k}(1-p)^{n-k}\right)\right] = \frac{t_1}{\left[-\ln\left(1-\dfrac{k}{n}\right)\right]^{1/\delta}} = \frac{3}{\left[\left[-\ln\left(1-\dfrac{8}{20000}\right)\right]\right]^{1/3.3}} = 32.11986564 \tag{6}$$

The estimate q^{ML} of q is obtained by substituting β^{ML} into (4) for q, giving (for the steam generator)

$$q^{\mathrm{ML}} = \exp\left[-\left(t_1 / \beta^{\mathrm{ML}}\right)^{\delta}\right] - \exp\left[-\left(t_2 / \beta^{\mathrm{ML}}\right)^{\delta}\right] = 0.0206396 \tag{7}$$

The corresponding ML point prediction for the number of future cracked heat exchanger tubes is

$$l_{\bullet}^{\mathrm{ML}} = nq^{\mathrm{ML}} = 412.79199 \cong 413 \text{ (cracked tubes)} \tag{8}$$

Prediction Intervals

Prediction of an unobserved random variable is a fundamental problem in statistics. Hahn and Nelson (1973), Patel (1989), and Hahn and Meeker (1991) provided surveys of methods for statistical prediction for a variety of situations on this topic. In the areas of reliability and life-testing, this problem translates to obtaining prediction intervals for lifetime distributions. Nordman and Meeker (2002) compared probability ratio (PR), simplified probability ratio (SPR) and likelihood ratio (LR) procedures proposed by Nelson (2000), assuming known the Weibull shape parameterδ. An approximate procedure based on a likelihood ratio (LR) statistic and suggested by Nelson (2000) is appropriate for more general values of p and q. This procedure can be viewed as a special case of more general likelihood-based prediction intervals described by Lejeune and Faulkenberry (1982) and Bjornstad (1990). A complete review on a number of different versions of predictive likelihood is given by Bjornstad (1990). In (Bjornstad 1996), partial concepts of likelihood in case of nuisance parameters and factorized likelihood are introduced. The general situation, which covers a wide variety of statistical problems like parametric inference and prediction in parametric, Bayes, empirical Bayes, and latent models, as well as missing-data problems, is discussed. In (Geweke and Amisano 2012), prediction and misspecified models are considered. Marginal or conditional composite likelihoods are reviewed by Varin et al. (2011). In (Varin and Vidoni 2006, 2009), using the notion of weighted composite likelihood, and in particular the notion of weighted pairwise likelihood, a useful surrogate for the true unknown predictive distribution function is introduced. This distribution can be considered for specifying predictors and prediction intervals. For applications involving failure times with censored data, methods presented by Lawless (1973), Nelson (1982), Mee and Kushary (1994), and Escobar and Meeker (1999) are useful. These prediction methods cannot be used for the limited amount of inspection data considered here.

Table 2 provides equal-tail 90% prediction intervals for the number of cracked tubes in the span of 3 to 10 years, based on the three procedures described in (Nordman and Meeker 2002). Two additional values of the Weibull shape parameter are used to evaluate the effect of misspecification of δ on the prediction limits. The PR and simplified PR (SPR) intervals are wider than the LR intervals.

It should be noted that Nelson (2000) had a very limited amount of inspection data (cumulative number of failures $k_{\bullet}$=8 in the interval [0,t_1=3.0 years]) and used hard computing, i.e., conventional computing, which requires a precisely stated analytic model to make statistical conclusions concerning the within-sample prediction of the number of future failures.

In this chapter, the above-mentioned amount of inspection data, where the ordered observations of failures$\left(X_1 \leq \ldots \leq X_k\right)$in the interval $[0, t_1]$ are unknown, is transformed into the kth order statistic $X_{k=k_{\bullet}}$ and it is assumed that $X_k = t_1$. However, if $X_k < t_1$,then our assumption $\left(X_k = t_1\right)$ purposely includes some imprecision into a computing process for the within-sample prediction of the future number of failures. Taking into account the definition of soft computing presented in (Li et al. 1998). "Every computing process that purposely includes imprecision into the calculation on one or more

Table 2. 90% prediction intervals $[\underset{\sim}{l}_\bullet, \tilde{l}_\bullet]$ for the number $l_\bullet$ of future cracked heat exchanger tubes

Procedure	Shape Parameter, δ		
	3.0	3.3	3.6
PR $[\underset{\sim}{l}_\bullet, \tilde{l}_\bullet]$	[140, 524]	[205, 756]	[297, 1090]
SPR $[\underset{\sim}{l}_\bullet, \tilde{l}_\bullet]$	[142, 521]	[206, 753]	[298, 1087]
LR $[\underset{\sim}{l}_\bullet, \tilde{l}_\bullet]$	[148, 487]	[216, 700]	[311, 1001]

levels and allows this imprecision either to change (decrease) the granularity of the problem, or to 'soften' the goal of optimization at some stage, is defined as to belonging to the field of soft computing", which is perhaps the most suitable among the possible alternative definitions, we can conclude that the computing process for the within - sample prediction of the future number of failures, proposed in this chapter, can be defined as to belonging to the field of soft computing. Undoubtedly, there exist situations where the calculation results obtained through a soft computing process may be more suitable for decision making than the calculation results obtained through a conventional computing process. The situations, which are discussed in this chapter, belong to such ones.

NOVEL APPROACHES TO PREDICTION OF A FUTURE NUMBER OF FAILURES

In this chapter, we use frequentist approaches to within-sample prediction via previous and future order statistics when the time-to-failure follows the two-parameter Weibull distribution indexed by scale and shape parameters β and δ, respectively. We consider the case when the parameter β is unknown, but the shape parameter δ is known.

The pivotal quantity averaging approach (PQAA) proposed here for constructing point prediction and simple prediction limits emphasizes pivotal quantities relevant for eliminating unknown parameters from the problems and represents a special case of the method of invariant embedding of sample statistics into a performance index applicable whenever the statistical problem is invariant under a group of transformations, which acts transitively on the parameter space (Nechval et al., 2003, 2010, 2012, 2016).

Prediction of a Future Number of Failures via Order Statistic X_k and PQAA

It is assumed that the order statistic $X_k = t_1$, where $k = 8$. Table 3 illustrates a situation of within-sample prediction based on the order statistic X_k and pivotal quantity averaging approach (PQAA).

Point Prediction

The probability density function of the kth order statistic is given by

Table 3. The situation of within-sample prediction of the number $l_\bullet$ of future failures via the order statistic X_k and PQAA

What Is Known?
Units at start: n=20000
First inspection was at the age $t_1 = 3.0$ years
Cumulative number of failures in the interval $[0, t_1]$ was $k_\bullet = 8$
Future inspection will be at the age $t_2 = 10.0$ years
What Is Unknown?
The ordered observations of failures $(X_1 \leq \ldots \leq X_k)$ in the interval $[0, t_1]$ are unknown
What Is Assumed?
It is assumed that the kth order statistic $X_k = t_1$, where k=$k_\bullet$
What Should Be Found?
Point prediction for the number $l_\bullet$ of future failures in the interval $(t_1, t_2]$
Prediction interval $[\underset{\sim}{l}_\bullet, \tilde{l}_\bullet]$ for the number $l_\bullet$ of future failures in the interval $[t_1, t_2]$

$$g_\theta(x_k) = \frac{1}{\mathrm{B}\left(k, n-k+1\right)}\left[F_\theta\left(x_k\right)\right]^{k-1}\left[1 - F_\theta\left(x_k\right)\right]^{n-k} f_\theta(x_k), \quad x_k > 0 \tag{9}$$

It can be shown that

$$g_\theta(x_k)dx_k = \frac{1}{\mathrm{B}(k, n-k+1)}\left[F_\theta(x_k)\right]^{k-1}\left[1 - F_\theta(x_k)\right]^{n-k} dF_\theta(x_k)$$

$$= \frac{1}{\mathrm{B}(k, n-k+1)} v^{k-1}\left(1-v\right)^{n-k} dv = g_{k,n}(v)dv, \quad v \in (0,1), \tag{10}$$

where

$$V = F_\theta(X_k) \tag{11}$$

is a pivotal quantity or simply pivot. Then it follows from (4) that

$$q = \exp\left[-\left(t_1/\beta\right)^{\delta}\right] - \exp\left[-\left(t_2/\beta\right)^{\delta}\right] = \left[1-v\right]^{\left(t_1/x_k\right)^{\delta}} - \left[1-v\right]^{\left(t_2/x_k\right)^{\delta}} = q\left(v, x_k\right) \tag{12}$$

Using the pivotal quantity averaging approach (Nechval *et al.* 2003, 2010, 2012, 2016), we have that

$$q(x_k) = E\left\{q\left(V, x_k\right)\right\} = \int_0^1 q(v, x_k) g_{k,n}(v) dv = 1 - F_{(x_k,\delta)}(t_1) - \left[1 - F_{(x_k,\delta)}(t_2)\right]$$

$$= \frac{\mathrm{B}(k, n-k+1+(t_1/x_k)^{\delta}}{\mathrm{B}(k, n-k+1)} - \frac{\mathrm{B}(k, n-k+1+(t_2/x_k)^{\delta}}{\mathrm{B}(k, n-k+1)} = 0.9996 - 0.978989 = 0.020611. \tag{13}$$

Thus, the corresponding point prediction for the number of future cracked heat exchanger tubes based on (X_k & PQAA) is

$$l_{\bullet}^{X_k \& \text{ PQAA}} = nq(x_k) = 412.2296 \cong 412 \text{ (cracked tubes)} \tag{14}$$

Statistical Inference

The corresponding point prediction for the number of future cracked heat exchanger tubes obtained via the ML procedure of Nelson (2000) is $l_{\bullet}^{\text{ML (Nelson)}} = 413$ cracked tubes. The corresponding point prediction for the number of future cracked heat exchanger tubes obtained via the proposed in this paper procedure based on the order statistic X_k and PQAA is $l_{\bullet}^{X_k \& \text{ PQAA}} = 412$ cracked tubes. Two additional values of the Weibull shape parameter are used to evaluate the effect of misspecification of δ on the point prediction (Table 4).

Thus, in the same situation, these procedures provided practically the same results of point prediction.

Table 4. Point prediction for the number $l_{\bullet}$ of future cracked heat exchanger tubes

Procedure of Point Prediction for $l_{\bullet}$	Shape Parameter, δ		
	3.0	3.3	3.6
ML (Nelson) $l_{\bullet}^{\text{ML}}$	[286]	[413]	[593]
(X_k & PQAA) $l_{\bullet}^{X_k \& \text{ PQAA}}$	[286]	[412]	[592]

Prediction Intervals

In order to find an $(1-\alpha)$ prediction interval $(\underset{\sim}{l}_{\bullet}, \tilde{l}_{\bullet})$ for the number $l_{\bullet}$ of future cracked tubes in the span of 3 to 10 years, we construct, at first, an equal-tail $(1-\alpha)$ confidence interval for the unknown parameter β, which is based on X_k. The obvious choice for a pivot V is given by

$$V = 1 - \exp\left[-\left(X_k / \beta\right)^{\delta}\right] \tag{15}$$

which has the distribution (10). Consider a probability

$$P\left(1 - \exp[-a] < 1 - \exp\left[-\left(X_k / \beta\right)^{\delta}\right] < 1 - \exp\left[-b\right]\right)$$

$$= P\left(1 - \exp[-a] < V < 1 - \exp[-b]\right) \tag{16}$$

and choose a and b so that

$$P\left(1 - \exp[-a] < V < 1 - \exp[-b]\right) = \int_{1-\exp[-a]}^{1-\exp[-b]} g_{k,n}(v)dv = 1 - \alpha. \tag{17}$$

Since it follows from (17) that

$$P\left\{X_k^{\delta} / b < \beta^{\delta} < X_k^{\delta} / a\right\} = 1 - \alpha, \tag{18}$$

the equal-tail $(1-\alpha)$ confidence interval for the unknown scale parameter β, which is based on X_k, is given by

$$\left[\underset{\sim}{\beta}, \tilde{\beta}\right] = \left[\left(X_k^{\delta}/b\right)^{1/\delta}, \left(X_k^{\delta}/a\right)^{1/\delta}\right] = \left[\left(X_k/b\right)^{1/\delta}, \left(X_k/a\right)^{1/\delta}\right] \tag{19}$$

Taking into account (4) and (19), we have that

$$q\left(\underset{\sim}{\beta}\right) = \exp\left[-\left(t_1/\underset{\sim}{\beta}\right)^{\delta}\right] - \exp\left[-\left(t_2/\underset{\sim}{\beta}\right)^{\delta}\right] \tag{20}$$

$$q\left(\tilde{\beta}\right) = \exp\left[-\left(t_1/\tilde{\beta}\right)^{\delta}\right] - \exp\left[-\left(t_2/\tilde{\beta}\right)^{\delta}\right] \tag{21}$$

Thus, in this case, the $(1-\alpha)$ prediction interval for the number of future cracked heat exchanger tubes, based on the equal-tail confidence interval of level $(1-\alpha)$ for the unknown scale parameter β, is given by

$$\left[\underset{\bullet}{l},\tilde{l}\right]=\left[nq\left(\tilde{\beta}\right),nq\left(\underset{\sim}{\beta}\right)\right] \tag{22}$$

If $1-\alpha=0.9$, it follows from (17) that

$$a=0.0002,\quad b=0.00066 \tag{23}$$

it follows from (19) and (23) that

$$\left[\underset{\sim}{\beta},\tilde{\beta}\right]=\left[\frac{3}{0.00066^{1/3.3}},\frac{3}{0.0002^{1/3.3}}\right]=\left[27.59916,\quad 39.62974\right]; \tag{24}$$

it follows from (20), (21) and (24) that

$$q\left(\underset{\sim}{\beta}\right)=0.99934-0.96553=0.03381 \tag{25}$$

$$q\left(\tilde{\beta}\right)=0.99980-0.98943=0.01037 \tag{26}$$

it follows from (22), (25) and (26) that

$$\left[\underset{\bullet}{l},\tilde{l}\right]=\left[207.473,\quad 676.216\right]\cong\left[207,\quad 676\right] \tag{27}$$

Statistical Inference

Table 5 provides 90% prediction intervals for the number of cracked tubes in the span of 3 to 10 years, based on the three procedures described in (Nordman and Meeker 2002) and (X_k & PQAA) procedure described above. Two additional values of the Weibull shape parameter are used to evaluate the effect of misspecification of δ on the prediction limits. The PR, simplified PR (SPR) and LR intervals are wider than the (X_k & PQAA) intervals.

Table 5. 90% prediction intervals $[\underset{\sim}{l}_\bullet, \tilde{l}_\bullet]$ for the number $l_\bullet$ of future cracked heat exchanger tubes

Shape Parameter, δ	Prediction Procedure of Interval $[\underset{\sim}{l}_\bullet, \tilde{l}_\bullet]$			
	PR $[\underset{\sim}{l}_\bullet, \tilde{l}_\bullet]$	SPR $[\underset{\sim}{l}_\bullet, \tilde{l}_\bullet]$	LR $[\underset{\sim}{l}_\bullet, \tilde{l}_\bullet]$	(X_k & PQAA) $[\underset{\sim}{l}_\bullet, \tilde{l}_\bullet]$
$\delta = 3.0$	[140, 524]	[142, 521]	[148, 487]	[144, 470]
Width of prediction interval	384	379	339	326
$\delta = 3.3$	[205, 756]	[206, 753]	[216, 700]	[207, 676]
Width of prediction interval	551	547	484	469
$\delta = 3.6$	[297, 1090]	[298, 1087]	[311, 1001]	[299, 969]
Width of prediction interval	793	789	690	670

Prediction of a Future Number of Failures via Order Statistics ($X_k \leq X_l$) and PQAA

Some Results of Constructing Prediction Limits on Future Order Statistics

Theorem 1. Let $X_1 \leq ... \leq X_n$ be the n ordered observations in a sample (current or future) of size n from a probability distribution (continuous or discrete) with density function $f_\theta(x)$, distribution function $F_\theta(x)$, where θ is a known parameter (in general, vector). Then a lower $(1-\alpha)$ prediction limit $\underset{\sim}{h}$ on the lth future order statistic X_l, $l \in \{1, \ldots, n\}$, in a set of n ordered observations $X_1 \leq \ldots \leq X_n$ is given by

$$\underset{\sim}{h} = \arg\left[P_\theta\left\{X_l > \underset{\sim}{h} \mid n\right\} = 1-\alpha\right] = \arg\left[P_\theta\left\{X_l \leq \underset{\sim}{h} \mid n\right\} = \alpha\right]$$

$$= \arg\left[F_\theta\left(\underset{\sim}{h}\right) = \frac{l}{l + (n-l+1)q_{2(n-l+1),2l;1-\alpha}}\right] \tag{28}$$

where $q_{2(n-l+1),2l;1-\alpha}$ is the quantile of order $1-\alpha$ for the F distribution with $2(n-l+1)$ and $2l$ degrees of freedom.

(Observe that an upper $1-\alpha$ prediction limit $\tilde{h}$ on the lth future order statistic X_l from a set of n ordered observations $X_1 \leq \ldots \leq X_n$ may be obtained from a lower $(1-\alpha)$ prediction limit by replacing $1-\alpha$ by α.)

Proof. Suppose an event occurs with probability p per trial. It is well-known that the probability P of its occurring l or more times in n trials is termed a cumulative binomial probability, and is related to the incomplete beta function $I_x(a, b)$ as follows:

$$P = \sum_{j=l}^{n} \binom{n}{j} p^j (1-p)^{n-j} = I_p\left(l, n-l+1\right) \tag{29}$$

It follows from (29) that

$$P_\theta\left\{X_l \le h \mid n\right\} = \sum_{j=l}^{n} \binom{n}{j} \left[F_\theta(h)\right]^j \left[1 - F_\theta(h)\right]^{n-j}$$

$$= I_{F_\theta(h)}(l, n-l+1) = \frac{1}{\mathrm{B}(l, n-l+1)} \int_0^{F_\theta(h)} u^{l-1}(1-u)^{(n-l+1)-1} du$$

$$= \frac{\left(\frac{2(n-l+1)}{2l}\right)^{2(n-l+1)/2}}{\mathrm{B}\left(\frac{2l}{2}, \frac{2(n-l+1)}{2}\right)} \int_0^{F_\theta(h)} u^{\frac{2(n-l+1)+2l}{2}} \left(\frac{1-u}{u} \frac{2l}{2(n-l+1)}\right)^{2(n-l+1)/2-1} \frac{-2l}{2(n-l+1)} \left(-\frac{du}{u^2}\right)$$

$$= \frac{\left(\frac{2(n-l+1)}{2l}\right)^{2(n-l+1)/2}}{\mathrm{B}\left(\frac{2(n-l+1)}{2}, \frac{2l}{2}\right)} \int_{\frac{1-F_\theta(h)}{F_\theta(h)} \frac{2l}{2(n-l+1)}}^{\infty} x^{2(n-l+1)/2-1} \left(1 + \frac{2(n-l+1)}{2l} x\right)^{-[2(n-l+1)+2l]/2} dx$$

$$= \int_{\frac{1-F_\theta(h)}{F_\theta(h)} \frac{2l}{2(n-l+1)}}^{\infty} f_{2(n-l+1),2l}(x) dx, \tag{30}$$

where

$$x = \frac{1-u}{u} \frac{2l}{2(n-l+1)} \tag{31}$$

$$f_{2(n-l+1),2l}(x) = \frac{1}{\mathrm{B}\left(\frac{2(n-l+1)}{2}, \frac{2l}{2}\right)} \left(\frac{2(n-l+1)}{2l}\right) \left(\frac{2(n-l+1)}{2l} x\right)^{2(n-l+1)/2-1}$$

$$\times \left(1 + \frac{2(n-l+1)}{2l} x\right)^{-[2(n-l+1)+2l]/2}, \quad x > 0 \tag{32}$$

is the probability density function of an F distribution with $2(n-l+1)$ and $2l$ degrees of freedom. Thus,

$$\begin{aligned} P_\theta\{Y_l \le h \mid n\} &= P_\theta\left\{F_{2(n-l+1),2l} > \frac{1-F_\theta(h)}{F_\theta(h)}\frac{2l}{2(n-l+1)}\right\} \\ &= 1 - P_\theta\left\{F_{2(n-l+1),2l} \le \frac{1-F_\theta(h)}{F_\theta(h)}\frac{2l}{2(n-l+1)}\right\} \end{aligned} \tag{33}$$

where $F_{2(n-l+1),2l}$ has the F distribution with $2(n-l+1)$ and $2l$ degrees of freedom. It follows from (33) that

$$\underset{\sim}{h} = \arg\left[P_\theta\left\{Y_l > \underset{\sim}{h} \mid n\right\} = 1-\alpha\right] = \arg\left[P_\theta\left\{Y_l \le \underset{\sim}{h} \mid n\right\} = \alpha\right]$$

$$= \arg\left(P_\theta\left\{F_{2(n-l+1),2l} \le \frac{1-F_\theta\left(\underset{\sim}{h}\right)}{F_\theta\left(\underset{\sim}{h}\right)}\frac{2l}{2(n-l+1)}\right\} = 1-\alpha\right) = \arg\left(P_\theta\left\{F_{2(n-l+1),2l} \le q_{2(n-l+1),2l;1-\alpha}\right\} = 1-\alpha\right) \tag{34}$$

Since (from (34))

$$\frac{1-F_\theta\left(\underset{\sim}{h}\right)}{F_\theta\left(\underset{\sim}{h}\right)}\frac{2l}{2(n-l+1)} = q_{2(n-l+1),2l;1-\alpha}, \tag{35}$$

we have that

$$F_\theta\left(\underset{\sim}{h}\right) = \frac{l}{l+\left(n-l+1\right)q_{2(n-l+1),2l;1-\alpha}}. \tag{36}$$

This ends the proof.

Corollary 1.1: An upper $1-\alpha$ prediction limit $\tilde{h}$ on the lth future order statistic X_l, $l\in\{1, \ldots, n\}$, in a set of n ordered observations $X_1 \le \ldots \le X_n$ is given by

$$\tilde{h} = \arg\left[P_\theta\left\{Y_l > \tilde{h} \mid n\right\} = \alpha\right] = \arg\left[P_\theta\left\{Y_l \le \tilde{h} \mid n\right\} = 1-\alpha\right]$$

$$= \arg\left[F_\theta\left(\tilde{h}\right) = \frac{l}{l+\left(n-l+1\right)q_{2(n-l+1),2l;\alpha}}\right] \tag{37}$$

Corollary 1.2: If the lower $(1-\alpha)$ prediction limit $\underset{\sim}{h}$ on an unknown order statistic X_l in a set of n future ordered observations $X_1 \le \ldots \le X_n$ is prespecified, then the number l of this statistic is determined as

$$l \equiv \underset{\sim}{l}_{1-\alpha} = \arg\left[F_\theta\left(\underset{\sim}{h}\right) = \frac{l}{l+(n-l+1)q_{2(n-l+1),2l;1-\alpha}}\right]$$

$$\cong \arg\min_l \left[F_\theta\left(\underset{\sim}{h}\right) - \frac{l}{l+(m-l+1)q_{2(n-l+1),2l;1-\alpha}}\right]^2, \tag{38}$$

where l is non-negative integer.

Corollary 1.3: If the upper $1-\alpha$ prediction limit $\tilde{h}$ on an unknown order statistic X_l in a set of n future ordered observations $X_1 \le \ldots \le X_n$ is prespecified, then the number l of this statistic is determined as

$$l \equiv \tilde{l}_{1-\alpha} = \arg\left[F_\theta\left(\tilde{h}\right) = \frac{l}{l+(n-l+1)q_{2(n-l+1),2l;\alpha}}\right]$$

$$\cong \arg\min_l \left[F_\theta\left(\tilde{h}\right) - \frac{l}{l+(n-l+1)q_{2(n-l+1),2l;\alpha}}\right]^2 \tag{39}$$

where l is non-negative integer.

Upper Prediction Limit with a Prescribed Confidence Level (1−α)

Table 6 illustrates a situation of within-sample prediction based on the order statistics ($X_k \le X_l$) and the pivotal quantity averaging approach (PQAA). It is assumed that $X_k = t_1$ and $X_l \le t_2$.

Let us assume that

$$\tilde{h} = t_2 \tag{40}$$

then

$$F_\theta\left(\tilde{h}\right) = F_\theta\left(t_2\right) = 1 - \exp\left[-\left(\frac{t_2}{\beta}\right)^\delta\right] \tag{41}$$

Table 6. The situation of within-sample prediction of the number $l_\bullet$ of future failures via the order statistics ($X_k \le X_l$) and PQAA

What Is Known?
Units at start: n=20000
First inspection was at the age $t_1 = 3.0$ years
Cumulative number of failures in the interval $[0, t_1]$ was $k_* = 8$
Future inspection will be at the age $t_2 = 10.0$ years
What Is Unknown?
The ordered observations of failures $(X_1 \le ... \le X_k)$ in the interval $[0, t_1]$ are unknown
The ordered observations of failures $(X_{k+1} \le ... \le X_l)$ in the interval $[t_1, t_2]$ are unknown
What Is Assumed?
It is assumed that the kth order statistic $X_k = t_1$, where k=$k_\bullet$
It is assumed that the lth order statistic $X_l \le t_2$, where l=$k_\bullet$+ $l_\bullet$
What Should Be Found?
Prediction of the number l of a future lth order statistic $X_l \le t_2$ with a prescribed confidence level (1−α) via ($X_k \le X_l$ & PQAA) approach
Upper prediction limit for the number of future failures $\tilde{l}_{\bullet 1-\alpha}$ = l−$k_\bullet$ in the interval $(t_1, t_2]$ with the prescribed confidence level 1−α

Since the parameter β is unknown, (41) is transformed to

$$1-\exp\left[-\left(\frac{t_2}{\beta}\right)^{\delta}\right]=1-\exp\left[-\left(\frac{t_2}{x_k}\frac{x_k}{\beta}\right)^{\delta}\right]=1-\exp\left[-\left(\frac{t_2}{x_k}\right)^{\delta}\left(\frac{x_k}{\beta}\right)^{\delta}\right]$$

$$=1-\left(\exp\left[-\left(\frac{x_k}{\beta}\right)^{\delta}\right]\right)^{(t_2/x_k)^{\delta}}=1-[1-v]^{(t_2/x_k)^{\delta}}=F_{(x_k,\delta)}(v,t_2). \tag{42}$$

Using the pivotal quantity averaging approach, we obtain

$$F_{(x_k,\delta)}(t_2) = E\left\{F_{(x_k,\delta)}\left(V,t_2\right)\right\} = \int_0^1 F_{(x_k,\delta)}\left(v,t_2\right)g_{k,n}(v)dv = 1 - \frac{\mathrm{B}\left(k, n-k+1+\left(t_2/x_k\right)^{\delta}\right)}{\mathrm{B}(k, n-k+1)} \tag{43}$$

where the unknown parameter β is eliminated from the problem. Then, using (39), where $F_{(x_k,\delta)}(t_2)$ is substituted into (39) for $F_\theta(\tilde{h})$, we have the following numerical results, which are presented in Table 7.

Prediction of a Future Number of Failures via Anticipatory Likelihood Function

Table 8 illustrates a situation of within-sample prediction based on the cumulative number of failures ($k_\bullet$=8) and anticipatory likelihood function (ALF).

It is assumed that the failure times are independent observations from a Weibull distribution with unknown scale parameter β and given (known) shape parameter δ.

Table 7. Upper prediction limit $\tilde{l}_{\bullet 1-\alpha}$ for the number $l_\bullet$ of future cracked heat exchanger tubes with a prescribed confidence level ($1-\alpha$) via the order statistics ($X_k \leq X_l$) and PQAA

Confidence Level ($1-\alpha$)	Shape Parameter, δ		
	3.0	3.3	3.6
0.9 $\tilde{l}_{\bullet 1-\alpha}$	[265]	[387]	[562]
0.95 $\tilde{l}_{\bullet 1-\alpha}$	[259]	[380]	[553]
0.99 $\tilde{l}_{\bullet 1-\alpha}$	[248]	[366]	[537]

Table 8. The situation of within-sample prediction for the number $l_\bullet$ of future cracked heat exchanger tubes via the cumulative number of failures ($k_\bullet$ = 8) and anticipatory likelihood function (ALF)

What Is Known?
Units at start: $n = 20000$
First inspection was at the age $t_1 = 3.0$ years
Cumulative number of failures in the interval $[0, t_1]$ was $k_\bullet = 8$
Future inspection will be at the age $t_2 = 10.0$ years
What Should Be Found?
Point prediction for the number $l_\bullet$ of future failures in the interval $\left[t_1, t_2\right]$

Point Prediction

Under Weibull distribution model (1), the anticipatory likelihood function is given by

$$L\left(\beta, l_\bullet \mid k_\bullet, \delta\right)$$

$$= \frac{n!}{k_\bullet! l_\bullet! \left(n - k_\bullet - l_\bullet\right)!} \left[1 - \exp\left(-\left[\frac{t_1}{\beta}\right]^\delta\right)\right]^{k_\bullet} \left[\exp\left(-\left[\frac{t_1}{\beta}\right]^\delta\right) - \exp\left(-\left[\frac{t_2}{\beta}\right]^\delta\right)\right]^{l_\bullet} \left[\exp\left(-\left[\frac{t_2}{\beta}\right]^\delta\right)\right]^{n-k_\bullet-l_\bullet} \quad (44)$$

Then ALF estimates β^{ALF} and $l_\bullet^{\mathrm{ALF}}$ must be found numerically by maximizing $\ln[L(\beta, l_\bullet \mid k_\bullet, \delta)]$,

$$\left(\beta^{\mathrm{ALF}}, l_\bullet^{\mathrm{ALF}}\right) = \arg\max_{\beta, l_\bullet} \ln\left[L\left(\beta, l_\bullet \mid k_\bullet, \delta\right)\right], \quad (45)$$

where

$$\ln\left[L\left(\beta, l_\bullet \mid k_\bullet, \delta\right)\right] = \ln\Gamma(n+1) - \ln\Gamma\left(k_\bullet + 1\right) - \ln\Gamma\left(l_\bullet + 1\right) - \Gamma\left(n - k_\bullet - l_\bullet + 1\right)$$

$$+ k_\bullet \ln\left[1 - \exp\left(-[t_1/\beta]^\delta\right)\right] + l_\bullet \ln\left[\exp\left(-[t_1/\beta]^\delta\right) - \exp\left(-[t_2/\beta]^\delta\right)\right] + \left(n - k_\bullet - l_\bullet\right)\ln\left[\exp\left(-[t_2/\beta]^\delta\right)\right] \quad (46)$$

The corresponding point prediction for the number $l_\bullet$ of future cracked heat exchanger tubes obtained via the ALF is given in Table 9. Two additional values of the Weibull shape parameter are used to evaluate the effect of misspecification of δ on the point prediction.

Statistical Inference

For comparison, the results of point prediction for the number $l_\bullet$ of future cracked heat exchanger tubes obtained via corresponding prediction procedures described in this chapter, are given in Table 10. Two additional values of the Weibull shape parameter are used to evaluate the effect of misspecification of δ on the point prediction.

Table 9. Point prediction for the number $l_\bullet$ of future cracked heat exchanger tubes via the ALF

Point Prediction Procedure	**Shape Parameter, δ**		
	3.0	**3.3**	**3.6**
ALF $l_\bullet^{\mathrm{ALF}}$	[268]	[388]	[558]

Table 10. Point prediction for the number $l_\bullet$ of future cracked heat exchanger tubes via corresponding prediction procedures described in this chapter

Prediction Procedures		Shape Parameter, δ		
		3.0	3.3	3.6
ML (Nelson) $l_\bullet^{\mathrm{ML}}$		[286]	[413]	[593]
(X_k & PQAA) $l_\bullet^{X_k \,\&\, \mathrm{PQAA}}$		[286]	[412]	[592]
($X_k \le X_l$ & PQAA)	$1-\alpha = 0.90$ $\tilde{l}_{\bullet 1-\alpha}$	[265]	[387]	[562]
	$1-\alpha = 0.95$ $\tilde{l}_{\bullet 1-\alpha}$	[259]	[380]	[553]
	$1-\alpha = 0.99$ $\tilde{l}_{\bullet 1-\alpha}$	[248]	[366]	[537]
ALF $l_\bullet^{\mathrm{ALF}}$		[268]	[388]	[558]

Thus, the procedures ($X_k \le X_l$ & PQAA), where the confidence level $1-\alpha = 0.90$, and ALF provided practically similar results of point prediction in the same situations.

NUMERICAL EXAMPLE

Let us assume that a steam generator with n = 20000 tubes, of which $k_\bullet$ = 8 have failed by the inspection at the age t_1 = 3.0 years and $l_\bullet$ = 380 have failed by the inspection at the age t_1 = 10.0 years, is considered. Due to stress and corrosion, the tubes develop cracks over time. Cracks are detected during planned inspections. The cracked tubes are subsequently plugged to remove them from service. To develop efficient inspection and plugging strategies, plant management can use a prediction of the number of tubes that will need plugging by a specified future time. The failure-time of tubes is modeled with a two-parameter Weibull distribution indexed by scale and shape parameters β and δ, respectively. The case is considered when the scale parameter β and the shape parameter δ are unknown. It is necessary to find a point prediction as well as upper prediction limit for the number $m_\bullet$ of future tubes that will need plugging by a future inspection at age t_3 = 20.0 years. Table 11 illustrates the above situation for within-sample prediction via ALF and ($X_1 \le X_m$ & PQAA) approaches.

Point Prediction for a Future Number of Failures via ALF Approach

Under the Weibull distribution model (1), the anticipatory likelihood function is given by

$$L\left(m_\bullet, \beta, \delta \mid k_\bullet, l_\bullet\right) = \frac{n!}{k_\bullet! l_\bullet! m_\bullet! (n - k_\bullet - l_\bullet - m_\bullet)!} \left[1 - \exp\left(-\left[{t_1}/{\beta}\right]^{\delta}\right)\right]^{k_\bullet}$$

Table 11. The situation for within-sample prediction of the number $m_\bullet$ of future failures via ALF and ($X_l \le X_m$ & PQAA) approaches

What Is Known?
Units at start: n=20000
First inspection was at the age of $t_1 = 3$ years
Second inspection was at the age of $t_2 = 10$ years
Cumulative number of failures in the interval $[0, t_1]$ was $k_\bullet = 8$
Cumulative number of failures in the interval $(t_1, t_2]$ was $l_\bullet = 387; \quad 413; \quad 425$ (for analysis)
Future inspection will be at the age of $t_3 = 20$ years
What Is Unknown?
The ordered observations of failures $(X_1 \le \ldots \le X_k)$ in the interval $[0, t_1]$ are unknown
The ordered observations of failures $(X_{k+1} \le \ldots \le X_l)$ in the interval $(t_1, t_2]$ are unknown
The future ordered observations of failures $(X_{l+1} \le \ldots \le X_m)$ in the interval $(t_2, t_3]$ are unknown
The parameters β and δ of the Weibull distribution are unknown
What Is Assumed?
It is assumed that the lth order statistic $X_l = t_2$, where $l = k_\bullet + l_\bullet$
It is assumed that the mth order statistic $X_m \le t_3$, where $m = m_\bullet + k_\bullet + l_\bullet$
What Should Be Found?
Point prediction for the number $m_\bullet$ of future failures in the interval $(t_2, t_3]$ via ALF approach
Prediction of the number m of a future mth order statistic $X_m \le t_3$ with a prescribed confidence level $(1-\alpha)$ via ($X_l \le X_m$ & PQAA) approach
Upper prediction limit $\tilde{m}_{\bullet 1-\alpha} = m - k_\bullet - l_\bullet$ for the number of future failures in the interval $(t_2, t_3]$ with the prescribed confidence level $(1-\alpha)$

$$\times\left[\exp\left(-\left[{t_1}/{\beta}\right]^{\delta}\right)-\exp\left(-\left[{t_2}/{\beta}\right]^{\delta}\right)\right]^{l_\bullet}\left[\exp\left(-\left[{t_2}/{\beta}\right]^{\delta}\right)-\exp\left(-\left[{t_3}/{\beta}\right]^{\delta}\right)\right]^{m_\bullet}\left[\exp\left(-\left[{t_3}/{\beta}\right]^{\delta}\right)\right]^{n-k_\bullet-l_\bullet-m_\bullet} \tag{47}$$

Then the ALF estimates of $m_\bullet$, β and δ must be found numerically by maximizing $\ln L\left(m_\bullet,\beta,\delta \mid k_\bullet,l_\bullet\right)$, i.e.,

$$\left(m_\bullet^{\mathrm{ALF}},\beta^{\mathrm{ALF}},\delta^{\mathrm{ALF}}\right)=\arg\max_{m_\bullet,\beta,\delta}\ln\left[L\left(m_\bullet,\beta,\delta \mid k_\bullet,l_\bullet\right)\right] \tag{48}$$

where

$$\ln\left[L\left(m_\bullet,\beta,\delta \mid k_\bullet,l_\bullet\right)\right]=\ln\Gamma(n+1)-\ln\Gamma\left(k_\bullet+1\right)-\ln\Gamma\left(l_\bullet+1\right)-\ln\Gamma\left(m_\bullet+1\right)-\ln\Gamma\left(n-k_\bullet-l_\bullet-m_\bullet+1\right)$$

$$+k_\bullet\ln\left[1-\exp\left(-\left[{t_1}/{\beta}\right]^{\delta}\right)\right]+l_\bullet\ln\left[\exp\left(-\left[{t_1}/{\beta}\right]^{\delta}\right)-\exp\left(-\left[{t_2}/{\beta}\right]^{\delta}\right)\right]$$

$$+m_\bullet\ln\left[\exp\left(-[t_2/\beta]^{\delta}\right)-\exp\left(-[t_3/\beta]^{\delta}\right)\right]+(n-k_\bullet-l_\bullet-m_\bullet)\ln\left[\exp\left(-[t_3/\beta]^{\delta}\right)\right], \tag{49}$$

$m_\bullet$ is non-negative integer, β and δ are non-negative.

It follows from (48) that

$$\left(m_\bullet^{\mathrm{ALF}},\beta^{\mathrm{ALF}},\delta^{\mathrm{ALF}}\right)=\left(3445,\ 31.83634,\ 3.301018\right) \tag{50}$$

Upper Prediction Limit for a Future Number of Failures via ($X_l \le X_m$&PQAA) Approach

It follows from (39) that the number m of the mth order statistic $X_m \le t_3$ with a prescribed confidence level $(1-\alpha)$ can be found as follows:

$$m\equiv m_{1-\alpha}=\arg\left[P_\theta\left\{X_m\le t_3 \mid n\right\}=1-\alpha\right]=\arg\left[P_\theta\left\{X_m> t_3 \mid n\right\}=\alpha\right]$$

$$= \arg\left[F_\theta(t_3) = \frac{m}{m+(n-m+1)q_{2(n-m+1),2m;\alpha}}\right] \cong \arg\min_m \left[F_\theta(t_3) - \frac{m}{m+(n-m+1)q_{2(n-m+1),2m;\alpha}}\right]^2, \tag{51}$$

where

$$F_\theta(t_3) = 1-\exp\left[-\left(\frac{t_3}{\beta}\right)^\delta\right] = 1-\exp\left[-\left(\frac{t_3}{x_l}\frac{x_l}{\beta}\right)^\delta\right] = 1-\exp\left[-\left(\frac{t_3}{x_l}\right)^\delta\left(\frac{x_l}{\beta}\right)^\delta\right]$$

$$= 1-\left(\exp\left[-\left(\frac{x_l}{\beta}\right)^\delta\right]\right)^{(t_3/x_l)^\delta} = 1-[1-v]^{(t_3/x_l)^\delta} = F_{(x_l,\delta)}(v,t_3) \tag{52}$$

Assuming that

$$\delta \cong \delta^{\mathrm{ALF}} = 3.301018 \tag{53}$$

and

$$F_\theta(t_3) \cong F_{(x_l,\delta)}(t_3) = E\left\{F_{(x_l,\delta)}(V,t_3)\right\} = \int_0^1 F_{(x_l,\delta)}(v,t_3)g_{l,n}(v)dv = 1-\frac{\mathrm{B}\left(l, n-l+1+\left(t_3/x_l\right)^\delta\right)}{\mathrm{B}\left(l,n-l+1\right)} \tag{54}$$

where

$$g_{l,n}(v) = \frac{1}{\mathrm{B}\left(l,n-l+1\right)} v^{l-1}\left(1-v\right)^{n-l}, \quad v \in (0,1) \tag{55}$$

prediction of the number m of the mth order statistic $X_m \le t_3$ with a prescribed confidence level $(1-\alpha)$ can be found via ($X_l \le X_m$ & PQAA) approach as follows:

$$m^{(X_l \le X_m \text{ \& PQAA})} \cong \arg\min_m \left[1-\frac{\mathrm{B}\left(l,n-l+1+\left(t_3/x_l\right)^\delta\right)}{\mathrm{B}(l,n-l+1)} - \frac{m}{m+(n-m+1)q_{2(n-m+1),2m;\alpha}}\right]^2 \tag{56}$$

where m is non-negative integer. Table 12 illustrates the results of within-sample prediction for the number $m_\bullet$ of future failures via ALF and ($X_l \leq X_m$ & PQAA) approaches.

Statistical Inference

Statistical analysis of the results presented in Table 12 allows us to propose the following methodology for predicting the number of future failures when both parameters β and δ of the Weibull distribution (1) are unknown.

Step 1: Using the ALF approach, the estimate of the unknown parameter δ is determined from (48):

$$\delta^{\text{ALF}} = \arg \max_{m_\bullet, \beta, \delta} \ln\left[L\left(m_\bullet, \beta, \delta \mid k_\bullet, l_\bullet\right)\right] \tag{57}$$

where $\ln\left[L\left(m_\bullet, \beta, \delta \mid k_\bullet, l_\bullet\right)\right]$ is given by (49).

Step 2: Using the ($X_l \leq X_m$ & PQAA) approach, prediction of the number m of the mth order statistic $X_m \leq t_3$ with a prescribed confidence level $(1-\alpha)$ is determined from (56):

$$m \equiv \tilde{m}_{1-\alpha} \cong \arg \min_{m} \left[1 - \frac{\mathrm{B}\left(l, n-l+1+\left(t_3 / x_l\right)^{\delta^{\text{ALF}}}\right)}{\mathrm{B}(l, n-l+1)} - \frac{m}{m+(n-m+1) q_{2(n-m+1), 2m; \alpha}}\right]^2, \tag{58}$$

where δ^{ALF} is substituted into (56) for δ.

Step 3: The upper prediction limit for the number of future failures in the interval $(t_2, t_3]$ with the prescribed confidence level $(1-\alpha)$ is determined by

$$\tilde{m}_{\bullet 1-\alpha} = m - k_\bullet - l_\bullet . \tag{59}$$

It should be noted that the statistical conclusions concerning predictions (based on previous inspections) of heat exchanger tubes cracked in subsequent intervals $(t_4, t_5]$, $(t_5, t_6]$, $(t_6, t_7]$, and so on, can be obtained through the ALF and ($X_l \leq X_m$ & PQAA) approaches in the same manner as described above. Since prediction of the number of future failures is carried out through the sequential order statistics, it is adaptive.

Table 12. The results of within-sample prediction for the number $m_\bullet$ of future failures via ALF and ($X_l \leq X_m$ & PQAA) approaches

Approach	Prediction of Number m for X_m	Confidence Level $1-\alpha$ for $x_m\left(P_\theta\left\{X_m \leq t_3 \mid n\right\} = 1-\alpha\right)$	Significance Level α for $x_m\left(P_\theta\left\{X_m > t_3 \mid n\right\} = \alpha\right)$	Prediction: Point $m_\bullet^{\mathrm{ALF}}$ Limit $\tilde{m}_{\bullet 1-\alpha}$
ALF	$l_\bullet = 387;\ \left(m_\bullet^{\mathrm{ALF}}, \beta^{\mathrm{ALF}}, \delta^{\mathrm{ALF}}\right) = \left(3000, 33.70408, 3.222629\right)$			
	3395	0.505	0.495	$m_\bullet^{\mathrm{ALF}} = 3000$
$X_l \leq X_m$ & PQAA	$l = 395;\ \delta^{\mathrm{ALF}} = 3.222629$			
	3323	0.9	0.1	$\tilde{m}_{\bullet 0.9} = 2928$
	3302	0.95	0.05	$\tilde{m}_{\bullet 0.95} = 2907$
	3263	0.99	0.01	$\tilde{m}_{\bullet 0.99} = 2868$
ALF	$l_\bullet = 413;\ \left(m_\bullet^{\mathrm{ALF}}, \beta^{\mathrm{ALF}}, \delta^{\mathrm{ALF}}\right) = \left(3304, 32.38295, 3.277393\right)$			
	3725	0.51	0.49	$m_\bullet^{\mathrm{ALF}} = 3304$
	$l = 421;\ \delta^{\mathrm{ALF}} = 3.277393$			
$X_l \leq X_m$ & PQAA	3649	0.9	0.1	$\tilde{m}_{\bullet 0.9} = 3228$
	3628	0.95	0.05	$\tilde{m}_{\bullet 0.95} = 3207$
	3587	0.99	0.01	$\tilde{m}_{\bullet 0.99} = 3166$
ALF	$l_\bullet = 425;\ (m_\bullet^{\mathrm{ALF}}, \beta^{\mathrm{ALF}}, \delta^{\mathrm{ALF}}) = (3445,\ 31.83634,\ 3.301018)$			
	3878	0.525	0.475	$m_\bullet^{\mathrm{ALF}} = 3445$
$X_l \leq X_m$ & PQAA	$l = 433;\ \delta^{\mathrm{ALF}} = 3.301018$			
	3804	0.9	0.1	$\tilde{m}_{\bullet 0.9} = 3371$
	3782	0.95	0.05	$\tilde{m}_{\bullet 0.95} = 3349$
	3740	0.99	0.01	$\tilde{m}_{\bullet 0.99} = 3307$

CONCLUSION AND FUTURE WORK

We have illustrated the within-sample prediction methodology of the number of future failures, which is based on the ALF and ($X_l \leq X_m$ & PQAA) approaches, for the two-parameter Weibull distribution. Application to other (such as, say, log-location-scale) distributions could follow directly.

It will be noted that in the literature only the case is considered when the scale parameter β of the Weibull distribution is unknown, but the shape parameter δ is known. As a rule, in practice the Weibull shape parameter δ is not known. Instead it is estimated subjectively or from relevant data. Thus, its value is uncertain. This δ uncertainty may contribute greater uncertainty to the construction of prediction limits for a future number of failures. Therefore, in this chapter, we consider the case when both parameters β and δ are unknown.

In literature, for this situation, usually a Bayesian approach is used. Bayesian methods are not considered here. We note, however, that although subjective Bayesian prediction has a clear personal probability interpretation, it is not generally clear how this should be applied to non-personal prediction or decisions. Objective Bayesian methods, on the other hand, do not have clear probability interpretations in finite samples.

It should be noted that the calculation results obtained in this chapter through the soft computing process proved to be more suitable for decision making than those obtained through the conventional computing process.

It is expected that the ideas presented in this chapter will give interesting and novel contributions to statistical theory and its applications. The methodology described here can be extended in several different directions to handle various problems that arise in practice.

REFERENCES

Bjornstad, J. F. (1990). Predictive likelihood: A review. *Statistical Science*, *5*(2), 242–254. doi:10.1214/ss/1177012175

Bjornstad, J. F. (1996). On the generalization of the likelihood function and the likelihood principle. *Journal of the American Statistical Association*, *91*, 791–806.

Escobar, L. A., & Meeker, W. Q. (1999). Statistical prediction based on censored life data. *Technometrics*, *41*(2), 113–124. doi:10.1080/00401706.1999.10485632

Geweke, J., & Amisano, G. (2012). Prediction and misspecified models. *The American Economic Review*, *102*(3), 482–486. doi:10.1257/aer.102.3.482

Hahn, G. J., & Meeker, W. Q. (1991). *Statistical Intervals: A Guide for Practitioners*. New York: Wiley. doi:10.1002/9780470316771

Hahn, G. J., & Nelson, W. (1973). A survey of prediction intervals and their applications. *Journal of Quality Technology*, *5*, 178–188.

Lawless, J. F. (1973). On estimation of safe life when the underlying distribution is Weibull. *Technometrics*, *15*(4), 857–865. doi:10.1080/00401706.1973.10489118

Lejeune, M., & Faulkenberry, G. D. (1982). A simple predictive density function. *Journal of the American Statistical Association, 77*(379), 654–657. doi:10.1080/01621459.1982.10477866

Li, X., Ruan, D., & van der Wal, A. J. (1998). Discussion on soft computing at FLINS96. *International Journal of Intelligent Systems, 13*(2-3), 287–300. doi:10.1002/(SICI)1098-111X(199802/03)13:2/3<287::AID-INT10>3.0.CO;2-4

Mee, R., & Kushary, D. (1994). Prediction limits for the Weibull distribution utilizing simulation. *Computational Statistics & Data Analysis, 17*(3), 327–336. doi:10.1016/0167-9473(92)00072-Y

Nechval, N. A., Berzins, G., & Danovics, V. (2016). Optimization of statistical decisions for age replacement problems via a new pivotal quantity averaging approach. *American Journal of Theoretical and Applied Statistics, 5*, 21–28. doi:10.11648/j.ajtas.s.2016050201.14

Nechval, N.A., Nechval, K.N., & Vasermanis, E.K. (2003). Effective state estimation of stochastic systems. *Kybernetes (An International Journal of Systems & Cybernetics), 32*, 666–678.

Nechval, N. A., Purgailis, M., Berzins, G., Cikste, K., Krasts, J., & Nechval, K. N. (2010). Invariant embedding technique and its applications for improvement or optimization of statistical decisions. In K. Al-Begain, D. Fiems & W. Knottenbelt (Eds.), Analytical and Stochastic Modeling Techniques and Applications. LNCS (Vol. 6148, pp. 306–320). Berlin, Heidelberg: Springer-Verlag. doi:10.1007/978-3-642-13568-2_22

Nechval, N. A., Purgailis, M., Nechval, K. N., & Bruna, I. (2012). Optimal prediction intervals for future order statistics from extreme value distributions. In *Proceedings of the World Congress on Engineering WCE '12, LNECS* (Vol. 3, pp. 1340–1345).

Nelson, W. (1982). *Applied Life Data Analysis*. New York: Wiley. doi:10.1002/0471725234

Nelson, W. (2000). Weibull prediction of a future number of failures. *Quality and Reliability Engineering International, 16*(1), 23–26. doi:10.1002/(SICI)1099-1638(200001/02)16:1<23::AID-QRE283>3.0.CO;2-Q

Nordman, D. J., & Meeker, W. Q. (2002). Weibull prediction for a future number of failures. *Technometrics, 44*(1), 15–23. doi:10.1198/004017002753398191

Patel, J. K. (1989). Prediction intervals–a review. *Communications in Statistics. Theory and Methods, 18*(7), 2393–2465. doi:10.1080/03610928908830043

Varin, C., & Vidoni, P. (2006). Pairwise likelihood inference for ordinal categorical time series. *Computational Statistics & Data Analysis, 51*(4), 2365–2373. doi:10.1016/j.csda.2006.09.009

Varin, C., & Vidoni, P. (2009). Pairwise likelihood inference for general state space models. *Econometric Reviews, 28*(1-3), 170–185. doi:10.1080/07474930802388009

Chapter 12
Selection of Appropriate Turbulance Model in Fuel Bundle of Nuclear Energy:
In Context With Inter-Subchannel Mixing of Coolant for Single Phase Flow

Shashi Kant Verma
National Institute of Technology Raipur, India

S. L. Sinha
National Institute of Technology Raipur, India

D. K. Chandraker
BARC, India

ABSTRACT

This chapter presents an overview of various types of turbulence model and their effect on thermal-hydraulic characteristics of nuclear fuel bundle, both past and present using Computational Fluid Dynamic (CFD) approach. It includes the mathematical definition related to fuel bundle in nuclear energy. The various types of geometrical arrangement like Pressurized Water Reactor (PWR), Boiling Water Reactor (BWR), etc., are stressed. The solution procedures that are applicable to the various reactor types are introduced here and presented in detail for different types of turbulence models. Study of these characteristics enables the student to appreciate the effect of the different types of turbulence models on turbulent mixing and related phenomena. Finally, recommendations of turbulence model for rod bundle are finalized. The inclusion of related references provides a starting point for the interested reader / researchers /industrialists.

DOI: 10.4018/978-1-5225-3035-0.ch012

INTRODUCTION

The basic mode of operation of the most nuclear reactors is that the coolant flowing through the core of the reactor removes heat produced by nuclear fission and conveys it to a steam generator. The steam drives a turbine coupled to a generator, which produces electricity. Fuel rod bundle is the heart of the nuclear reactor system. Over the precedent 20 years, momentous improvements have been ended in industrial Computational Fluid Dynamics (CFD) codes, in computing supremacy and in parallel-computing. These improvements have facilitated the utilization of CFD as a universal practice in numerous sectors, its use in the nuclear diligence is rising. In industrial computational fluid dynamics codes, there is more than one turbulence model built in. It is the addict accountability to decide one of those models, appropriate for the dilemma considered. Even though considerable progresses completed in the pasture, the use of CFD in predicting single-phase flows in rod bundles tranquil faces a few challenges due to difficulties in precisely predicting the turbulent structures such as Secondary flows, Vortex shedding, and Flow pulsations that donate to inter-channel mixing (Krauss and Meyer, 1996; Baglietto and Ninokata, 2005). The mixing of cooling fluid in rod bundles from one subchannel to another through the gaps between the rods reduces the temperature differences in the coolant as well as beside the perimeter of the rods. The observable fact of natural mixing was earliest intensively investigated in the 1960s and leftovers a theme of research up to the current period. Universal features of these simulations are the computations performed by the researchers by means of k-ε turbulence model. Evaluation of the simulation grades with experimental data was rarely acceptable. Models based on the Reynolds-stress transport equations can offer qualitatively accurate solutions. As a probable justification, secondary flow was deduced from measurements by numerous experimentalists (Trupp and Azad, 1975; Trippe and Weinberg, 1979b; Seale, 1979; Rehme, 1987) and fairly recently its subsistence has been supported by analytical marks (Kim and Kim, 2004). Turbulent mixing is recognized as a transport method which is caused by the lateral velocity variation in the gap among two sub-channels. In this means, mass, momentum or energy can be elated in lateral way. This procedure has a diffusive nature. For the explanation of this method, a diffusion coefficient and a gradient of the transportable flow variable is requisite. Normally, our endeavour is informative, we would identify the reader's attention to the reality that turbulence models have to be certain based on theoretical considerations and / or enough information obtained from measurements. The significance of this revision can be traced back to the deletion of heat in a reactor. To sustain the reactor at a safe and constant temperature and, at the same time, operate at highest efficiency, the stability and control of the heat elimination is of great significance. The secondary flow and cross-channel flow of water in different geometry promotes the heat deletion in the radial direction. This chapter will concentrate on the analysis of cross-flow in rod-bundle geometry. Although the theoretical and computational approach is very interesting, this learning was solely performed on computational basis. Due to dissimilar profile of the sub-channels, the thermal hydraulic descriptions of the coolant flowing through the sub-assembly are also unlike which requests a thorough and profound understanding.

In the boiling disaster, the rod exterior is in touch with merely steam and heat transfer deteriorates which leads to increase in the cladding temperature and threatens the reliability of the fuel rod. In the PWR, generally the boundary to CHF is accessible with the exodus from Nucleate Boiling Ratio (DNBR), which is the ratio of heat flux required to cause the Departure from Nucleate Boiling (DNB) to the confined heat flux. Design safety limits are utilized to make certain the safety confines of significant parameters, which are likely to by no means surpass in any modes of process. The safety margin is distinct as the distinction (or ratio) among the limiting value for the parameter and the actual value of the parameter on the plant. In several cases, neither the limiting value nor the actual value can be resolved precisely. In these cases, a dictatorial recognition standard is set for the parameter, which is usually lower than the estimated value for the safety limit to take into account the uncertainty in the assessment. The safety margin is then defined as the dissimilarity among these receiving criteria and the authentic value according to International Atomic Energy Agency (IAEA). For an easy parameter like the highest cladding temperature, the case is clear-cut as the cladding material is known to start a speedy chemical reaction with steam (primary to cladding injure) at a certain temperature (1200°C). Then it is adequate for the proprietor of the plant to exhibit that the actual value of the temperature is lesser than this temperature and with an adequate margin of safety. If a conservative analysis does not illustrate adequate margin, a so-called best-estimate analysis can be used to added reveal the adequate margin as per International Atomic Energy Agency (IAEA).

A significant constraint in heat transfer design is the Critical Heat Flux (CHF) which is definite as the limiting heat flux between boiling with high heat transfer coefficient and boiling with low heat transfer coefficient. Small heat transfer from the nuclear core to the coolant leads to high temperatures of the fuel rods and accelerated oxidation of the cladding, and can eventually lead to release the radioactivity if a cladding collapse occurs. Significant feature of the fuel element design is the design of spacer grids, which grasp the fuel rods in place and an appropriate distance (spacing) away from each other to keep away from mechanical stress due to vibrations etc. Heat transfer and the CHF can be amplified with the use of spacer mixing vanes. Full scale CHF experiments are extremely costly. Computational Fluid Dynamics (CFD) can be used in the design process for simulating several dissimilar spacer configurations before full scale experiments. CFD provides thorough 3D flow description, which assist the trendy to recognize the flow experience and hold the design process. The invent parameters can be distorted and simulations run with no costs distant from computational costs. Several models are implemented into CFD for the simulation. Experimental data are essential for these models (experimental correlations) and to authenticate the CFD results.

Figure 1. the criteria for safety margins is illustrated

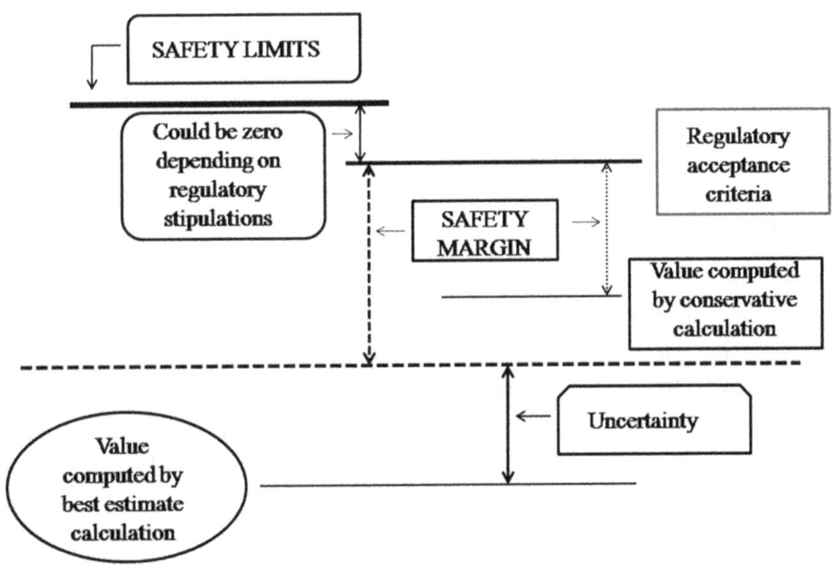

BACKGROUND

In a nuclear reactor core, fuel rods are supported by grid structures referred to as spacer grids. These spacer grids afford bear to the fuel rods maintaining them at a predetermined distance from each other, reducing the vibrations that would take place if the rods were not supported. As one of the key flow obstructions in the core, spacer grids have been established to influence the coolant flow hydrodynamics and heat transfer description. There have been studies in regard to a single-phase coolant travelling along the fuel rods and through the spacer grids (Rehme, 1973; Dominguez-Ontiveros and Hassan, 2009). However, there is modest data accessible to offer as to what occurs when two phases travel through the spacer grid. To solve the realistic intricate engineering problem, it is essential to calculate the solutions of the Navier-Stokes equations numerically with the help of super computers. This is a branch of science called as Computational Fluid Dynamics (CFD). Before presenting the diverse CFD approaches, it is vital to introduce the universal phenomena in fluid motion i.e. the concept of turbulence.

Turbulence is not a fluid property but belongings of flow itself (Saxena, 2014). Turbulent flow can be merely defined as a flow that is disordered in time and space. But precisely turbulent flows may have fairly dissimilar dynamics, may be three dimensional or infrequently quasi two dimensional, and may exhibit ordered rotational structures called as eddies. The most useful property of turbulent flows is that they are able to merge transported quantities much more quickly than if only molecular diffusion processes were concerned. This could be a required property because it enhances the mixing of physical and chemical properties within a flow. Turbulence is a ubiquitous phenomenon found in nature and industries. From an industrial point of analysis, its applications are diverse such as flow around aircraft, vehicles from submarines, turbulent flow generated by combustion, flow in a nuclear reactor. Thus, turbulence denotes a state of fluid in which all properties (velocity, pressure, density, etc.) fluctuate endlessly in an irregular, chaotic, non-repeating and unpredictable manner (Hinze, 1975). Reynolds (Reynolds, 1895) was the first to suggest a standard for differentiation between laminar and turbulent flows in his classic dye visualization. He observed that a dimensionless parameter called as Reynolds number, Re regulates the onset of turbulence. This dimensionless parameter is defined as ratio of inertial to viscous force and mathematically given as follows:

$$Re = \frac{l^* u^*}{\vartheta}$$

where u^* is the velocity scale, l^* is the characteristic length of the flow domain and ϑ the kinematic viscosity of fluid. Reynolds showed that above a critical value of Reynolds number ($Re_c = 2100$ for fluid flow in pipe) the topology and dynamics of the flow changes considerably and the flow becomes turbulent. Turbulence can be decomposed into rotational structures called eddies that are characterized by very dissimilar scales of length, time and speed. Numerous turbulent structures of very different sizes coexist in an identical flow and possess a precise dynamics of space-time scales that distinguish them. The largest structures have the characteristic length of the domain L and enclose most of the energy. Large eddies take out energy from the mean motion, become unstable and breakup, transferring their energy to lesser size eddies. The small eddies go through a analogous break-up process, and move their energy to lesser and lesser eddies until this energy is dissipated by viscosity. This is called energy cascade (Richardson, 1992). Computational Fluid Dynamics involves guess of fluid flow, heat transfer and associated phenomena by solving the mathematical equations that govern these processes using a numerical method. The main steps involved in a CFD study are as follows:

1. Conception of mesh to symbolize the required geometry;
2. Provide suitable boundary and initial conditions for the problem;
3. Selection of proper turbulence models depending on the required level of accuracy;
4. Solving the equations of mass, momentum and energy all over the domain of interest.

Since the growth of computers, numerical simulation has a significant place in research and in industry. It predicts numerous results, both qualitative and quantitative exclusive of setting up experiments that can be costly. If we seem at a turbulent fluid flow from a statistical point of view, a mean field appears on which fluctuations of dissimilar scales are superimposed. The problem of numerical simulation of turbulent flows is to assess with more or less accuracy, the average and fluctuations of these fields for a wide range of scales present in the flow. The numerical simulations can be separated in to three chief families depending on the level of accuracy and extent of modelling used. They are Direct Numerical Simulation (DNS), Large Eddy Simulation (LES) and Reynolds averaged Navier-Stokes simulation (RANS). These three simulation techniques are used in a different way depending on the needs and available computational resources. Figure 2 represents a digest of differences among the three approaches. The stages from recognition to authentication of the vortex flow in rod bundles and compound channels are concise under as shown in Table 1. The character of this flow is ultimately fit acknowledged by the scientific commune and the calculation of the flow in rod bundles should be on the true way now.

MAIN FOCUS OF THE CHAPTER

This chapter gives an overview of various types of turbulence model and their effect on thermal-hydraulic characteristics of nuclear fuel bundle, both past and present using Computational Fluid Dynamic (CFD) approach. This chapter presents the mathematical definition related to fuel bundle in nuclear energy. The various types of geometrical arrangement like PWR, BWR etc. are stressed. The solution procedures that are applicable to the various reactor types are introduced here and presented in detail for different types of turbulence models. Study of these characteristics enables the student to appreciate the effect of the different types of turbulence models on turbulent mixing and related phenomena. Finally recommendations of turbulence model for rod bundle are finalized. The many references provide a starting point for the interested reader.

Waters (1961) calculated the quantity of fluid mixing among flow channels within 7-rod bundle fuel elements. In that learning, a salt solution was introduced into the rod bundle, and the fluid mixing was resolute by measuring the salt concentration. Hanson and Todreas (1977) considered the fluid mixing in a hexagonal 61-pin, wire-wrapped rod bundle. In their learning, salt was injected in the bundle, and fluid mixing was studied by investigating the salt tracer dispersion. Meyer (2010) published a literature revision on large scale vortices in rod bundles. He summarised a variety of experimental and numerical hard work that resulted in the discovery of large scale flow fluctuations. These large-scale patterns are particularly chief in tight lattice bundles (small P/D). Zimmermann (2015) performed Systematic CFD investigations of the flow in a 5×5 rod bundle geometry containing a single spacer grid with split-type mixing vanes. He used a cubic non-linear k-ε turbulence model to simulate the typical rod bundle flow

Figure 2. Represents a digest of differences among the three approaches

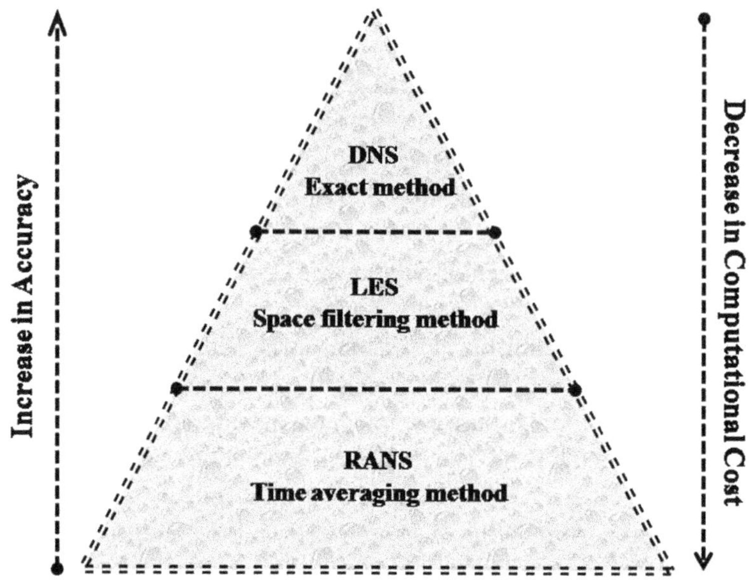

effects such as secondary flow and non-isotropic turbulence. He concluded that for the detailed analysis of the forces and flows in the gap showed that mixing vane induced swirling flow in the sub-channels muscularly influence the inter subchannel cross-flow. A new model was developed based on the CFD simulation results to predict the forced cross-flow between the sub-channels. In general, single-phase flow in a mixture of rod bundles and channel arrays has been deliberated with numerous diverse techniques. Natural mixing has been characterised by straight measuring mixing of a scalar quantity (i.e. tracer) of a new kind (radioactive, chemical, thermal) or ultimately by measuring turbulent flow (LDA, PIV, hot-wire anemometry). Ylonen (2013) found that, the turbulent Schmidt number (Sc) is supposed to be stable and has an evasion value of 0.9. Nevertheless, numerous researchers have done that the value of the turbulent Schmidt number is extremely case reliant. He presented his result by dimensionless mixing scalars and the results with useful turbulence models, RSM and k-ε as shown in Figure 8. Divide tint map ranges are used for experiments and computational results due to differences in the magnitudes of dimensionless mixing scalars. Figure 11 showed the allocation of the tracer at 50 mm downstream from the spacer. The result of the applied turbulence model is roughly unimportant at this point as the distributions look approximately the identical. Computational results are too qualitatively exceptionally comparable to the experimental results. On the other hand, quantitatively the dimensionless mixing scalar distributions vary visibly, which is evident in the unlike scale of the colour bars. Also, Walker et al. (2010) described the outcome of applied Sc_t when they considered mixing in T-junctions. They done that Sc_t had to be decreased to rate 0.1 in order to increase turbulent diffusion subsequent to a T-junction. They also concluded based on prior studies and found that the optimum value for turbulent Schmidt number ranges from 0.1 to 2.2. Based on these conclusions, the sensitivity study with Sc_t was conducted and the value of turbulent Schmidt number was varied in the range 0.1 to 1.5, and the pure convection case was also modelled. Baglietto and Ninokata (2005) found that the inclusion of adequate anisotropy modelling enables to accurately reproduce the wall shear stress distribution and velocity field in tight lattice fuel bundles. They were evaluating the influence of secondary flow and the significance of the anisotropy modelling for the dissimilar turbulence closures. They used different type of grid like Coarse (6048 nodes), Fine (15,888) and Refined (26,108) for the y^+ value at the near wall cell as 3.5, 1.1 and 0.7 respectively. Yu et al. (2012) have done research in subchannels for a 37-element simulated Canada Deuterium Uranium (CANDU) fuel bundle. Results from this study, obtained for different mass flow rates, revealed the presence of strong turbulent swirling flow in many bundle subchannels around the entrance region. They were adopted different meshing scheme as shown in Figure 3. A coarse uniform mesh is used for the sub-region far away from the bundle. A non-uniform mesh is used for the sub-region adjacent to the fuel bundle. A magnification of 1.1 is applied to cells from the upstream endplate to the boundary between the two pipe flow sub-regions. The adjacent regions are then coupled using the non-conformal meshing scheme. The endplate and bundle flow regions are meshed with the finest grid sizes. The whole computational domain contains about 6.3 million cells.

Podila and Rao (2016) said that proper mesh generation constitutes a key Factor affecting the numerical stability and the accuracy of the computation results as shown in figure 5. They found that for a intricate geometry, the time for mesh generation accounts for a huge portion of the simulation time as it is an iterative or trial-and-error process, and the mesh generation has to be frequent several times to attain a desired resolution. Moreover, generation of unstructured meshes requires larger quantity of memory per core especially when the mesh routines are not parallelised. For the rod bundle problem in hand, meshing was computationally challenging. They had explained that the Two-equation and Reynolds stress models (RSM) were assessed beside the measured turbulence intensities downstream of the split-vane spacer grid. The mesh was set on at the split-vane spacer grid area in an endeavour to exactly predict the increase in turbulence intensities downstream of the split-vane spacer grid region. The assessment of the CFD predictions was made with experiments from KAERI (Korean Atomic Energy Research Institute) that was provided to participants of the blind benchmark exercise. Finally, they had summarized that the three turbulence models tested all under predicted the turbulence intensities downstream of the split-vane spacer grid, with the Realizable k-ε model predicting the maximum values nearer to the experiments.

To establish the extent of the increase predicted by each of the models tested, the turbulence intensity at axial locations just downstream of the split-vane spacer grid was evaluated in Figure 3 (a). The data pulling out to observe the increase in turbulence intensity among two locations was facilitated by creating lines just downstream position of the split-vane spacer grid. The CFD predictions accessible in Figures 3 (a) symbolize the variation of turbulence intensities at all the three subchannels. As seen in Figures 3 (a) at the downstream location, the Realizable k-ε model predicted (up to 30%) higher turbulence intensities than the $k-\omega$SST model did. The vane area is connected with flow separation and re-circulation which the $k-\omega$SST model with high y+ wall treatment futile to arrest thereby foremost to lower predicted turbulent intensities. Dissimilar in the $k-\omega$SST model, the dissipation term in the Realizable k-ε model was derived from an exact equation for the transport of the mean square vortices fluctuation (STAR-CCM+ user manual, 2015). These positions need to be enhanced to arrest the flow physics downstream of spacers compared to the $k-\omega$SST model.

Zhang et al. (2017) applied structured hexahedral meshes at the bundle channels away from the grids, while tetrahedral meshes are used at the grids where geometry is quite complex as shown in Figure 4. The sensitivity analyses of mesh size were performed on the basis of maximum mesh size. They used different maximum mesh size like 0.5, 1.0, 2.0 and 4.0 mm for the mesh number 11283877, 8086472, 7208496, and 6702348 respectively. The solution times were 350, 264, 236 and 178 min respectively.

The flow characteristics at the specific area were represented by the streamlines at the outlet of spacer grid and mid span mixing grid as shown in Figure 9 and Figure 10 respectively. It was seen that at the

Table 1. Important footprints for recognition to authentication of the vortex flow in rod bundles and compound channels

S. No.	Published Year	Researcher	Turbulence model	Remarks
1.	2005	Baglietto and Ninokata	Quadratic *k-ε* model	This model explains gifted potential in this esteem is adjusted in its coefficients, and the familiar model is useful to fully developed flow in an infinite triangular array, with a variety of Reynolds numbers.
2.	2007	Chang and Tavoularis	URANS	The flow restricted large-scale coherent structures, which pretentious sturdily the local velocity fluctuations, close to the gaps among rods or among rods and the surrounding wall.
3.	2010	Ikeno and Kajishima	Large eddy simulation (LES)	This revision suggests that simulating the physics in the flow, as well as predicting the mean flow by means of Turbulence model is one of the mainly significant roles of CFD.
4.	2012	Yu et al.	*k-ε*, LES	Occurrence of swirling flow inside the bundle.
5.	2012	Liu et al.	*k–ω* (shear stress transport) SST, realizable *k-ε* (RKE) and RSM	The near-wall turbulence modelling was acknowledged as a momentous kindness when choosing a proper turbulence model for the simulations. The sensation of the SST *k–ω* turbulence model is endorsed to a near-wall treatment, which relies on damping of the turbulent viscosity in low Reynolds numbers regions rather than standard wall treatment as implemented for the turbulence models.
6.	2014	Cinosi et al.	RANS	A RANS model by means of any of *k-ε, k–ω* and Reynolds-stress turbulence models was enough for the forecast of mean velocity profiles, but they all three misjudge the time-averaged turbulent velocity components.
7.	2014	Podila et al.	Reynolds stress model (RSM), Standard *k-ε*	The URANS RSM solution method along with a mesh count of 15.31 million cells below predicted the turbulence intensity due to the generated vortices at the back of the spacers.
8.	2016	Podila and Rao	Realizable *k-ε*, *k–ω* SST and Reynolds stress models	The designed turbulence intensities were established to be reliant on the selection of the turbulence model used, the mesh density on the spacers as well as on the domain length upstream of the spacers.
9.	2016	Chen et al.	RNG *k-ε*	The RNG *k-ε* turbulence model can guess the turbulence flow superior than the SKE model and the RKE model.
10.	2016	Miku and Tiselj	Wall-resolved LES using WALE model	Large Eddy Simulation (LES) using WALE model is set up as an appealing accurate tool for simulations of the flow through the rod bundle, if adequately fine mesh is applied.
11.	2016	Batta and Class	RANS–CFD	Prediction of pressure drop across spacer grids by RANS–CFD using well resolved meshes in the spacer grid region yields more accurate results than any available correlation.
12.	2016	Merzari et al.	LES, *k–ωSST*	No turbulence model is capable to hold the flow in direct wake of the wire, when compared with LES. *k–ω*SST reproduces exactly the behaviour in the instantaneous proximity of the wire (when not in the wake) while all *k–ε* models miscalculate it.
13.	2016	Kang and Hassan	Steady Reynolds-Averaged Navier–Stokes (RANS)	Steady RANS come close to could not confine unsteady, large-scale cross-flow fluctuations and qualitative cross-flow blueprint modify due to the tangentially restrained test section.
14.	2017	Zhang et al.	*k–ωSST*	The upstream grid has no remarkable impact on the flow field formed at its downstream grids if the cross flow is sufficiently developed.
15.	2017	Jeong et al.	RANS	Authors presents a suitable way for a practical RANS based CFD methodology which is relevant to valid scale 217-pin wire-wrapped fuel assembly of KAERI (Korea Atomic Energy Research Institute) PGSFR (Prototype Gen-IV Sodium-cooled Fast Reactor).
16.	2017	Mikuž & Tiselj	URANS	Turbulent flow in rod bundles URANS (as well as steady-state RANS) simulations predicted mean velocity profiles fairly well.

Figure 3. Graphic of meshing scheme (Yu et al., 2012)

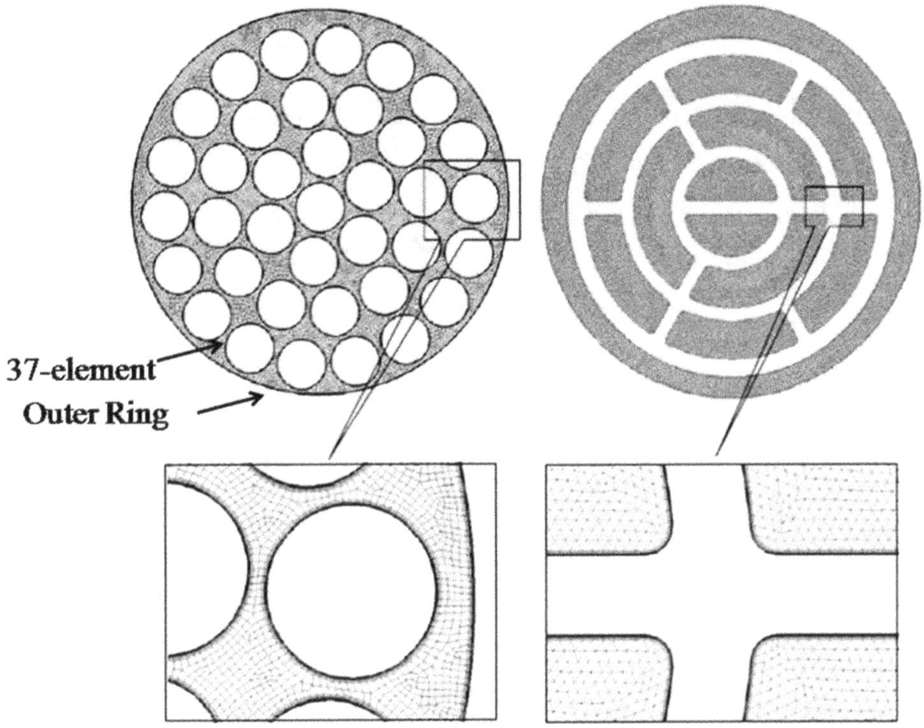

Figure 4. Exploded examination of the domain and boundary meshes for particular bundle and endplate subchannels in a cross section (Yu et al., 2012)

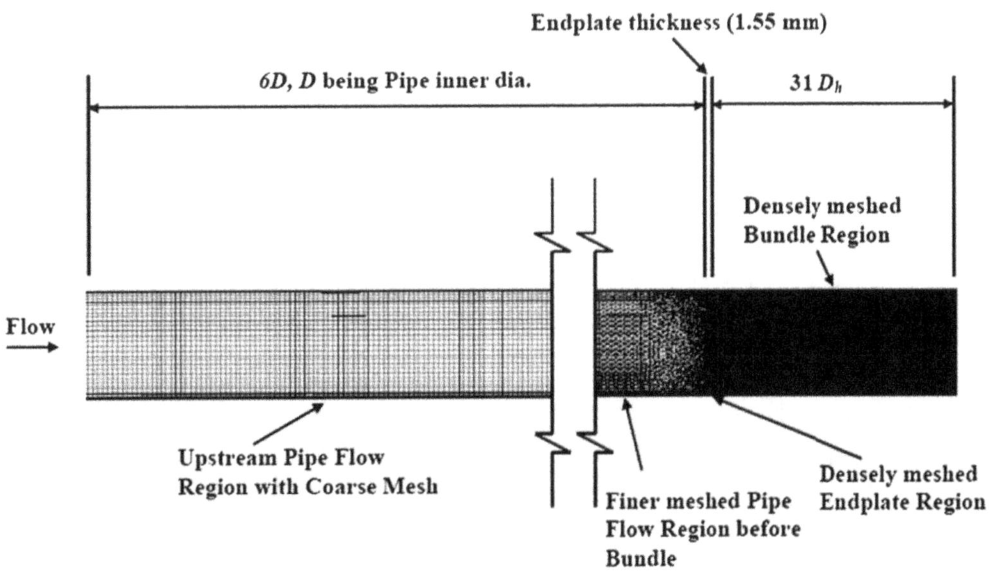

Figure 5. Mesh used for the 5-By-5 rod bundle simulations (Podila and Rao 2016)

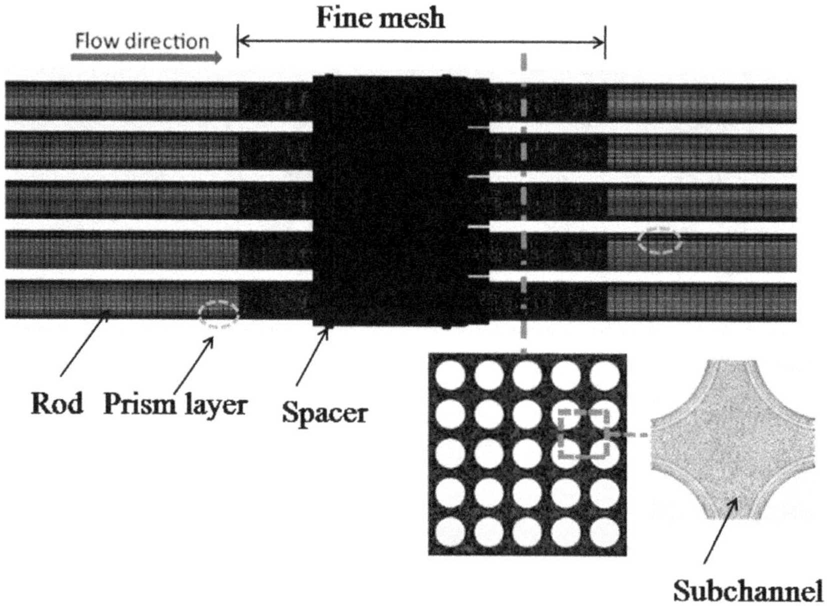

Figure 6. CFD predictions of the turbulence intensities using two-equation models at locations just downstream of the split-vane spacer grid (Podila and Rao, 2016)

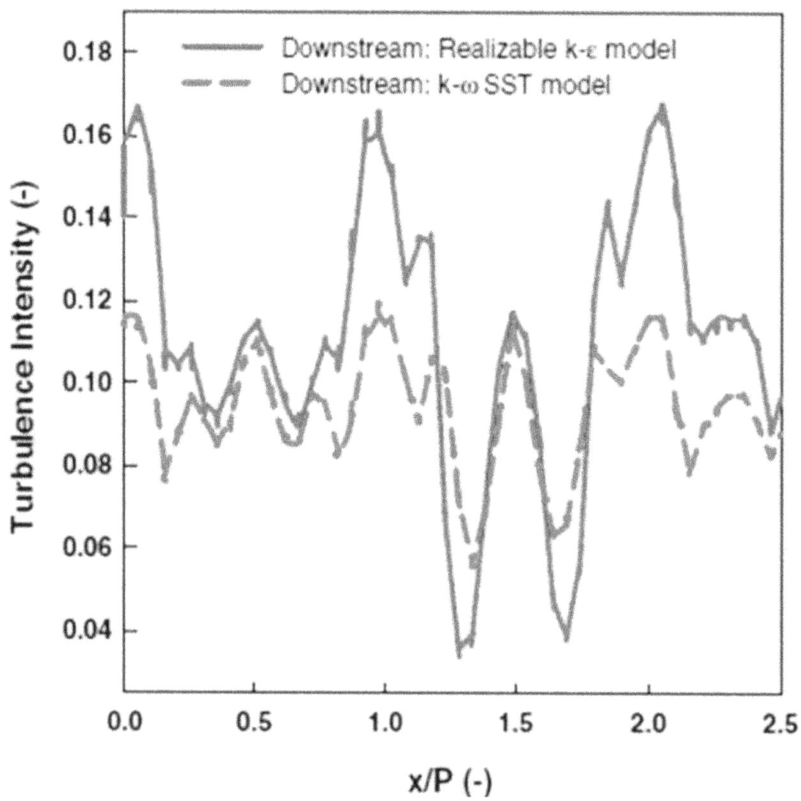

Figure 7. Applied mesh for pressurized water reactor (Zhang et al., 2017)

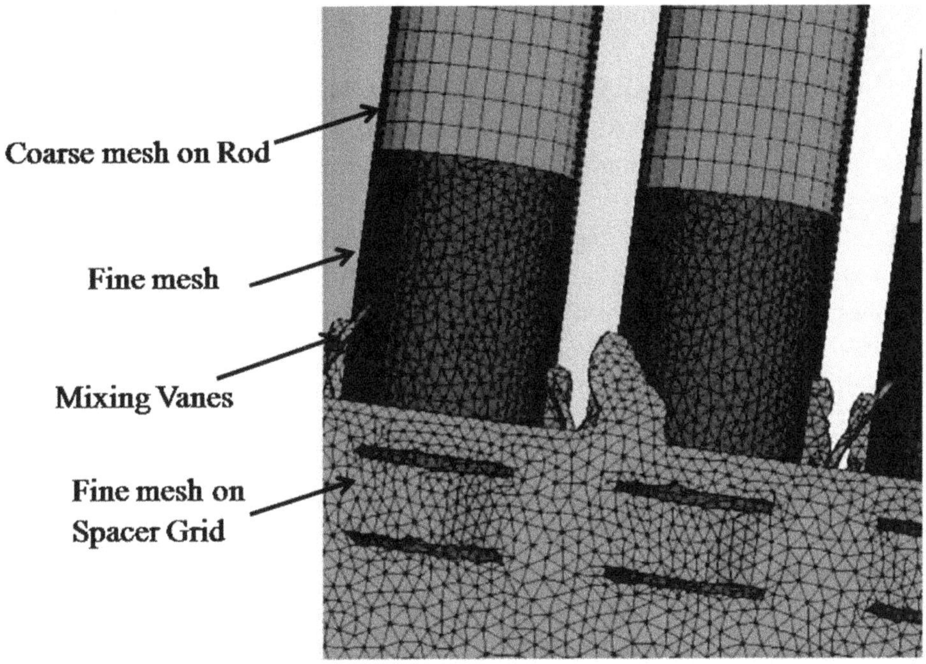

Figure 8. Cross-section used for comparative analysis (Zhang et al., 2017)

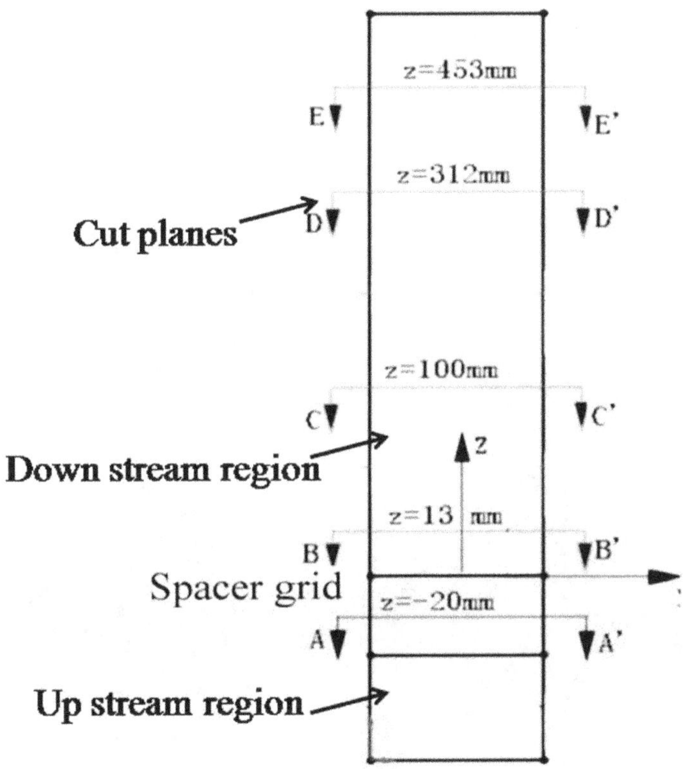

outlet of the spacer grid due to cross flow created among the neighbouring channels, comparatively big complete vortexes with their major axis being steady to the stretching direction of the mixing vane which were normally visible.

In addition, they found that, because of the dimple and spring, numerous smaller vortexes are likely to come into view close to the outsized vortex. As for the mid span mixing grid, due to an analogous structure of the mixing vane as mentioned above, no much variation of the streamline distribution at its exit can be observed (Figure 10).

Jeong et al. (2017) studied by adapting an innovative grid generation method using a Fortran based in-house code with a General Grid Interface (GGI) function in a general-purpose commercial CFD code, CFX. They suggested that, the RANS based CFD methodology can be successfully extended to the real scale 217-pin wire-wrapped fuel bundles of KAERI PGSFR.

Mikuž & Tiselj (2017) performed with the Shear Stress Transport (SST) turbulence model and automatic wall-treatment using OpenFOAM, an open-source CFD code. Results of URANS simulations are compared with the measurements of the MATiS-H experiment, which was performed at Korean Atomic Energy Research Institute (KAERI) in 2011–2012.

Issues, Controversies, Problems

In industrial computational fluid dynamic codes, there is more than one turbulence model built in. It is the client accountability to select one of those models, appropriate for the difficulty considered. In the preceding decade, numerous computations were offered using computational fluid dynamics for

Figure 9. Streamline at the outlet of spacer grid (Zhang et al., 2017)

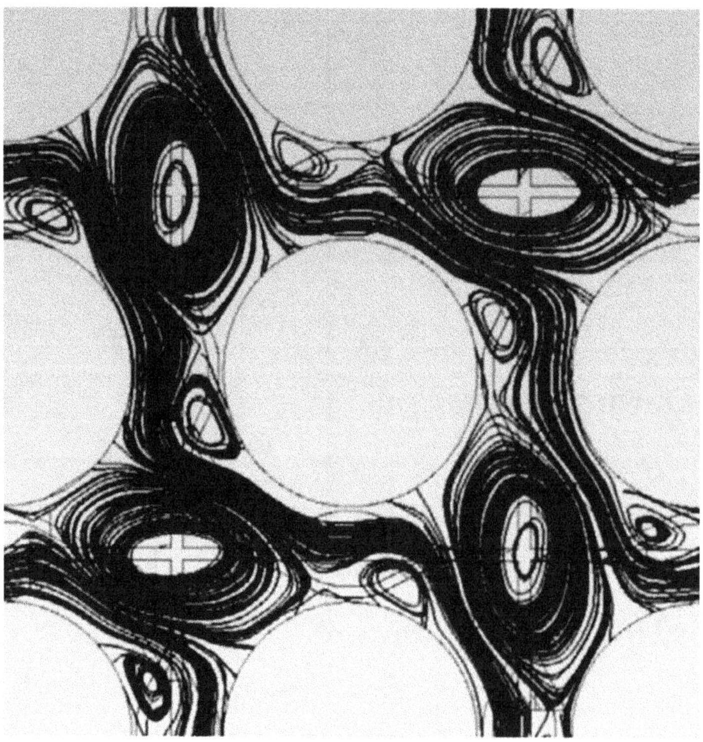

Figure 10. Streamline at the outlet of mid span mixing grid (Zhang et al., 2017)

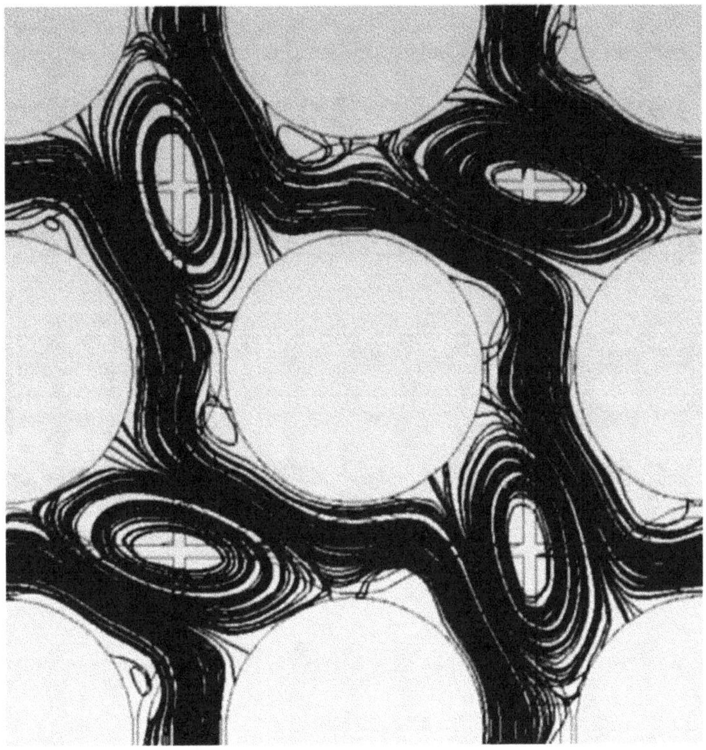

the simulation of a variety of effort of the nuclear trade. A familiar characteristic in an amount of those simulations is that they were done by means of the standard $k–\varepsilon$ turbulence model with no mitigating the preference of the model. The simulation results were hardly ever acceptable. In this chapter, we shall think the flow in a fuel rod bundle as a case study and argue why the claim of the standard $k–\varepsilon$ model unable to present logical results in this condition (Hazi, 2005). It is also showed that a turbulence model based on the Reynolds stress transport equations can give qualitatively truthful results. Normally, our plan is instructive; we would like to call the reader's concentration to the fact that turbulence models have to be certain based on speculative considerations and / or sufficient in sequence obtained from measurements.

In the years precedent numerous authors ended a challenge to simulate the hydrodynamics of fuel rod bundles by CFD techniques (Lestinen and Gango, 1999; Bergeron et al., 1999; Kriventsev and Ninokata, 1999; Rautaheimo et al., 1999; Tzanos, 2001, 2002, 2004). A universal characteristic of these simulations is that the researchers done the computations using $k–\varepsilon$ turbulence model. Evaluation of the simulation results with experimental data was rarely acceptable. Tzanos (2004) has given simulation results comparing the recital of variants of the $k–\varepsilon$ models. His final recommendation is that the models have to be enhanced in instruct to achieve realistic simulation results for fuel rod bundle simulations. However, the reasons of the malfunction of the standard $k–\varepsilon$ model and its clones were not elucidated. At this juncture, it is explained that the relevance of the standard $k–\varepsilon$ model is not reliable for rod bundle flows, fundamentally since the assumptions useful for the duration of the improvement of these models do not grasp in such flows. By means of the standard $k–\varepsilon$ model the results are not only inexact, but those are qualitatively incorrect. Additionally, it is revealed that former models, e.g., models based on the Reynolds-stress transport equations can give qualitatively truthful solutions.

Figure 11. Dimensionless mixing scalar distributions, from distances 50-250 mm (Ylonen, 2013)

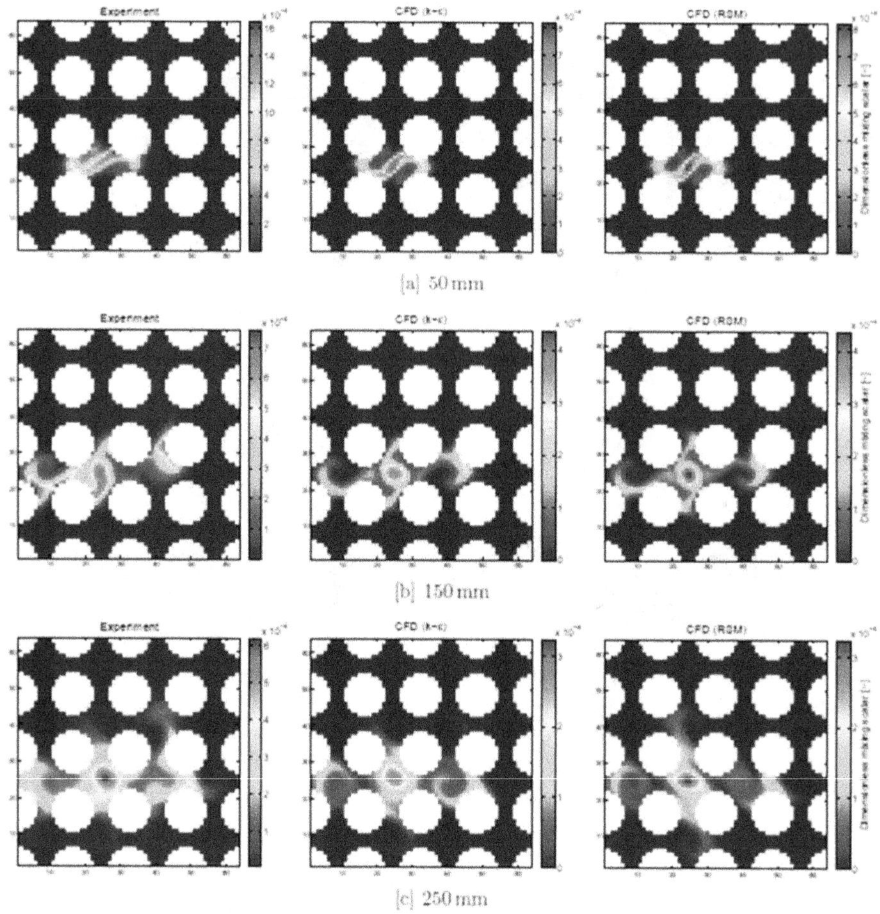

SOLUTIONS AND RECOMMENDATIONS

CFD category experiments are vital in the advance and validation of numerical methods and models worn to revision the spacer grid possessions. Rod bundle correlations are proprietary in nature. Empirical correlations are not reliable for new design of a rod bundle. So far, no CHF data generated for a rod bundle in BARC (power requirement ~ 6 MW and fuel simulator design). Mechanistic modelling of CHF is expected to provide good prediction. For AHWR with a new spacer design, establishing a methodology for the evaluation of thermal margin is necessary. Pre-test calculations can be performed and based on those results; probable candidates for a novel spacer grid design can be meticulous for the adiabatic experiments. The information would be then used mutually with post-test calculations to find the best candidate(s) for further expensive heated bundle experiments.

FUTURE RESEARCH DIRECTIONS

The spacer modeling is important to ascertain the maximum allowable power that can be produced in a fuel bundle. Accurate prediction of spacer effect provides higher thermal margin and hence leads to improve economy. The thermal-hydraulic aspects of spacing devices for water-cooled nuclear fuel bundles have been intentional in several laboratories. The grades explain that spacing devices can have great possessions on CHF. The possessions are typically encouraging but pessimistic things on CHF are reasonable with ill-designed spacing strategy. Bundle pressure drops are typically improved drastically by the occurrence of spacers. Development and validation of dry-out modeling code FIDOM-Rod yields a good prediction of CHF for Indian AHWR. CFD proved to be useful for spacer effect analysis for CHF.

CONCLUSION

The numerical simulation provides an exclusive nearby into the mixing concept in a rod bundle. A methodology for the dry-out analysis in BWR assemblies has been established and a computer code, FIDOM-Rod developed and validated. It was revealed that from previous investigation, the k-ε model could not replicate the fundamental description of such flows. On the erstwhile dispense, it had been confirmed that the Reynolds-stress model could be a superior entrant for the exact modelling of rod bundles. It is momentarily discussed that these conclusions could be drained a priori, studying the experimental grades and the assumptions useful throughout the expansion of the turbulence models. A simple spacer model for AHWR has been proposed. The minimum critical power ratio of the AHWR core was evaluated to be 1.44. (Required typical MCPR is ~1.1 for BWR). Due to adequate thermal margin, the possibility of power up rating of AHWR exists (at least by 15-20% considering 10% prediction uncertainty). Limited elevated resolution concentration dimensions shown momentous differences among the tested spacer grid designs. Cross-mixing presentation depended particularly sturdily on the deal of the mixing vanes. The inference of lateral velocities presented supplementary in order on the flow deeds downstream of the spacer grid.

REFERENCES

Baglietto, E., & Ninokata, H. (2005). A turbulence model study for simulating flow inside tight lattice rod bundles. *Nuclear Engineering and Design*, *235*(7), 773–784. doi:10.1016/j.nucengdes.2004.10.007

Bhattacharya, A., Yu, S. D., & Kawall, G. (2012). Numerical simulation of turbulent flow through a 37-element CANDU fuel bundle. *Annals of Nuclear Energy*, *40*(1), 87–105. doi:10.1016/j.anucene.2011.10.017

Chang, D., & Tavoularis, S. (2007). Numerical simulation of turbulent flow in a 37–rod bundle. *Nuclear Engineering and Design*, *237*(6), 575–590. doi:10.1016/j.nucengdes.2006.08.001

Cheng, S., Chen, H., & Zhang, X. (2017). CFD analysis of flow field in a 5×5 rod bundle with multi grid. *Annals of Nuclear Energy*, *99*, 464–470. doi:10.1016/j.anucene.2016.09.053

Cinosi, N., Walker, S. P., Bluck, M. J., & Issa, R. (2014). CFD simulation of turbulent flow in a rod bundle with spacer grids (MATIS-H) using STAR-CCM+. *Nuclear Engineering and Design*, *279*, 37–49. doi:10.1016/j.nucengdes.2014.06.019

Combest, D. P., Ramachandran, P. A., & Dudukovic, M. P. (2011). On the Gradient Diffusion Hypothesis and Passive Scalar Transport in Turbulent Flows. *Industrial & Engineering Chemistry Research*, *50*(15), 8817–8823. doi:10.1021/ie200055s

Dominguez-Ontiveros, E., & Hassan, Y. A. (2014). Experimental Study of a Simplified 3x3 Rod Bundle Using DPTV. *Nuclear Engineering and Design*, *279*, 50–59. doi:10.1016/j.nucengdes.2014.04.037

Hanson, A. S., & Todreas, N. (1977). *Fluid mixing studies in a hexagonal 61-pin, wire wrapped rod bundle (tech. report)*. doi:10.2172/5159920

Ikeno, T., & Kajishima, T. (2010). Analysis of dynamical flow structure in a square arrayed rod bundle. *Nuclear Engineering and Design*, *240*(2), 305–312. doi:10.1016/j.nucengdes.2008.07.012

Ikeno, T., & Kajishima, T. (2010). Analysis of dynamical flow structure in a square arrayed rod bundle. *Nuclear Engineering and Design*, *240*(2), 305–312. doi:10.1016/j.nucengdes.2008.07.012

International Atomic Energy Agency (IAEA). (2003). IAEA-TECDOC-1332: Safety margins of operating reactors (tech. report).

Jae-Ho, J., Min-Seop, S., & Kwi-Lim, L. (2017). RANS based CFD methodology for a real scale 217-pin wire-wrapped fuel assembly of KAERI PGSFR. *Nuclear Engineering and Design*, *313*, 470–485. doi:10.1016/j.nucengdes.2017.01.007

Krauss, T., & Meyer, L. (1996). Characteristics of turbulent velocity and temperature in a wall channel of a heated rod bundle. *Experimental Thermal and Fluid Science*, *12*(1), 75–86. doi:10.1016/0894-1777(95)00076-3

Meyer, L. 2010. From discovery to recognition of periodic large scale vortices in rod bundles as source of natural mixing between sub channels, - A Review. *Nuclear Engineering and Design*, *240*(6), 1575 – 1588. doi:10.1016/j.nucengdes.03.014

Mikuž, B., & Tiselj, I. (2016). Wall-resolved Large Eddy Simulation in grid-free 5 x 5 rod bundle of MATiS-H experiment. *Nuclear Engineering and Design*, *298*, 64–77.

Blaž, M., & Iztok, T. (2017). URANS prediction of flow fluctuations in rod bundle with split-type spacer grid. *International Journal of Heat and Fluid Flow*, *64*, 10–22. doi:10.1016/j.ijheatfluidflow.2017.01.008

Podila, K., & Rao, Y. (2016). CFD modelling of turbulent flows through 5×5 fuel rod bundles with spacer-grids. *Annals of Nuclear Energy*, *97*, 86–95. doi:10.1016/j.anucene.2016.07.003

Podila, K., Rao, Y. F., Krause, M., & Bailey, J. (2014). A CFD simulation of 5 X 5 rod bundles with split-type spacers. *Progress in Nuclear Energy*, *70*, 167–175. doi:10.1016/j.pnucene.2013.08.012

Rehme, K. (1973). Pressure drop correlations for fuel element spacers. *Nuclear Technology*, *17*, 15.

Rehme, K. (1987). The structure of turbulent flow through rod bundles. *Nuclear Engineering and Design*, *99*, 141–154. doi:10.1016/0029-5493(87)90116-6

Reynolds, O. (1895). On the dynamical theory of incompressible viscous fluids and the determination of the criterion. *Philosophical Transactions of the Royal Society of London*, *186*(0), 123–164. doi:10.1098/rsta.1895.0004

Saxena, A. (2014). Thermal-hydraulic numerical simulation of fuel sub-assembly for Sodium-cooled Fast Reactor [Ph.D. Thesis]. CEA Cadarache, Aix-Marseille University.

Seale W. J. (1979). Turbulent Diffusion of Heat between Connected Flow Passages, Part 1: Outline of Problem and Experimental Investigation. *Nuclear Engineering and Design*, *54*(2), 183-195. (79)90166-310.1016/0029-5493

STAR-CCM+. (2015). Retrieved from http://www.cd-adapco.com/products/STAR-CCMplus/

Trippe, G., & Weinberg, D. (1979b). Non-isotropic eddy viscosities in turbulent flow through rod bundles. In S. Kakac & D. B. Spalding (Eds.), *Turbulent Forced Convection in Channels and Bundles* (Vol. 1, p. 505). Hemisphere, Washington.

Trupp, A. C., & Azad, R. S. (1975). The structure of turbulent flow in triangular array rod bundles. *Nuclear Engineering and Design*, *32*(1), 47–84. doi:10.1016/0029-5493(75)90090-4

Walker, C., Manera, A., Niceno, B., Simiano, M., & Prasser, H.-M. (2010). Steady-state RANS-simulations of the mixing in a T-junction. *Nuclear Engineering and Design*, *240*(9), 2107–2115. doi:10.1016/j.nucengdes.2010.05.056

Experiments and CFD Code Applications to Nuclear Reactor Safety (XCFD4NRS).

Waters, E. D. (1961). *Fluid mixing experiments with a wire wrapped 7-rod bundle fuel assembly*. Annular Space.

Ylonen, A. T. (2013). High-resolution flow structure measurements in a rod bundle [Ph.D. Thesis]. Lappeenranta University of Technology.

Zimmermann, M. (2015). Development and Application of a Model for the Cross-Flow Induced by Mixing Vane Spacers in Fuel Assemblies [Ph.D. Thesis]. Karlsruhe Institute of Technology.

KEY TERMS AND DEFINITIONS

Boiling Water Reactor (BWR): A common type of light water reactor, where water is allowed to boil in the core thus generating steam directly in the reactor vessel.

Computational Fluid Dynamics (CFD): CFD is anxious with obtaining numerical solution to fluid flow, heat transfer, mass transfer, chemical reactions and associated phenomena effort using computers.

Coolant: The liquid circulated through the reactor core to remove heat generated by the core. It can be a liquid (usually water or molten sodium) or a gas (usually carbon dioxide or helium).

Critical Heat Flux (CHF): The CHF is defined as the maximum heat flux from the fuel rod to the coolant before the boiling crisis occurs.

Light Water Reactor (LWR): A common nuclear reactor cooled and usually moderated by ordinary water. It is a generic designation including BWR and PWR types.

Pressurised Water Reactor (PWR): The most common type of light water reactor (LWR), it uses water at very high pressure in a primary circuit and steam is formed in a secondary circuit.

Spacer: It is a device to maintain the rod bundle configuration for coolant flow.

Turbulent Mixing: Due to turbulent fluctuations, there is a lateral movement of fluid particles. The same mass that is transported over the gap is replaced by an equal mass from the adjacent sub-channel, so there is no net mass transfer in single-phase flow.

Chapter 13
Summability Techniques and Their Applications in Soft Computing

Smita Sonker
National Institute of Technology Kurukshetra, India

Alka Munjal
National Institute of Technology Kurukshetra, India

ABSTRACT

Summability methods are a useful tool in dealing with the problems in the soft computing like in filtering of the signals and for stabilizing the systems. Signals can be in the form of various types of series (Infinite Series, Fourier series, etc.) and hence, summability theory is applicable in finding the error of approximation and degree of approximation of such signals. In this chapter, the authors gave an introductory discussion on summability theory and approximation of the signals. Further, they explained about the stability of the frequency response of the system. Also, they used the Fourier approximation in the soft computing models (multilayer perceptrons; radial basis function (RBF) or regularization networks, and fuzzy logic models) and found the output data of requirement.

INTRODUCTION

In this modern world, the soft computing becomes very effective in developing the functional and robust intelligent systems. The theory of the soft computing is developed by mathematical and logical frameworks considering the natural attributes in learning and reasoning. It enables the computing devices with more reasonable and logical for an effective and intelligent functioning. The soft computing exists in every area of our life which uses the technology like computer systems, transport vehicles, appliances, or the things with hidden digital electronics.

Summability techniques are very much applicable in soft computing as a tool to filter the signals in the form of series (Infinite Series, Fourier series, wavelets etc.) (Bachmann et al., 2012) using the various summability methods (Zygmund, 1988) such as matrix summability, Cesàro and Generalized

DOI: 10.4018/978-1-5225-3035-0.ch013

Cesàro summability, Hölder summability, Harmonic summability, Riesz's and Riesz's typical means summability, Nörlund and Generalized Nörlund summability, indexed summability, Abel summability, Euler summability, Borel summability, Hausdorff summability, Banach summability etc. (Dutta and Rhoades, 2016). In this chapter, we will discuss the stability of the system using the summability and absolute summability methods.

Kumar et al. (2010) determined the new approach for Generating Parametric Orthogonal Wavelets in which scaling function filter represent as a product of two Laurent polynomials. In 2011, they used Discrete Wavelet Transform in place of Fourier transform and improve the results. Also, Kumar et al. (2016) determined the Wavelet Variance, Covariance and Correlation Analysis of BSE and NSE Indexes Financial Time Series. In 2016, Sonker and Munjal (2016a) determined a theorem on generalized absolute Cesàro summability with the help of sufficient conditions for infinite series and in (2016b), they used the concept of triangle matrices for obtaining the minimal set of sufficient conditions of infinite series to be bounded. In 2017, Sonker and Munjal (2017a) obtained boundness conditions of absolute summability $\phi - | A |_k$ factors. In this way by using the advanced summability method, we can improve the quality of the filters.

Summability Theory

In broad, summability theory is the theory of the assignment of limits, which is fundamental in analysis, function theory, topology and functional analysis. In mathematical analysis, a summability method is a famous formulation for convergence of a divergent series. Summability is a field in which we study the non-convergent series/ integrals and assigns a value (number) to it.

For instance, followings are some interesting cases, for which the assignment of limits can be done,

- Real or complex sequences for the limit process '$n \to \infty$',
- Series (convergence of series),
- Sequences and series of functions like power series, Fourier series, etc.,
- Limit of a function at a point (continuity, continuous extension),
- Differentiation of functions,
- Integration of functions.

In 1890, Cesàro (1890) deals with the sum of some divergent series which later known as Cesàro summation. In specialized mathematical contexts, values can be objectively assigned to certain series whose sequence of partial sums diverges. This is to make meaning of the divergence of the series.

For example, Cesàro summation assigns Grandi's series $1 - 1 + 1 - 1 + 1 - \ldots$ the value 1/2.

There exist three types of summability,

1. Ordinary Summability
2. Absolute Summability
3. Strong Summability

- **Ordinary Summability:** Let $\sum u_n$ be an infinite series of real numbers with a sequence of partial sums $\{s_n\}$. Let $T \equiv (a_{n,k})$ be an infinite matrix with real or complex constants. Then the sequence-to-sequence transformation,

$$t_n = \sum_{k=0}^{n} a_{n,k} s_k, \quad n = 0,1,2,\ldots$$

defines the matrix transform of the sequence $\{s_n\}_{n=1}^{\infty}$. Here the column vector of the t_n is the product of the matrix T with the column vector of the s_n. The sequence $\{s_n\}$ or the series $\sum u_n$ is said to be matrix summable to s, if $\lim_{n\to\infty} t_n = s$.

- **Absolute Summability:** The infinite series $\sum_{n=0}^{\infty} a_n$ with the sequence of the partial sum $\{s_n\}$ is absolute summable by the method A (A-summable) to the limit s, if it is A-summable to s, i.e. if $\lim_{n\to\infty} t_n = s$ and if the sequence $\{t_n\}$ is of bounded variation:

$$\sum_{n=1}^{\infty} \left| t_n - t_{n-1} \right| < \infty$$

- **Absolute Summability Of Index q:** The infinite series $\sum_{n=0}^{\infty} a_n$ with the sequence of the partial sum $\{s_n\}$ is absolute summable with the index q by the method A (A-summable) to the limit s, if it is A-summable to s, i.e.

$$\sum_{k=1}^{n} k^{q-1} \left| t_k - t_{k-1} \right|^q < \infty, \quad \text{as } n \to \infty,$$

and if $t_n \to s$, as $n \to \infty$. It is denoted by $\left| A, q \right|$.

- **Strong Summability:** The infinite series $\sum_{n=0}^{\infty} a_n$ with the sequence of the partial sum $\{s_n\}$ is strong summable with index q by the method A (A-summable) to the limit s, if it is A-summable to s, i.e. if $\sum_{k=1}^{n} k^q \left| t_k - t_{k-1} \right|^q = O(n)$, as $n \to \infty$, and if $t_n \to s$, as $n \to \infty$. It is denoted by $\left[A, q \right]$.

The following inclusion relations hold,

$$\left| A, q \right| \subset \left[A, q \right] \subset \left(A \right).$$

- **Methods for Summability:** These are some methods of summability,
  - $(C, 1)$ means when $a_{n,k} = \frac{1}{n+1}, \quad 0 \le k \le n.$
  - Harmonic means when $a_{n,k} = \frac{1}{(n-k+1)\log n}, \quad 0 \le k \le n.$
  - (C,δ) means when $a_{n,k} = \frac{E_{n-k}^{\delta-1}}{E_n^{\delta}}, \quad 0 \le k \le n.$
  - Nörlund means when $a_{n,k} = \frac{p_{n-k}}{P_n}, \quad 0 \le k \le n.$
  - Riesz mean when $a_{n,k} = \frac{p_k}{P_n}, \quad 0 \le k \le n.$
  - General Nörlund mean (N, p, q) when $a_{n,k} = \frac{p_{n-k}q_k}{R_n}$ where $R_n = \sum_{k=0}^{n} p_k q_{n-k}.$
  - **Deferred Cesàro Mean:** Agnew define the Deferred Cesàro mean of the sequence $x = (x_k)$ by $\left(D_{p,q}\right)_n = \frac{1}{q(n) - p(n)} \sum_{k=p(n)+1}^{q(n)} x_k$ where $q(n)$ and $p(n)$ are sequences of positive natural numbers satisfying $q(n) < p(n)$ and $\lim_{n\to\infty} q(n) = \infty.$

Trigonometric Fourier Series

Let *f* be a 2π-periodic function, integrable over the interval $\left[-\pi, \pi\right]$. Let the Trigonometric Fourier series associated with *f* be

$$s_n\left(f;x\right) = \frac{a_0}{2} + \sum_{k=1}^{n}\left(a_k coskx + b_k sinkx\right), \ \forall n \ge 1 \quad \text{with} \quad s_0\left(f;x\right) = \frac{a_0}{2}$$

denotes the $(n + 1)^{th}$ partial sums, called trigonometric polynomials of nth degree, of the Fourier series of *f*. The conjugate series of the Fourier series of *f* is defined by $\sum_{n=1}^{\infty}\left(b_n cosnx - a_n sinnx\right) = \sum_{n=0}^{\infty} v_n$ and its nth partial sum is defined as

$$\tilde{s}_n\left(f;x\right) = \sum_{k=1}^{n}\left(b_k coskx - a_k sinkx\right), \quad \forall n \ge 1 \ \text{ and } \ \tilde{s}_n\left(f;x\right) = 0$$

where

$$a_k = \frac{1}{\pi}\int_{-\pi}^{\pi} f(x)\cos kx\, dx, \quad k = 0,1,2,\ldots$$

and

$$b_k = \frac{1}{\pi}\int_{-\pi}^{\pi} f(x)\sin kx\, dx, \quad k = 1,2,3,\ldots$$

are called the Fourier coefficients of f. The sequence of partial sums of series $\sum_{k=0}^{\infty} u_k(x)$, given by

$s_n\left(f;x\right) = \frac{a_0}{2} + \sum_{k=1}^{n}\left(a_k \cos kx + b_k \sin kx\right)$, is a trigonometric polynomial of order n.

- **Convergence of Fourier Series:** The key to the convergence of the Fourier series of the integrable function f is the smoothness of f. The smoother f is (i.e., the more continuous f and its derivatives are), the better the convergence of its Fourier series to f. If f is merely continuous at a point, its Fourier series may diverge, but if f is piecewise smooth and continuous, series converges uniformly and absolutely to f. the convergence is also faster if the function is smoother. The essential content is that if f is continuously differentiable k times, then its Fourier coefficients a_n and b_n go to 0 faster than $1/n_k$.

Approximation of Functions

The approximation theory generally selects a closely matched target function among the well-defined class function. The requirement of approximating the function is very popular in engineering and physical sciences. This branch of analysis investigates the certain known signals and approximated it to a desirable signal using the specific class of functions (polynomial and Fourier series) and the elementary properties (limit values, continuity and integrability etc.). Depending on the structure of the domain and co-domain of the problem, several techniques of approximation may be applicable.

In Soft computing, approximating functions are used to describe the behavior of the activity of a neural network, Support Vector Machine (SVM) Network, Radial Basis Function (RBF) network etc. with the help of a Fourier truncated series and a polynomial approximating function. The approximation can be done by two ways: Polynomial approximation and Fourier approximation.

According to the Weierstrass theorem, a continuous real-valued bounded function can be approximated with the help of the sup-norm if the order of the polynomial function is infinity while a discontinuous even function can be approximated by a norm other than the sup-norm. Taylor series expansion is used for the polynomial approximation of the function and the quality of the approximation depends on the number of terms taken. Of course, for a function to have a Taylor series, it must be infinitely differentiable in some interval, and this is a very restrictive condition. But in Fourier approximation, sines and cosines functions are used and serve as much more versatile elements than powers of any variable.

Approximations of a real-valued continuous function $f(x)$ by an approximation function $T(x)$ (by Fourier series, orthogonal series) have a finite number of the terms. Our main focus is to solve the engineering problem (a set of sparse or noisy training data points) by approximation of the continuous function with the help of summability theory with minimized error and to find the degree of approximation. We will give some theoretical explanation (with the help of theorems) which is required for the degree of approximation and error minimization of data.

The major items in an approximation problem are the type of the approximating function applied and measure the goodness of an approximation. This is also known as the question of choosing the *form* and *norm*. The choice of the approximating function (form) is more important than the choice of a measure of goodness, that is, a distance function or norm that measures the distance between f and f_a. Unfortunately, there is no theoretical method of determining which out of many possible approximating functions will yield the best approximation. Before examining the approximation properties of models, we can consider some basic shortcomings of classical approximation schemes. On the other hand, there are only a few feasible functions in use or under investigation. The most popular functions are tangent hyperbolic, a few radial basis functions (notably Gaussians and multi-Quadrics), polynomial functions, and three standard membership functions applied in fuzzy models (triangle, trapezoid, and singleton). These functions are called activation, basis and, membership functions in multilayer perceptrons; radial basis function (RBF) or regularization networks, and fuzzy logic models. These models are, together with support vector machines, the most popular soft modeling (Herrero et al., 2015) and learning functions. Their mathematical forms follow.

- **Multilayer Perceptron:** A multilayer perceptron is a representation of the non-linear basis function expansion (approximation)

$$o = f_a\left(x, w, v\right) = \sum_{i=1}^{N} w_i \varphi_i\left(x, v_i\right)$$

where $\varphi_i\left(x, v_i\right)$ is a set of given functions.

Radial Basis Function (RBF) Network

A radial function ϕ generally satisfies the property $\phi\left(\mathrm{x}\right) = \phi\left(\mathrm{x}\right)$. The RBF is a real-valued function which depends only on the distance from the origin or a centre c so that $\phi\phi\left(\mathrm{x}\right) = \phi\left(\mathrm{x}\right)$; or $\phi\left(\mathrm{x},\ \mathrm{c}\right) = \phi\left(\mathrm{x} - \mathrm{c}\right)$. Sums of these functions are commonly used to approximate functions. A RBF network is a representative of a linear basis function expansion

$$o = f_a\left(x, w\right) = \sum_{i=1}^{N} w_i \varphi_i\left(x\right).$$

- **Fuzzy Logic Model (Kecman, 2001):** A fuzzy logic model, like an RBF network, can be a representative of a linear or non-linear basis function expansion:

$$o = f_a(x, c, r) = \frac{\sum_{i=1}^{N} G(x, c_i) r_i}{\sum_{i=1}^{N} G(x, c_i)}.$$

Algebraic Polynomial and Truncated Fourier Series Expansion:

1. An algebraic polynomial expansion:

$$o = f_a(x) = \sum_{i=0}^{N} w_i x^i$$

2. A truncated Fourier series expansion:

$$o = a_0 + a_1 sin(x) + b_1 cos(x) + a_2 sin(2x) + \ldots + a_n sin(nx) + b_n cos(nx)$$

The major question to be answered is the choice of the norm (the distance between the data and the approximating function ($f_a(x, w)$). This choice is less important than, the choice of form $f_a(x, w)$. If $f_a(x, w)$ is compatible with an underlying function $f(x)$ that produced the training data points, then almost any reasonable measure will lead to an acceptable approximation to $f(x)$. If $f_a(x, w)$ is not compatible with $f(x)$, none of the norms can improve bad approximations to $f(x)$. However, in many practical situations, one norm of the approximation is naturally preferred over another.

Norm

A norm on V is a function p: $V \rightarrow R\ \forall\ a \in F$ and all $u, v \in V$ with the following properties:

1. $p(a\,v) = |a|\,p(v)$,
2. $p(u + v) \leq p(u) + p(v)$,
3. If $p(v) = 0$, then v is the zero vector,
4. $p(v) \geq 0$ (non-negativity).

A well-known L_p norm is most commonly used tool in the study of Fourier series. In letting $p \rightarrow \infty$, the L_p norm becomes the essential upper bound and L_p behavior formally becomes Lipschitz behavior. The results on Fourier series can be generalize by using the more general classes of functions in place of power function

- **L_0 Norm:** The first norm is a L_0-norm. L_0 -norm of vector *x* is

$$x_0 = \sqrt[0]{\sum_i x_i^0}$$

that is a total number of non-zero elements in a vector.

- **L_1 Norm (Absolute Value Norm):** It is sum of absolute difference between two vectors or matrices,

$$x_1 = \sum_i \left| x_i - x_{i-1} \right|.$$

- **L_2 Norm (Euclidean Norm):** A most popular application of Euclidean norm is in the signal processing field as the Mean-Squared Error (MSE) measurement. It is a sum of squared difference

$$x_1 - x_{22} = \sqrt{\sum_i \left(x_{1_i} - x_{2_i} \right)^2}.$$

- **L_p Norm:** It is given by

$$x_1 - x_{2p} = \sqrt[p]{\sum_i \left(x_{1_i} - x_{2_i} \right)^p}, \qquad 1 \leq p < \infty.$$

- **L_∞ Norm:** It is the maximum entries' magnitude of the vector

$$x_\infty = max\left(\left| x_i \right| \right)$$

Norm of Approximation

The norm of the approximation measures the match of a specific approximation $f_a(x)$ to the given set of noisy data. It is a positive scalar to measures the error, length, size, distance, etc. Here a norm usually represents an error. The L_p (Holder) norm is the most common mathematical class of norms to measure the discrete data set. The L_p norm is a *p*-norm of an error vector *e*. The choice of the measure of the goodness of the approximation depends on the simplicity of approximation and the algorithms of the optimization. The L_2 norm is the best one for data corrupted with normally distributed (Gaussian) noise. In this case, it is known that the estimated parameters or weights obtained in L_2 norm correspond to the maximum-likelihood estimates. The L_1 norm is much better than the Euclidean norm for data that have outliers because the L_1 norm tends to ignore such data. The Chebyshev norm is ideal in the case of exact data with errors in a uniform distribution.

Deficiency of Polynomial Approximations

The example below shows the deficiency of polynomial approximations. They behave badly in the proximity of domain boundaries even though they are the perfect interpreter of giving training data points.

Example 1: Approximate function $f(x) = 1/(1 + 25x^2)$ defined over the range [- 1, + 1] and sampled at 21 equidistant points x_i, $i = 1, 2, \ldots, 21$, (i.e., $P = 21$) by polynomials of the twelfth, sixteenth, and twentieth orders.

For this particular function (see Figure 1) it is found that for any point $x \neq x_i$, where $|x| > 0.75$, the error $|f_a(x) - f(x)|$ of the approximation increases without bound as the order of the approximating polynomials n increases. This is true even though $f_a(x_i) = f(x_i)$ when the order of the polynomial, $n = P - 1$, where P is the number of training data. In this example, $f_a(x_i) = f x_i)$ for a 20th order polynomial, which means that both L_1 and L_2 norms are equal to zero, wrongly suggesting perfect errorless approximation over the whole domain of input variable x.

When considering an error over a whole range, a more satisfactory objective is to make the maximum error as small as possible. This is the *minimax* or a Chebyshev type of approximation and the function $f_a(x)$ is chosen so that L_∞ is minimized. It is in this context that the Chebyshev polynomials have found a wide range of applications.

There are also many other polynomials, notably a class of orthogonal polynomials, that can be used for approximations. All of them have similar deficiencies in the sense that as the number of training points P increases, the approximation improves near the center of the interval but shows pronounced oscillatory behavior at the ends of the interval.

Both the polynomials and trigonometric sums have similar disadvantages in that they cannot take sharp bends followed by relatively flat behavior. Both functions are defined over the whole domain (i.e., they are globally acting activation functions), and they generally vary gently. Such characteristics can be circumvented by increasing the degree of these functions, but this has a consequence wild behavior of the approximator close to boundaries (Figure 1).

Truncated Fourier Approximation

In truncated Fourier approximation, Sines and Cosines functions are used. It serves as much more versatile elements than powers of any variable. Sines and Cosines are not only helpful in an approximation of non-analytic but also in the wildly discontinuous functions. In signal analysis, Fourier approximation has assumed important new dimensions due to their wide range of applications. In obtaining the trigonometric functions on an analog computer, a set of completely new problems must be solved. One of these is "frequency response" which has to do with the highest frequency of interest contained in the input signal. For example, the sine of an angle can be obtained by means of a suitably wound potentiometer which produces the proper voltage as a result of a shaft rotation proportional to the angle. Obviously, a scheme of this kind is only suitable for slowly changing angles, i.e., frequencies less than 10 cps.

Figure 1. Polynomial approximations of a function 1 /(1 + 25x²)

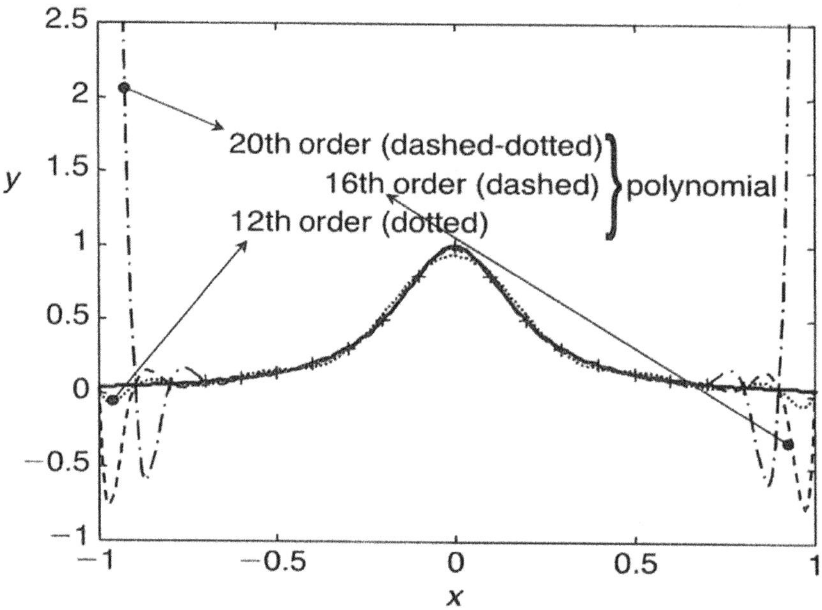

Fourier approximations based upon the summation of a series of orthogonal polynomials. The trigonometric polynomials are dense in the set of continuous functions on the unit circle (with the uniform norm). Thus, it is a demonstrate way for a function to approximated by a trigonometric polynomial.

Trigonometric Fourier Series Approximation With the Help of Summability Method (Cesàro Sub-Methods)

Let f be 2π-periodic (bounded, integrable) and $f \in L_p\,[0, 2\pi] = L_p$ for $p \geq 1$. Then,

$$s_n\left(f;x\right) = \frac{1}{2}a_0 + \sum_{k=1}^{n}(a_k \sin kx + b_k \cos kx) \equiv \sum_{k=1}^{n}u_k\left(f;x\right)$$

is the partial sum of the first $(n+1)^{\text{th}}$ terms of the Fourier series of $f \in L_p\left(p \geq 1\right)$ at a point x.

Then, it is known that

$$s_n\left(f;x\right) - f\left(x\right) = \frac{2}{\pi}\int_0^{\pi}\varnothing_x\left(t\right)\frac{sin\left(n+1/2\right)t}{2sin\left(t/2\right)}dt$$

where

$$\varnothing_x(t) = \frac{1}{2}\{f(x+t) + f(x-t) - 2f(t)\}.$$

A measurable 2π-periodic function w: $[0, 2\pi] \rightarrow [0, \infty]$ is called a weight function if the set $w^{-1}(\{0,\infty\})$ has the Lebesgue measure zero and $f \in L_w^p([0,2\pi])$ is the weighted Lebesgue space of all measurable 2π-periodic functions if

$$f_{p,w} = \left(\int_0^{2\pi} |f(x)|_p \, w(x)\, dx\right)^{1/p} < \infty$$

A weight function w belongs to the Muckenhoupt class $A_p(1 < p < \infty)$, if

$$sup\left(\frac{1}{|I|}\int_I w(x)\, dx\right)\left(\frac{1}{|I|}\int_I [w(x)]^{-1/(p-1)}\, dx\right)^{p-1} < \infty$$

where the supremum is over all intervals I with length $|I| \leq 2\pi$.

Let $f \in L_w^p$ and $w \in A_p(1 < p < \infty)$. Then modulus of continuity of the function f is defined by

$$\Omega(f,\delta)_{p,w} = \sup_{|h|\leq\delta} |\Delta_h(f)|_{p,w}, \quad \delta > 0,$$

where

$$\Delta_h(f(x)) = \frac{1}{h}\int_0^h |f(x+t) - f(x)|\, dt$$

For $0 < \alpha \leq 1$, the Lipschitz class $Lip(\alpha,p,w)$ is give as,

$$Lip(\alpha,p,w) = \{f \in L_w^p : \Omega(f,\delta)_{p,w} = O(\delta^\alpha), \delta > 0\}.$$

Let $T \equiv (a_{n,k})$ be a lower triangular matrix with non-negative entries. For each n, the sequence $\{a_{n,k}\}$ is either non-increasing or non-decreasing in k, $0 \leq k \leq n$, the matrix T is said to have monotonic rows.

A positive sequence $c = \{c_n\}$ is called almost monotone increasing (or decreasing) if there exists a constant $K = K(c)$, such that for all $n \geq m$,

$$c_n \geq K c_m, \ (or\, c_n \leq K c_m).$$

Such sequences will be denoted by $c\in$ AMDS (or $c\in$ AMIS).

If $e\in$ AMIS (or $e\in$ AMDS), then the sequence e is almost monotone increasing (or decreasing) mean sequence, denoted by $e\in$ AMIMS (or $e\in$ AMDMS) and have the following relations, AMIS $\subset$ AMIMS and AMDS $\subset$ AMDMS.

Let E be an infinite subset of N and consider E as strictly increasing sequence of positive integers, say $E=\left\{\lambda\left(n\right)\right\}_{n=1}^{\infty}$. The summability method is defined as,

$$\left(C_{\lambda}\left(x\right)\right)_{n}=\frac{1}{\lambda\left(n\right)}\sum_{k=1}^{\lambda(n)}x_{k}$$

where $\left\{x_{k}\right\}$ is a sequence of real or complex numbers and $n = 1, 2, \ldots$. It is clear that C_{λ} is regular for any λ. Thus, the C_{λ}-method yields a subsequence of the Cesàro method C_{1}. C_{λ} is obtained by deleting a set of rows from Cesàro matrix.

The authors considered approximation of $f\in Lip\left(\alpha,p\right)$ with generalized matrix mean

$$\tau_{n}^{\lambda}\left(f;x\right)=\int_{k=0}^{\lambda(n)}a_{\lambda(n),k}s_{n}^{\lambda}\left(f\right)$$

where

$$s_{n}^{\lambda}\left(f;x\right)=\frac{1}{\pi}\int_{0}^{2\pi}f\left(x+t\right)D_{n}^{\lambda}\left(t\right)dt,$$

$$D_{n}^{\lambda}\left(t\right)=\frac{sin\left(\lambda\left(n\right)+1/2\right)t}{2sin\left(t/2\right)},$$

$$\sigma_{n}^{\lambda}\left(f;x\right)=\frac{1}{\lambda\left(n\right)+1}\sum_{m=0}^{\lambda(n)}s_{m}\left(f;x\right).$$

In this case, $a_{n,k}=\dfrac{1}{\lambda\left(n\right)+1}$ for $0\leq k\leq n$ and $a_{n,k}=0$ for $k>n$.

In addition to this, if $\lambda\left(n\right)=n$ for $\sigma_{n}^{\lambda}\left(f;x\right)$, then it coincides with Cesàro method of order one i.e. C_{1}.

Error of Approximation

The computation of error function $E_n(f)$, defined by

$$E_n(f) = Min\left\|T_n(x) - f(x)\right\|, \quad n > 0$$

It is called the error estimation of signals using summability techniques. If $E_n(f) > 0$ it means that our approximation is bigger than the true value. If $E_n(f) < 0$, then the approximation is too large. Oftenly, we are only interested (or can only find) in the magnitude of the error |E|. By taking the approximate value near to the exact value, we can reduce the error.

Hence the only reasonable way to measure the error of an electronic device that generates trigonometric functions is on the basis of percent of full scale.

This means that if the device operates between zero and 100 volts and it is rated as having an error of 1% of full scale and the following can happen:

1. The output may be in error by 1 volt at 100 volts
2. The output may be in error by 1 volt at 10 volts
3. The output may be in error by 1 volt at 1 volt, etc.

On the basis of percentage, we could say that the equipment gives a result that is in error by 1% at 100 volts, 10% at 10 volts, and 100% at 1 volt. To summarize, a trigonometric approximation is "best" from

Figure 2.

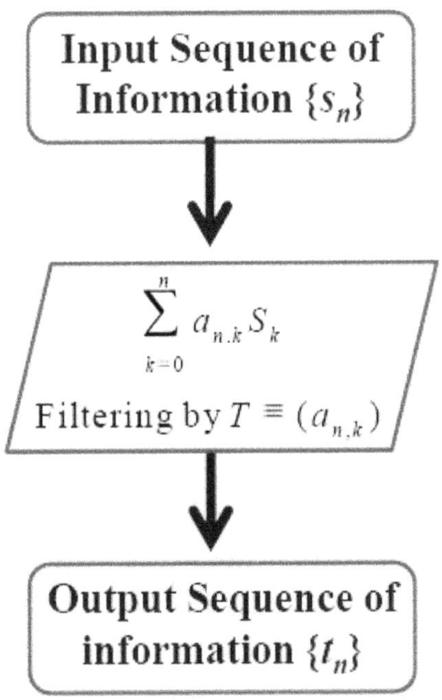

the viewpoint of the analog computer programmer if it allows the resolution of the highest frequencies of interest with a minimum full-scale error and with the most reliability.

The signals (functions) can be approximated with the help of trigonometric Fourier series with the Cesàro submethods. The degree of approximation $E_n(f)$ of a function $Lip(\alpha,p,w)$-space by a trigonometric polynomial $T_n(x)$ of degree *n* is given by

$$E_n(f)=\min_{T_n} f(x)-T_n(f)_{p,w}.$$

When approximation error $E_n(f)$ depends linearly upon the weights *w*, the error function $E_n(f)$, defined as the sum of error squares, is a hyper paraboloidal bowl with a guaranteed single (global) minimum.

Degree of Approximation by a General C_λ - Summability Method

Many researchers have studied the error estimation $E_n(f)$ through trigonometric-Fourier approximation for the situations in which the summability matrix dropped its monotonicity. In 2017, Sonker and Munjal (2017b) found the approximation of the function f ϵ Lip (α, p) using infinite matrices of Cesàro submethod. In this section, some theorems are explained which describe the degree of approximation of signals belonging to the class Lip(α, p) by a more general C_λ - method (summability method) which gives sharper estimations because all the estimations of the previous result are in terms of *n*, while our estimations are in terms of $\lambda(n)$; where $(\lambda(n))^{-\alpha}\leq n^{-\alpha}$ for $0<\alpha\leq 1$. The sharper estimates of infinite matrices are very much applicable in solid state physics which motivate us for investigation of perturbations of matrix valued functions.

Theorem 1: *Let* $f\in Lip(\alpha,p)$, *and* $T=(a_{\lambda(n),\,k})$ *be an infinite regular triangular matrix s.t.* $a_{n,k}\geq 0$

and $\sum_{k=1}^{n}a_{n,k}=1,\forall n,k$. *If, for p>1,* $0<\alpha<1$,

$\{a_{\lambda(n),k}\}\in AMDMS$ in k and

$(a_{\lambda(n),0})=O\left((\lambda(n)+1)^{-1}\right)$ satisfies,

Then

$$f-\tau_n^\lambda(f)_{p,w}=O\left(\lambda(n)^{-\alpha}\right)$$

Theorem 2: *Let* $f \in Lip(\alpha, p)$, *and let* $T = \left(a_{\lambda(n)}, k\right)$ be an infinite regular triangular

matrix s.t. $a_{n,k} \geq 0$ and $\sum_{k=1}^{n} a_{n,k} = 1, \forall n, k$. If, for p>1, $0 < \alpha < 1$,

$\left\{a_{\lambda(n),k}\right\} \in AMIMS$ in k and

$\left(a_{\lambda(n),0}\right) = O\left(\left(\lambda(n)+1\right)^{-1}\right)$ satisfies,

then

$$\left\| f - \tau_n^{\lambda}(f) \right\|_{p,w} = O\left(\lambda(n)^{-\alpha}\right)$$

In this Approximation, the sharpness of the estimation is improved by using the AMIMS (AMDMS) and more general summability method and with the help of trigonometric polynomials.

Lemmas

The following lemmas are used in the proof of theorem 1 and 2.

- **Lemma 1:** *If* $f \in Lip(\alpha, p)$, *for* $0 < \alpha \leq 1$ *and* $p > 1$, *then*

$$\left\| f - s_n(f) \right\|_{p,w} = O\left(n^{-\alpha}\right).$$

- **Lemma 2:** *Let matrix have AMS rows and satisfy*

$$\left(\lambda(n)+1\right) max\left\{a_{\lambda(n),0}, a_{\lambda(n),r}\right\} = O(1).$$

then, for $0 < \alpha < 1$

$$\sum_{k=0}^{\lambda(n)} a_{\lambda(n),k} (k+1)^{-\alpha} = O\left(\left(\lambda(n)+1\right)^{-\alpha}\right)$$

- **Proof:** Suppose that the rows of the matrix are *AMDS*. Then there exists a $K>0$ such that

$$\begin{aligned}\sum_{k=0}^{\lambda(n)} a_{\lambda(n),k}\left(k+1\right)^{-\alpha} &\le \sum_{k=0}^{\lambda(n)} K a_{\lambda(n),0}\left(k+1\right)^{-\alpha} \\ &= K a_{\lambda(n),0} \sum_{k=0}^{\lambda(n)} \left(k+1\right)^{-\alpha} \\ &= O\left(a_{\lambda(n),0}\left(\lambda(n)+1\right)^{1-\alpha}\right) \\ &= O\left(\left(\lambda(n)+1\right)^{1-\alpha}\right)\end{aligned}$$

A similar argument applies if the rows of the matrix are *AMIMS*.

- **Proof of Main Theorem:** If $p > 1, 0 < \alpha < 1$.

$$\begin{aligned}\tau_n^{\lambda}\left(f\right) - f &= \sum_{k=0}^{\lambda(n)} a_{\lambda(n),k} s_k^{\lambda}\left(f\right) - t_n^{\lambda} f + \left(t_n^{\lambda} - 1\right) f \\ &= \sum_{k=0}^{\lambda(n)} a_{\lambda(n),k}\left(s_k^{\lambda}\left(f\right) - f\right) + \left(t_n^{\lambda} - 1\right) f\end{aligned}$$

Using lemmas 1 and 2,

$$\begin{aligned}\tau_n^{\lambda}\left(f\right) - f_{p,w} &\le \sum_{k=0}^{\lambda(n)} a_{\lambda(n),k}\left(s_k^{\lambda}\left(f\right) - f\right)_{p,w} + \left|\left(t_n^{\lambda} - 1\right)\right| f_{p,w} \\ &= \sum_{k=0}^{\lambda(n)} a_{\lambda(n),k} O\left(\left(k+1\right)^{-\alpha}\right) + O\left(\left(\lambda(n)\right)^{-\alpha}\right) \\ &= O\left(\left(\lambda(n)+1\right)^{-\alpha}\right) + O\left(\left(\lambda(n)\right)^{-\alpha}\right) \\ &= O\left(\left(\lambda(n)\right)^{-\alpha}\right)\end{aligned}$$

Stability

The stability of the process of a production system must be a requirement for its smooth functioning. The innovations in the process of the products are necessary to remain on the market, but there is always a risk of losing the stability. The purpose of developing the soft computing models is to control the output loading and haulage process so that the process remains stable in the required period.

Control will consist of an adequate selection of process inputs in to get a required output values consistent with the pre-determined production plan. The process will remain stable if the production volume is consistent with the volume set in the production plan. In Mechanical Engineering, the input data is presented as a periodic signal which can be expressed as a harmonic summation of sinusoids. Similarly, a summation of sinusoids is used in presenting the electrical load signal. The drawback of electrical load data is boundlessness. It is a non-stationary with abrupt variations caused due to weather changes and other effects. This phenomenon results in the variation of the high-frequency component which may not be represented as a bounded and periodic spectrum. Summability methods are suitable for converting the unbounded data into a bounded data or filtering the data for stability.

A necessary and sufficient condition for a system to be BIBO (Bounded Input Bounded Output) stable is that the impulse response be absolutely summable, i.e.,

$$\text{BIBO stable} \Leftrightarrow \sum_{n=-\infty}^{\infty} \left|h(n)\right| < \infty$$

Summability techniques are trained to minimize the error. With the use of summability Technique, the output of the signals can be made stable, bounded and used to predict the behavior of the input data, the initial situation and the changes in the complete process.

Stability of the Frequency Response of the System

The discrete-time signals can be expressed in the different forms by using the continuous-time signals. The complex sinusoidal exponential sequences have an important role in expressing the discrete-time signals because the exponential complex sequence is the eigenfunction of linear time-invariant systems.

The impulse response of the moving-average system is

$$h(n) = \begin{cases} \dfrac{1}{M_1 + M_2 + 1}, & -M_1 \le n \le M_2 \\ 0, & otherwise \end{cases}$$

Therefore, the frequency response is

$$H\left(e^{jw}\right) = \sum_{n=-\infty}^{\infty} h(n) e^{-jwn}$$

The condition for stability is also a sufficient condition for the existence of the frequency response function. To see this, note that, in general,

$$\left|H\left(e^{jw}\right)\right| = \left|\sum_{n=-\infty}^{\infty} h(n) e^{-jwn}\right| \le \sum_{n=-\infty}^{\infty} \left|h(n) e^{-jwn}\right| \le \sum_{n=-\infty}^{\infty} \left|h(n)\right|$$

So, the general condition

$$\sum_{n=-\infty}^{\infty}\left|h(n)\right|<\infty$$

ensures that $H\left(e^{jw}\right)$ exists. It is very obvious that the existence condition of the frequency response is the same as the condition for dominance of the steady-state solution. Indeed, a complex exponential that exists for all *n* can be thought of as one that is applied at $n=-\infty$. The eigen-function property of complex exponentials depends on the stability of the system, since at finite *n*, the transient response must have become zero, so that we only see the steady-state response $H\left(e^{jw}\right)e^{jwn}$ for all finite *n*.

Thus, if *h*(*n*) is absolutely summable, then $H\left(e^{jw}\right)$ exists. Furthermore, in this case, the series can be shown to converge uniformly to a continuous function of *w*. Since a stable sequence is, by definition, absolutely summable, all stable sequences have Fourier transforms. It also follows, then, that any stable system will have a finite and continuous frequency response.

$$H\left(e^{jw}\right)=\frac{1}{M_1+M_2+1}\sum_{n=-M_1}^{M_2}e^{-jwn}$$

This can be expressed as

$$\begin{aligned}H\left(e^{jw}\right)&=\frac{1}{M_1+M_2+1}\frac{e^{jwM_1}-e^{-jw(M_2+1)}}{1-e^{-jw}}\\&=\frac{1}{M_1+M_2+1}\frac{e^{jw(M_1+M_2+1)/2}-e^{-jw(M_1+M_2+1)/2}}{1-e^{-jw}}e^{-jw(M_2-M_1+1)/2}\\&=\frac{1}{M_1+M_2+1}\frac{e^{jw(M_1+M_2+1)/2}-e^{-jw(M_1+M_2+1)/2}}{e^{jw/2}-e^{-jw/2}}e^{-jw(M_2-M_1)/2}\\&=\frac{1}{M_1+M_2+1}\frac{\sin\left[w\left(M_1+M_2+1\right)/2\right]}{\sin\left(w/2\right)}e^{-jw(M_2-M_1)/2}\end{aligned}$$

The magnitude of $H\left(e^{jw}\right)$ is plotted in Figure 3 for $M_1=0$ and $M_2=4$. Note that $H\left(e^{jw}\right)$ is periodic, as required for the frequency response of a discrete-time system. Note also that $\left|H\left(e^{jw}\right)\right|$ falls off at "high frequencies and $< S.H\left(e^{jw}\right)$. This attenuation of the high frequencies suggests that the system will smooth out rapid variations in the input sequence; in other words, the system is a rough approximation to a low- pass filter. This is consistent with what we would intuitively expect about the behavior of the moving-average system.

The sufficient condition for a Fourier transform representation is Absolute summability. To prove this, we calculated the Fourier transforms of the sequence $\left[1/\left(M_1 + M_2 + 1\right)\right]\left(u\left[n + M_1\right] - u\left[n - M_2 - 1\right]\right)$ and found it absolutely summable. Clearly, any finite length sequence is absolutely summable and thus has a Fourier transform representation. In the context of linear time-invariant systems, any FIR system will be stable and therefore will have a finite, continuous frequency response. When a sequence has infinite length, we must be concerned about the convergence of the infinite sum.

Applications

The approximation of the signals is widely used in the digital filter such as finite impulse response filters [FIR filters] and infinite impulse response filters [IIR filters] and these filters have numerous variety of applications like in speech and image processing, communications, radar, sonar and medical signal processing, etc. For an engineering problem having piecewise finitely differentiable data with jump discontinuities, there are more sophisticated ways to use the sampled data to obtain a more accurate solution using the approximations of the Fourier coefficients by summability techniques. With the help of a minimal set of sufficient conditions for a soft computing model, the error can be minimized by using summability methods so that the output data is filtered and according to our interests for observing the behavior and requirements of the modeled problem.

Figure 3. Magnitude of the frequency response of the moving-average system for the case $M_1 = 0$ and $M_2 = 4$

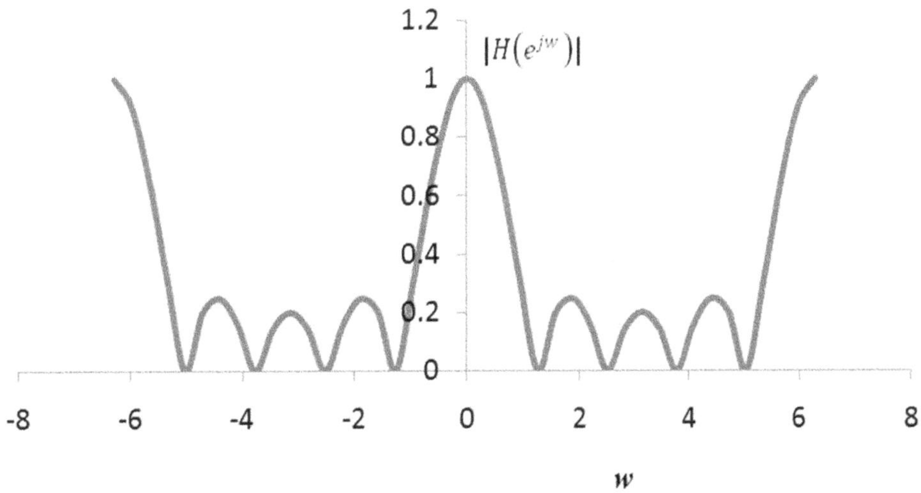

REFERENCES

Bachmann, G., Narici, L., & Beckenstein, E. (2012). *Fourier and wavelet analysis*. Springer Science & Business Media.

Cesàro, E. (1890). Sur la multiplication des séries. *Bulletin des Sciences Mathématiques, 14*(2), 114–120.

Dutta, H., & Rhoades, B. E. (Eds.). (2016). *Current Topics in Summability Theory and Applications*. Singapore: Springer. doi:10.1007/978-981-10-0913-6

Herrero, Á., Sedano, J., Baruque, B., Quintián, H., & Corchado, E. (2015). *Proceedings of the 10th International Conference on Soft Computing Models in Industrial and Environmental Applications*. Springer.

Kecman, V. (2001). *Learning and Soft Computing Support Vector Machines. In Neural Networks and Fuzzy Logic Models*. MIT Press.

Kumar, A., & Joshi, L. K. (2011). MODWT Based Time Scale Decomposition Analysis of BSE and NSE Indexes Financial Time Series. *Int. Journal of Math. Analysis, 5*(27), 1343–1352.

Kumar, A., Joshi, L. K., Pal, A. K., & Shukla, A. K. (2010). A New Approach for Generating Parametric Orthogonal Wavelets. *Journal of Wavelet Theory and Applications, 4*(1), 1–8.

Kumar, A., Pant, S., & Joshi, L. K. (2016). Wavelet Variance, Covariance and Correlation Analysis of BSE and NSE Indexes Financial Time Series. *International Journal of Mathematical. Engineering and Management Sciences, 1*(1), 26–33.

Sonker, S., & Munjal, A. (2016a). Sufficient conditions for triple matrices to be bounded. *Nonlinear Studies, 23*(4), 531–540.

Sonker, S., & Munjal, A. (2016b). Absolute Summability Factor $\varphi - |C, 1; \delta|_k$ of Infinite Series. *International Journal of Mathematical Analysis, 10*(23), 1129–1136. doi:10.12988/ijma.2016.6690

Sonker, S., & Munjal, A. (2017a). A note on boundness conditions of absolute summability ϕ-|A|k factors. In *Proceedings of the International Conference on Advances in Science and Technology* (pp. 208-210).

Sonker, S., & Munjal, A. (2017b). Approximation of the function $f \in Lip(\alpha, p)$ using infinite matrices of Cesàro submethod. *Nonlinear Studies, 24*(1), 113–125.

Zygmund, A. (1988). *Trigonometric series*. Cambridge University Press.

Chapter 14
Applying Multi-Objective Optimization Algorithms to Mechanical Engineering

Preeti Shivach
Graphic Era University, India

Lata Nautiyal
Graphic Era University, India

Mangey Ram
Graphic Era University, India

ABSTRACT

In today's scenarios, the utilization of simulation and optimization in the field of designing is achieving wider recognition in the various zones of commerce as the computational competences of computers upsurge day by day. The result is that the uses for numerical optimization have increased tremendously. Design process in engineering is a distinct practice of solving the problems where a group of recurrently indistinct objectives has to be well-adjusted deprived of violating any given circumstances. Consequently, it seems quite ordinary to consider a design process as an optimization process. The design process could be articulated as to allocate values to the system parameters to confirm that the state variables and the characteristics are as suitable as possible through an inclusive range of operating and environmental variables. This is a complex multi-objective optimization problem (MOOP). This chapter discusses the use of MOO algorithms in mechanical engineering.

INTRODUCTION

To improve the designing process of engineering there is a powerful tool called optimization. It is the act of finding the finest solution under particular conditions. It is more effective than conventional trail-and-error process of designing. In today's scenarios, the utilization of simulation and optimization in the field of designing is achieving wider recognition in the various zones of commerce as the

DOI: 10.4018/978-1-5225-3035-0.ch014

computational competences of computers upsurge day by day. The result is that the uses for numerical optimization have increased tremendously. There are many optimization approaches developed and used in the literature. These approaches are grounded on non-linear, linear and geometric programming etc. Most recent methods are neural network, genetic algorithms and stimulated annealing. These methods are also called non-traditional optimization methods. Engineering design problem generally consists of blend of numerical simulation, logical designs and list selection. Hence, non-gradient methods are sound suited to this type of problems.

Engineering design problems of real world can be categorized by more than one objective. Therefore, the designing problem in engineering can be viewed as a multi-objective problem. The design process requires a large number of goals to be satisfied and deal with various design variables. Some of these objectives may be conflicting to each other it means we work to achieve a particular objective there might be some objective which we are losing. For example, the automobile design could be realized as a multi-objective problem with two contradictory goals, namely, the minimization of weight and the maximization of the crash resistance. Though, consequences of reduction in the weight of the automobile is, increase in the crash resistance, and vice-versa.

Design variables are a factor that the designer or engineer might "alter" with the purpose of modifying the objects he is designing. The types of design variables are:

- Independent Variables
- Dependent Variables
- State Variables
- Operating Variables
- Environment Variables

Independent variables are the measures the engineer deals with precisely, such as physical properties and lubrication properties, etc. Dependent variables are factors the engineer can't precisely allocate values to but through independent parameters the designer deals with them. State variables are a transitional type of design parameters between independent and dependent. Operating variables are those factors that can be altered after the designing process. The last type of factors (also called external variables) is the environmental features that have impact on the design when used.

Therefore, the designing problem can be expressed as to allocate values or costs to the various design factors with the purpose of ensuring that the variables and the features are as suitable as promising through an inclusive span of environmental and operating variables. This is certainly a complex MOOP.

Single Objective Optimization

Single objective optimization problems (SOOP) are those problems that have only one goal to be achieved. It can be stated as:

Find a vector $X = \{x_1, x_2, x_3, \ldots, x_n\}^T$ that minimizes objective function f(x)

With the constraints; $g_j(x) \leq 0, j = 1, 2, 3, \ldots, m$

$h_l(x) \leq 0, l = 1, 2, 3, \ldots, p$

where, X is the vector of design variables.

The above problem is called a SOOP, since there is only one objective to be minimized.

Multi-Objective Optimization

When the number of objectives to be achieved is more than one, then the problem is called MOOP. These problems are solved by using multi-objective optimization methods (MOO). The concept of optimization was first proposed by Kuhn and Tukcer (1951). The procedure is referred as Vector Maximization. One of the earliest and simplest methods used to deal with multi-objectives is the formation of a single overall objective function combining individual objectives. This is usually done by a linear combination of the objectives by allocating weights to the individual objectives (Walley (1991), Salama et al., (1988)). This type of formulation assumes that the objectives are independent which might be unreal in some cases. Also, this procedure might not span the entire region of optimization.

Several realistic optimization problems typically have many contradictory objectives. In that MOOP, usually the solution that optimizes all objective functions concurrently does not exist. Hence, Pareto optimal solutions are introduced. These solutions are "efficient" in terms of all objective functions. In general there are numerous Pareto optimal solutions. Hence, we have to choose a decisive solution among these solutions, this process is known as "trade-off analysis". It is the most significant task in MOOP. The analysis is easy if there are only two or three objectives, but if number of objectives is greater than three then it is not possible to represent Pareto Frontier. To create local trade-off analysis interactive methods can be used, presenting a "certain" Pareto optimal solution.

The multi-objective optimization methods are generally classified into three key categories; priori articulation of preferences, post priori articulation of preferences and no articulation of preferences. General steps involved in generic process are as follows:

1. Formulate the problem
2. Computer possible design variables domain
3. Perform analysis (loop between step 1 to 3)
4. Compute Pareto-optimal set
5. Execute design synthesis

A MOOP can be defined as:

Find a vector $X = \{x_1, x_2, x_3 \ldots, x_n\}^T$ that minimizes objective functions $f_1(x), f_2(x), f_3(x) \ldots, f_k(x)$

- **Subject to Constraints:**

$g_j(x) \leq 0, j = 1, 2, 3, \ldots, m$ and $h_l(x) \leq 0, l = 1, 2, 3, \ldots, p$

where x_i is the design variable

As already discussed, the Pareto optimality concept is applied to MOOP. Essentially, $x_i \in S$ is supposed to be Pareto optimal if all other $x_j \in S$ have a greater value for at least one of f_i, with $i = 1, \ldots, k$, or have equal value for all the objectives. Generally, following are the definitions:

- A point x_i is called *weak* Pareto optimum iff no such x exists for which the condition $f_i(x_j) < f_i(x_i)$ is true, for all $i \in \{1, \ldots, n\}$.
- A point x_i is called *strict* Pareto optimum iff no such x exists for which the condition $f_i(x_j) \leq f_i(x_i)$ is true, for all $i \in \{1, \ldots, n\}$, with a minimum one strict inequality.

The image of the effective solutions is called Pareto front. The form of the Pareto front designates the kind of the trade-off between the diverse objectives. Figure 1 depicts the Pareto curve. In this curve, the points are termed non-inferior or non-dominated points.

Figure 2 depicts weak and strict Pareto Optima. Point *A*1 and *A*5 are weak Pareto optima; point *A*2, *A*3 and *A*4 are strict Pareto optima.

General Optimization Methods

General optimization approaches could be categorized into two broad classes; derivative and non-derivative methods, (refer Figure 3). Non-derivative methods are also called black-box methods because they don't require any derivative of objective function for getting optimal solution. These methods find a global optimal solution of a problem and don't suffer from native optima problem as gradient methods do.

DESGIN PROBLEMS IN ENGINEERING

In today's scenarios, the utilization of simulation and optimization in the field of designing is achieving wider recognition in the various zones of commerce as the computational competences of computers upsurge day by day. The result is that the uses for numerical optimization have increased tremendously. Analytical practices and numerical optimization are of great importance and can allow huge enhance-

Figure 1. Pareto Curve

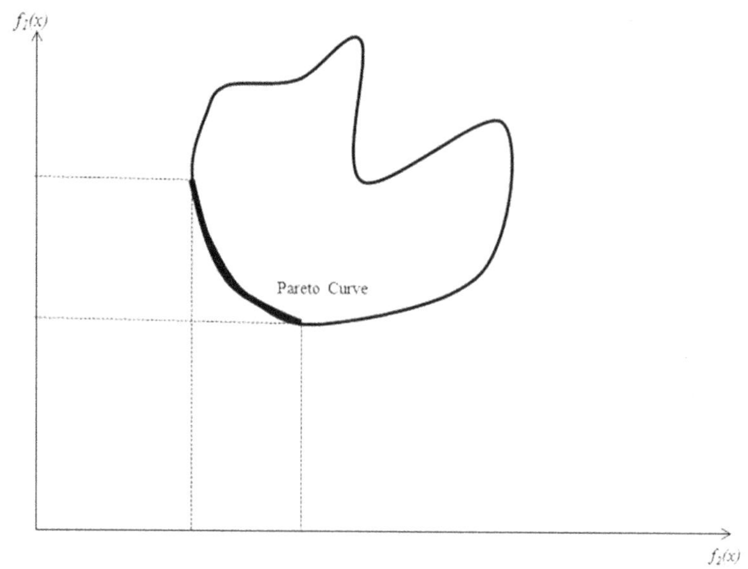

Figure 2. Weak and Strict Pareto Optima

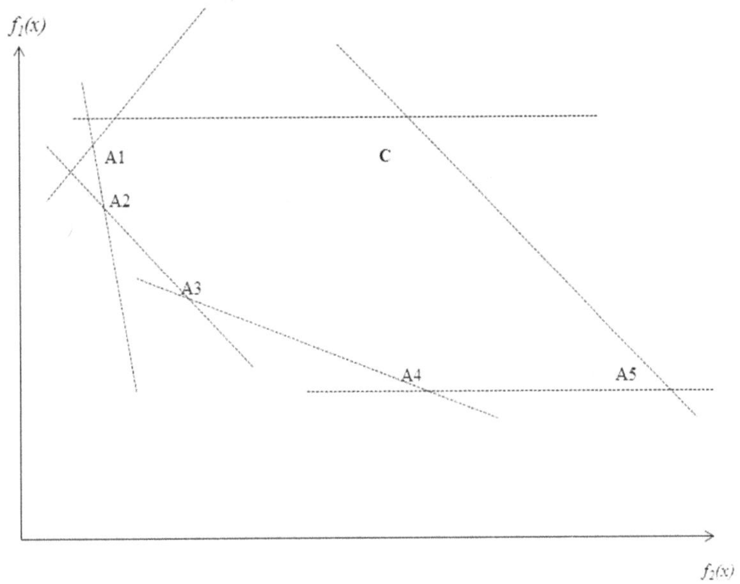

Figure 3. General Optimization Methods

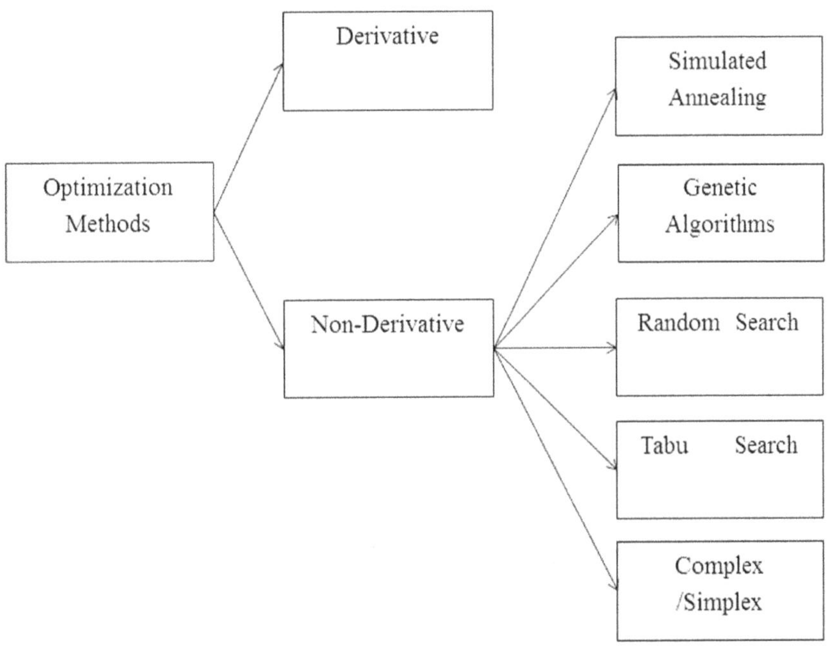

ments in the design field. The fact is that design problems in Engineering often contain large number of conflicting objectives. In several conditions, various objectives are combined into a single objective. After that the process of optimization is applied with one finest design as the outcome. There after result is intensely rely upon how the objectives are combined with each other.

Design process in engineering is a distinct practice of solving the problems where a group of recurrently indistinct objectives has to be well-adjusted deprived of violating any given circumstances. Consequently, it seems quite ordinary to consider a design process as an optimization process. Designing can be massively improved by implementing contemporary modeling, simulation and optimization techniques. The design process could be articulated as to allocate values to the system parameters to confirm that the state variables and the characteristics are as suitable as possible through an inclusive range of operating and environmental variables. There are number of researches that worked on optimization some of them are Kumar et al. (2016a), Kumar et al. (2017), Kumar et al. (2016b), Pant et al. (2015), Pant and Singh (2011), Pant et al. (2017a), Pant et al. (2015), Pant et al. (2017b) and Pant et al. (2017c)

WAYS TO PERFORM MULTI-OBJECTIVE OPTIMIZATION

As stated previously design problems of engineering are generally categorized by the existence of various contradictory objectives that has to be fulfilled by the design. Hence, it is acceptable to view the design problem MOOP. References to MOO could be located in (Hwang et al. (1980), Steuer (1986)) and engineering applications in (Eschenauer et al. (1990), Osyczka (1984)). There are number of methods exist that can be used to solve these problems. The MOOP can usually be solved in four diverse methods. The method used is determined by the condition when the decision-maker expressed his or her inclination to the diverse objectives. The decision maker may express his preference at four times; never, prior, during or after the real optimization process (refer Figure 4).

We will apply some of these methods on a single problem that is an electric power distribution problem. The goal of cost-effectively possible supply of electric power is to choose generator output

Figure 4. Multi-Objective Optimization Methods

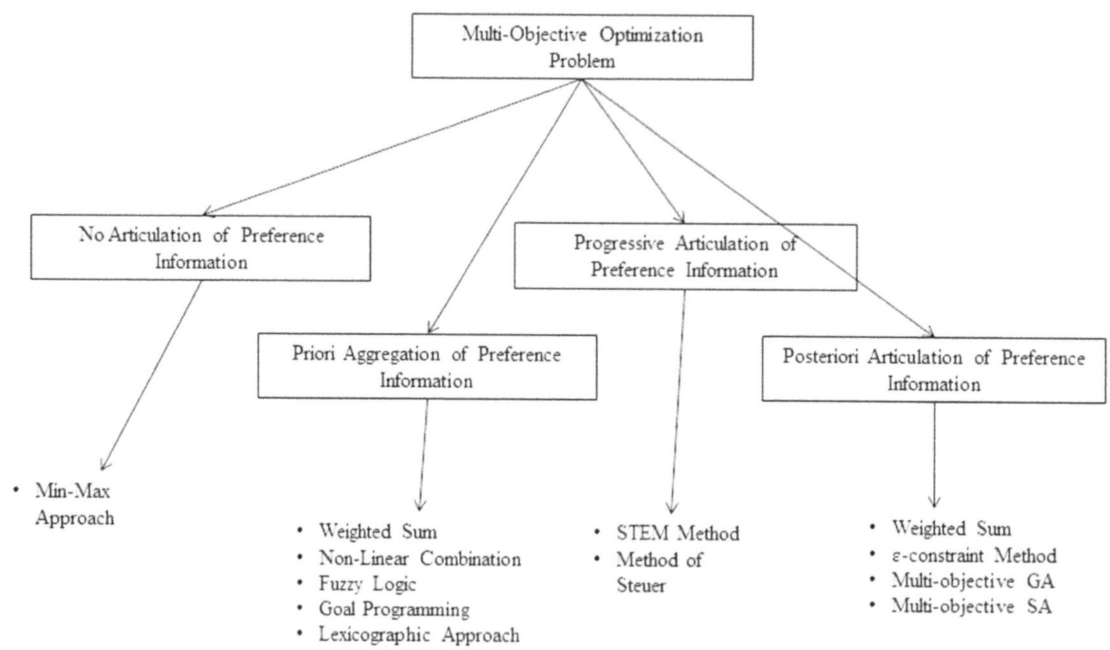

such that it meets the requirement at least operational cost and produce minimum atmospheric emission and waste. Therefore, goal is to minimize two conflicting objective functions, emission and fuel cost.

Actual power production of ith generator is represented by $X = [X_1, X_2, X_3 \ldots, X_n]^T$. The cost function is represented as follows,

$$F_c(P) = \sum_{i=1}^{n} a_i + b_i P_i + c_i P_i^2$$

The total emission $F_e(P)$ can be expressed as:

$$F_e(P) = \sum_{i=1}^{n} 10^{-2}\left(\alpha_i + \phi_i P_i + \gamma_i P_i^2\right) + \xi_i e^{(\lambda_i P_i)}$$

where $\alpha_i, \phi_i, \gamma_i, \xi_i, \lambda_i$ are emission characteristic coefficient of ith generator. Power output is constrained by inferior and superior limits,

$$P_i^{\inf} \le P_i \le P_i^{\sup}, i = 1, 2, 3 \ldots n$$

The total power generated should be equal to the sum of demand and loss of power in transmission lines, i.e.:

$$\sum_{i=1}^{n} P_i = P_D + P_{loss}$$

According to Abido and Lidiane (2003) for n=6 and $P_i^{inf}=10$, $P_i^{sup}=120$MW and total power is 283MW. Therefore, we can formulate the problem as:

Minimize $F(P) = [F_c(P), F_e(P)]$

Subject to constraints

$$h_i(P) = \sum_{i=1}^{n} P_i - 283 = 0$$

$$g_l(P) = P_l - 120£0, \qquad l = 1,\ 2,\ 3 \ldots 6$$

$$g_l(P) = 10 - P_l£0, \qquad l = 7,\ 8,\ 9 \ldots 12$$

No Preference Articulation

This type of method doesn't use any kind of preference information. Min-Max formulation and global criterion method fall under this category (Hwang et al., 1980; Osyczka, 1984; Steuer, 1986).

1. **Global Criterion Method:**

This approach is used to transform MOOP to a scalar optimization problem. The function for global criterion is described in a way so that the solution found is handy to best solution. Function is described as:

$$f(x) = \sum_{i=1}^{k} \left(\frac{f_i^0 - f_i(x)}{f_i^0} \right)^s$$

Some authors took s=1 whereas some used s=2. This method can also be applied using a family of L_p-metrices stated as:

$$L_p\left(f\right) = \left[\sum_{i=1}^{k} \left| f_i^0 - f_i\left(x\right) \right|^s \right]^{1/s}, \quad 1 \leq s \leq \infty$$

The exponent s gives various ways to calculate the distance. The value 1 is used for simple formulation and value 2 is used for Euclidean distance and ∞ for Tchebycheff norm.

Example: Power distribution problem can be solved with the following metrics:

L_{1}-Metric: $\min f(x) = \left| F_c^0 - F_c \right| + \left| F_e^0 - F_e \right|$

L_{2}-Metric: $\min f(x) = \left[\left(F_c^0 - F_c \right)^2 + \left(F_e^0 - F_e \right)^2 \right]^{1/2}$

Relative L_2-Metric: $\min f(x) = \left(\frac{F_c^0 - F_c}{F_c^0} \right)^2 + \left(\frac{F_e^0 - F_e}{F_e^0} \right)^2$

L_{3}-Metric: $\min f(x) = \left[\left(F_c^0 - F_c \right)^3 + \left(F_e^0 - F_e \right)^3 \right]^{1/3}$

Relative L_3-Metric: $\min f(x) = \left[\left(\frac{F_c^0 - F_c}{F_c^0} \right)^3 + \left(\frac{F_e^0 - F_e}{F_e^0} \right)^3 \right]^{1/3}$

Priori Articulation

Priori Articulation is most customary method used by decision makers for conducting MOO. In these methods prior to optimization, the objectives are accumulated to one single. There are various approaches used for this task; weighted sum, non-linear approach and goal programming etc.

1. **Weighted-Sum Approaches:**

This approach is perhaps the simplest and broadly used method. It is also called scalarization method. Steuer (1986) presented a good discussion on this approach. This method translates the MOOP into a scalar problem as follows:

$$f(x) = \sum_{i=1}^{k} w_i f_i(x) r_i$$

where, the sum of all weight should be equal to 1, r_i are the constants and w_i reflects the relative importance of particular objective. It gives best results when $r_i = 1/f_i^0$, where f_i^0 is the ideal solution of the problem.

Example: The problem of electric distribution can be formulated by this method as follows;

$$F(P) = w_1 \frac{F_c}{F_c^0} + w_2 \frac{F_e}{F_e^0}$$

2. **Goal Programming Method:**

This method was given by Charnes et al. (1955) and Charnes and Cooper (1961). This method tries to find precise target values of the objectives. For example:

$f_1(x) \geq u_1$

$f_2(x) = u_2$

$f_3(x) \leq u_3$

$x \in S.$

It is obvious that we have to discriminate two situations, i.e., if the joint amid the image set *C* and the *utopian set* (image of the permissible solutions), is null or not. In the earlier case, task converts into the task in which we need to discover a solution value of which is as handy as likely to the utopian set. This leads to constraints and extra variables introduction. For constraints of first type $f_1(x) \geq u_{1,}$ we announce a variable a_l- such that the constraint change to $f_1(x)+a_l^- \geq u_1$. For constraints of the second type $f_2(x) = u_2$, announce two variables a_2+ and a_2- such that the constraints convert $f2(x)+a_2^- - a_2^+ = u_2$. For constraints of the type $f_3(x) \leq u_3$, announce a variable a_3+ such that the constraints change to $f_3(x) - a_3^+ \leq u_3$.

Suppose, we symbolize the vector *a* of the added variables. A solution (*x, a*) is called a *strict Pareto-slack optimum* iff (*x', a'*), for every $x' \in S$, such that $a_i' \leq a_i$ with a minimum one strict inequality does not exist. A number of ways are there for enhancing the slack/surplus variables.

Progressive Articulation

Another name of these methods is interactive methods. These methods depend on prior information of preferences instantaneously when they explore the search space. They are usually applied in the arena of operational research. The decision maker acquires information when he/she encounters diverse possible solutions of the problem. Some of the advantages of interactive methods are:

- No prior information of preference is required,
- The decision maker gains a healthier consideration of the problem,
- There is a good chance of solution acceptance as the decision maker is concerned in search process.

The cons of these methods are:

- The solution completely depends on expression of the decision maker.
- An elevated attempt is needed from the decision maker.
- If decision maker modifies his/her preferences; the whole process has to be started again.

1. STEM Method

This approach was proposed by Benoyoun et al. (1971), and also called STEP method. In this method, preference information delivered by the decision maker is consumed to cut the search space. The optimization problem can be stated as:

$$\min\left\{\sum_{i=1}^{k}\left(w_i^h\left(f_i\left(x\right)-f_i^0\right)\right)^p\right\}^{1/p}$$

$$where,\quad w_i^h > 0 \text{ and } \sum_{i=1}^{k} w_i = 1$$

where, h is the run (iteration) number, p can have value between 1 and ∞. We get the objective vector $\vec{F}$ when the problem is solved and this vector is matched with best solution. If it is acceptable to some limits then the decision maker decides some relaxation on at least one of the goals. It means the search space is reduced and weight of that particular objective is assigned value 0. Again, we solve the problem, after this iteration there may be the case that the decision maker is satisfied with the solution or further reduce the search space. This way the algorithm proceeds.

Posteriori Articulation

These types of method don't depend on the preferences given by the decision maker because the solution is given to the decision maker after the whole process completes. This is also a negative point of these methods. Some methods also suffer from a huge computational load. Another dark side of these methods is that the decision maker has a large number of solutions to choose form. Some methods exist which perform this task of assessment (Morse (1980), Rosenman and Gero (1985)).

1. **Multi-Objective Genetic Algorithms (MOGA):** A great advantage of GA is that it operates on population of individuals. Therefore, it is appealing to create a policy in which the population takes the entire Pareto front in only one optimization run. Author in Fonseca and Fleming (1995) have categorized MOGAs in two classes as discussed below.

a. **Non-Pareto based Methods:** The very first MOGA was proposed by Schaffer (1985) named Vector Evaluating Genetic Algorithm (VEGA). Selection method of GA is used in VEGA to yield non-dominated individuals. Every distinct goal is labeled as the choice measure for a fraction of the population. This method produces a poor exposure of Pareto front. All of this type of techniques inclines to unite to a subclass of the Pareto-optimal frontier and leave a big part of the Pareto set uncharted. Diversity should be preserved so that the whole Pareto frontier is extracted. This act of preserving mixture will incline to increase vigor in problems by confirming that there is a genetic type for reproducing mechanisms to work upon (Grueninger, 1996; Harik, 1995).

b. **Pareto Based Approaches:** Goldberg (1989) proposed a non-dominated sorting to grade a search population bestowing to Pareto optimality. Non-dominated individuals are recognized first and assigned the rank 1 and separated. Then the non-dominated ones in the condensed population are categorized, given the rank 2, and then they are also separated. This procedure of recognizing individuals is continued till the entire population has been graded.

2. **ε-Constraints Method:** The ε *-constraints* method is suggested by Chankong and Haimes (1983). In this method, an objective is selected by the decision maker out of *n* to be minimized; value of the remaining objectives is must be less than or same as to a given target values. Let us assume $f_3(x)$ be the objective function chosen, then the problem $P(\varepsilon_3)$ will be:

$\min f_3(x)$

$f_i(x) \leq \varepsilon_i, \forall i \in \{1, \ldots, n\}\ \{3\}$

$x \in S.$

This expression can be stemmed by a more general result by Miettinen and Makela (1994):

If an objective j and a vector $\varepsilon = (\varepsilon_1, \ldots, \varepsilon_{j-1}, \varepsilon_{j+1}, \ldots, \varepsilon_n) \in R^{n-1}$ *exist, such that* x^* *is an optimal solution to the following problem* $P(\varepsilon)$:

$\min f_j(x)$

$f_i(x) \leq \varepsilon_i, \forall i \in \{1, \ldots, n\} \setminus \{j\}$

$x \in S,$

then x^* is a weak Pareto optimum."

One benefit of this approach is that it is able to attain effective points in a non-convex Pareto curve. For example, let us suppose we have two objective functions and $f_2(x)$ is selected to be minimized, i.e.

$\min f_2(x)$

$f_1(x) \leq \varepsilon_1$

$x \in S$

A situation is depicted in Figure 5, where, when $f_1(x) = \varepsilon_1, f_2(x)$ is an efficient point of the non-convex Pareto curve.

Hence, as suggested by Steuer (1986) the upper bounds (ε_i) can be adjusted by the decision maker to get weak Pareto optima which is also a disadvantage of ε-constraint method. Additionally, this method can't be applied to the problem in which there are more than two objectives. For these reasons, Erghott and Rusika (2005), recommended two alterations to enhance the method.

APPLICATIONS OF OPTIMIZATION

Designing process in engineering can be improved by using modeling and optimization techniques. There can be various criteria of a good design such as profit, efficiency, process safety and time etc. These objectives are conflicting, it means if we want to achieve optimal solution for one objective we have to compromise on one or more other objectives. Multi-objective optimization can be used in planning of string collection in steel manufacturing and it can also be used in long term scheduling of atomic power plants.

Kadar (2013) applied optimization to electric power generation systems. Author used some single-objective and some multi-objective optimization solution to the problem such as decision making, optimization of renewable resources. They have shown that in power systems there are various complex tasks to be maintained by computer control. Since the energy industry plays with large amounts of money, the optimization, moreover the profit optimization has high importance. At Óbuda University the developed some optimization solutions applying linear programming, constraint programming, weighting methods and rule based systems. A broad variety of optimization solutions is being used by businesses, governments, universities, and other groups. Many organizations are saving billions of dollars using this technique.

Figure 5. Non-Convex Pareto Curve

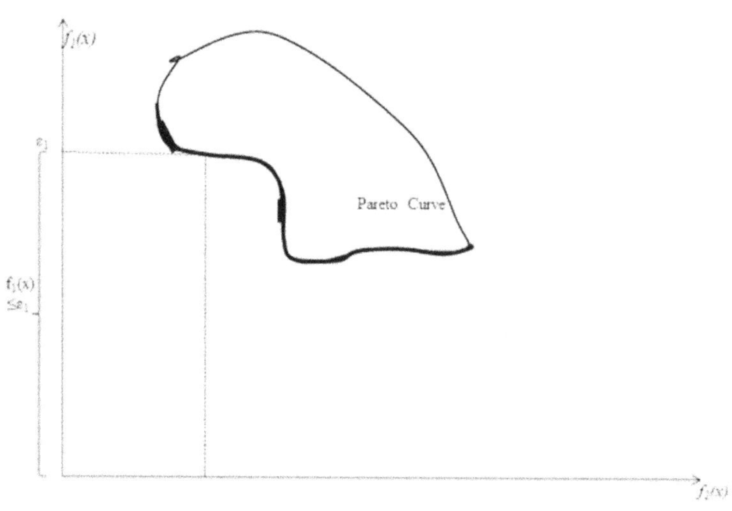

CONCLUSION

The design problems of mechanical engineering are multi-objective problems because there are number of objectives that have to be achieved by the design. These objectives may be conflicting to each other. It means when the designer chooses to achieve one goal he may compromise with the other objective. Hence, designing process of engineering can be solved by using multi-objective optimization algorithms. There are number of solutions available to this problem. This chapter discusses some of these algorithms and also applied them to a problem of electric power supply problem. Based on the time of preferences given by the decision maker the approaches are divided into four categories. First case is when there are no preferences given by the decision maker. Second case is when the decision maker provides preference information prior to the design. Third is the availability of the information after the process completes. Last case is that in which information is provided during the process.

REFERENCES

Abido, M. A. (2003). A niched Pareto genetic algorithm for multiobjective environmental/economic dispatch. *Electrical Power and Energy Systems*, *25*(2), 97–105. doi:10.1016/S0142-0615(02)00027-3

Benayoun, R., de Montgolfier, J., Tergny, J., & Laritchev, O. (1971). Linear programming with multiple objective functions: Step method (stem). *Mathematical Programming*, *1*(1), 366–375. doi:10.1007/BF01584098

Chankong, V., & Haimes, Y. Y. (1983). *Multiobjective Decision Making: Theory and Methodology*. New York: Elsevier Science Publishing.

Charnes, A., & Cooper, W. W. (1961). *Management models and industrial applications of linear programming*. New York: Wiley.

Charnes, A., Cooper, W. W., & Ferguson, R. (1955). Optimal estimation of executive compensation by linear programming. *Management Science*, *1*(2), 138–151. doi:10.1287/mnsc.1.2.138

Ehrgott, M., & Wiecek, M. (2005). Multiobjective programming. In J. Figueira, S. Greco, & M. Ehrgott (Eds.), *Multiple Criteria Decision Analysis. State of the Art Surveys* (pp. 667–722). Springer. doi:10.1007/0-387-23081-5_17

Eschenauer, H., Koski, J., & Osyczka, A. (1990). *Multicriteria Design Optimization*. Berlin: Springer-Verlag. doi:10.1007/978-3-642-48697-5

Fonseca, C., & Fleming, P. (1995). An overview of evolutionary algorithms in multiobjective optimization. *Evolutionary Computation*, *3*(1), 1–18. doi:10.1162/evco.1995.3.1.1

Goldberg, D. (1989). *Genetic Algorithms in Search and Machine Learning*. Reading Addison Wesley.

Grueninger, T., & Wallace, D. (1996). Multi-modal optimization using genetic algorithms (Technical Report 96.02). CAD lab, Massachusetts Institute of Technology, Cambridge, MA.

Harik, G. (1995). Finding multimodal solutions using restricted tournament selection. In *Proceeding of Sixth International Conference on Genetic Algorithms.*

Hwang, C., Paidy, S., Yoon, K., & Masud, A. S. M. (1980). Mathematical programming with multiple objectives: A tutorial. *Computers & Operations Research, 7*(1-2), 5–31. doi:10.1016/0305-0548(80)90011-8

Kadar, P. (2013). Application of Optimization Techniques in the Power System Control. *Acta Polytechnica Hungarica., 10*(5), 221–236.

Kuhn, H. W., & Tucker, A. W. (1995). Nonlinear Programming. In *Proceedings of the 2nd Berkley Symposium, Mathematics and Statistical Probability* (pp. 481-492).

Kumar, A., Pant, S., & Ram, M. (2016). System reliability optimization using grey wolf optimizer algorithm. *Quality and Reliability Engineering International.* doi:10.1002/qre.2107

Kumar, A., Pant, S., Ram, M., & Singh, S. B. (2017). *On Solving Complex Reliability Optimization Problem Using Multi-Objective Particle Swarm Optimization. In Mathematics Applied to Engineering* (pp. 115–131). Elsevier.

Kumar, A., Pant, S., & Singh, S. B. (2016). Reliability Optimization of Complex System by Using Cuckoos Search algorithm. In Mathematical Concepts and Applications in Mechanical Engineering and Mechatronics (pp. 95–112). Hershey, PA: IGI Global.

Miettinen, K., & Mäkelä, M. M. (1994). A Nondifferentiable Multiple Criteria Optimization Method Applied to Continuous Casting Process. In A. Fasano & M. Primicerio (Ed.), *Proceedings of the Seventh European Conference on Mathematics in Industry (ECMI 9)*, Stuttgart (pp. 255-262).

Morse, J. N. (1980). Reducing the size of the nondominated set: Pruning by clustering. *Computers & Operations Research, 7*(1-2), 55–66. doi:10.1016/0305-0548(80)90014-3

Osyczka, A. (1984). *Multicriterion Optimization in Engineering - with FORTRAN Programs.* Chichester: Ellis Horwood.

Pant, S., Anand, D., Kishor, A., & Singh, S.B. (2015). A Particle Swarm Algorithm for Optimization of Complex System Reliability. *International Journal of Performability Engineering, 11*(1), 33-42.

Pant, S., Kumar, A., Kishor, A., Anand, D., & Singh, S. B. (2015). Application of a Multi-Objective Particle Swarm optimization Technique to Solve Reliability Optimization Problem. In *Proceeding of IEEE Int. Conf. on Next generation Computing Technologies*, September 4-5 (pp. 1004-1007). doi:10.1109/NGCT.2015.7375272

Pant, S., Kumar, A., & Ram, M. (2017a). Reliability Optimization: A Particle Swarm Approach. Advances in Reliability and System Engineering. *Springer International Publishing, 2017*, 163–187.

Pant, S., Kumar, A., & Ram, M. (2017b). *Flower pollination algorithm development: a state of art review. International Journal of System Assurance Engineering and Management.* doi:10.1007/s13198-017-0623-7

Pant, S., Kumar, A., Singh, S. B., & Ram, M. (2017c). A Modified Particle Swarm Optimization Algorithm for Nonlinear Optimization. *Nonlinear Studies., 24*(1), 127–138.

Pant, S., & Singh, S. B. (2011). Particle Swarm Optimization to Reliability Optimization in Complex System. In *Proceedings of the IEEE Int. Conf. on Quality and Reliability*, Bangkok, Thailand, Sept 14-17 (pp. 211-215). doi:10.1109/ICQR.2011.6031711

Rosenman, M. A., & Gero, J. S. (1985). Reducing the pareto optimal set in multicriteria optimization (With application to Pareto optimal dynamic programing). *Engineering Optimization*, *8*(3), 189–206. doi:10.1080/03052158508902489

Salama, M., Garba, J., Demsetz, L., & Udwadia, F. (1988). Simultaneous Optimization of Controlled Structures. *Computational Mechanics*, *3*(4), 275–282. doi:10.1007/BF00368961

Schaffer, J. (1985). Multiple objective optimization with vector evaluated genetic algorithms. In *Proceeding of International Conference on Genetic Algorithms*, Pittsburgh.

Steuer, R. (1986). *Multiple criteria optimization: theory, computation and application.* New York: John Wiley & Sons, Inc.

Walley, P. (1991). *Statistical Reasoning with Imprecise Probabilities.* London: Chapman and Hall. doi:10.1007/978-1-4899-3472-7

Compilation of References

Abhishek, K., Datta, S., & Mahapatra, S. S. (2016). Multi-objective optimization in drilling of CFRP (polyester) composites: Application of a fuzzy embedded harmony search (HS) algorithm. *Measurement*, *77*, 222–239. doi:10.1016/j.measurement.2015.09.015

Abido, M. A. (2003). A niched Pareto genetic algorithm for multiobjective environmental/economic dispatch. *Electrical Power and Energy Systems*, *25*(2), 97–105. doi:10.1016/S0142-0615(02)00027-3

Ageron, B., Gunasekaran, A., & Spalanzani, A. (2012). Sustainable supply management: An empirical study. *International Journal of Production Economics*, *140*(1), 168–182. doi:10.1016/j.ijpe.2011.04.007

Ahi, P., & Searcy, C. (2013). A comparative literature analysis of definitions for green and sustainable supply chain management. *Journal of Cleaner Production*, *52*, 329–341. doi:10.1016/j.jclepro.2013.02.018

Ahi, P., & Searcy, C. (2014). A stochastic approach for sustainability analysis under the green economics paradigm. *Stochastic Environmental Research and Risk Assessment*, *28*(7), 1743–1753. doi:10.1007/s00477-013-0836-5

Ahmed, A. (2016). Deposition and Analysis of Composite Coating on Aluminum Using Ti – B_4C Powder Metallurgy Tools in EDM. *Materials and Manufacturing Processes*, *31*(4), 467–474. doi:10.1080/10426914.2015.1025967

Akkerman, R., Farahani, P., & Grunow, M. (2010). Quality, safety and sustainability in food distribution: A review of quantitative operations management approaches and challenges. *Spectrum (Lexington, Ky.)*, *32*(4), 863–904.

Ali, N., Othman, M. A., Husain, M. N., & Misran, M. H. (2014). A Review of Firefly Algorithm. *ARPN Journal of Engineering and Applies Science*, *9*(10), 1732–1736.

Amin, A. K. M. N., Lajis, M. A., & Radzi, H. C. D. M. (2009). The implementation of Taguchi method on EDM process of tungsten carbide. *European Journal of Scientific Research*, *26*, 609–617.

Amorima, F. L., & Weingaertner, W. L. (2005). The influence of generator actuation mode and process parameters on the performance of finish EDM of a tool steel. *Journal of Materials Processing Technology*, *166*(3), 411–416. doi:10.1016/j.jmatprotec.2004.08.026

Anitha, J., Das, R., & Pradhan, M. K. (2012). Comparison of Neural Network Learning Algorithms for Prediction of Surface Roughness in EDM. *Journal of Mechatronics and Intelligent Manufacturing*, *3*(1/2), 3.

Arora, S., & Sing, S. (2013). The Firefly Optimization Algorithm: Convergence Analysis and Parameter Selection. *International Journal of Computers and Applications*, *69*(3).

Asilturk, I., Neseli, S., & Ince, M. A. (2016). Optimization of Parameters affecting surface roughness of Co28Cr6Mo medical material during CNC lathe machining by using Taguchi and RSM Methods. *Measurement*, *78*, 120–128. doi:10.1016/j.measurement.2015.09.052

Attaran, M. (2012). Critical success factors and challenges of implementing RFID in supply chain management. *Journal of Supply Chain and Operations Management*, *10*(1), 144–167.

Bachmann, G., Narici, L., & Beckenstein, E. (2012). *Fourier and wavelet analysis*. Springer Science & Business Media.

Badia-Melis, R., Mishra, P., & Ruiz-García, L. (2015). Food traceability: New trends and recent advances. A review. *Food Control*, *57*, 393–401. doi:10.1016/j.foodcont.2015.05.005

Bagheri, A., & Zandieh, M. (2011). Bi-criteria flexible job-shop scheduling with sequence-dependent setup times—variable neighborhood search approach. *Journal of Manufacturing Systems*, *30*(1), 8–15. doi:10.1016/j.jmsy.2011.02.004

Bagheri, A., Zandieh, M., Mahdavi, I., & Yazdani, M. (2010). An artificial immune algorithm for the flexible job-shop scheduling problem. *Future Generation Computer Systems*, *26*(4), 533–541. doi:10.1016/j.future.2009.10.004

Baglietto, E., & Ninokata, H. (2005). A turbulence model study for simulating flow inside tight lattice rod bundles. *Nuclear Engineering and Design*, *235*(7), 773–784. doi:10.1016/j.nucengdes.2004.10.007

Bahram, A. (2009). Optimising the automated plasma cutting process by design of experiments. *International Journal of Rapid Manufacturing*, *1*(1), 19–40. doi:10.1504/IJRAPIDM.2009.028930

Baidya, A., Bera, U. K., & Maiti, M. (2013). Solution of multi-item interval valued solid transportation problem with safety measure using different methods. *Opsearch*, 1-22.

Baird, A. J., & Wilby, R. L. (1999). *Eco-hydrology: plants and water in terrestrial and aquatic environments*. Psychology Press.

Balinski, M. L. (1961). Fixed-cost transportation problems. *Naval Res. Logistics Q.*, *8*(1), 41–54. doi:10.1002/nav.3800080104

Baradaran, S., Fatemi Ghomi, S. M. T., Mobini, M., & Hashemin, S. S. (2010). A hybrid scatter search approach for resource-constrained project scheduling problem in PERT-type networks. *Advances in Engineering Software*, *41*(7-8), 966–975. doi:10.1016/j.advengsoft.2010.05.010

Bass, I. (2007). *Six sigma statistics with Excel and Minitab*. McGraw Hill.

Baykasoğlu, A., Hamzadayi, A., & Köse, S. Y. (2014). Testing the performance of teaching–learning based optimization (TLBO) algorithm on combinatorial problems: Flow shop and job shop scheduling cases. *Information Sciences*, *276*, 204–218. doi:10.1016/j.ins.2014.02.056

Bellman, R. E., & Zadeh, L. A. (1970). Decision-making in a fuzzy environment. *Management Science*, *17*(4), B-141–B-164. doi:10.1287/mnsc.17.4.B141

Benayoun, R., de Montgolfier, J., Tergny, J., & Laritchev, O. (1971). Linear programming with multiple objective functions: Step method (stem). *Mathematical Programming*, *1*(1), 366–375. doi:10.1007/BF01584098

Beri, N., Maheshwari, S., Sharma, C., & Kumar, A. (2008). Performance Evaluation of Powder Metallurgy Electrode in Electrical Discharge Machining of AISI D2 Steel Using Taguchi Method, *International Journal of Mechanical, Aerospace, Industrial, Mechatronic and Manufacturing Engineering*, *2*(2), 225–229.

Beri, N., Maheshwari, S., Sharma, C., & Kumar, A. (2014). Surface quality modification using powder metallurgy processed CuW electrode during electric discharge machining of Inconel 718. *Procedia Materials Science*, *5*, 2629–2634. doi:10.1016/j.mspro.2014.07.524

Bernardo, J. M., & Smith, A. F. (2001). *Bayesian theory*. Academic Press.

Beta, T., Nam, S., Dexter, J. E., & Sapirstein, H. D. (2017). Phenolic content and antioxidant activity of pearled wheat and roller-milled fractions. *LWT-Food Science and Technology*, *78*, 151–159.

Beulens, A. J., Broens, D. F., Folstar, P., & Hofstede, G. J. (2005). Food safety and transparency in food chains and networks Relationships and challenges. *Food Control*, *16*(6), 481–486. doi:10.1016/j.foodcont.2003.10.010

Bhanot, V. K., Beri, N., & Kumar, A. (2014). Machinability assessment of Superni-800 during EDM with powder metallurgy processed Cu-Ti electrode using the Taguchi method.

Bhattacharya, A., Yu, S. D., & Kawall, G. (2012). Numerical simulation of turbulent flow through a 37-element CANDU fuel bundle. *Annals of Nuclear Energy*, *40*(1), 87–105. doi:10.1016/j.anucene.2011.10.017

Bhoskar, T., Kukarni, O. K., Kulkarni, N. K., Patekar, S. L., Kakandikar, G. M., & Nandedkar, V. M. (2015). Genetic Algorithm and Its Applications to Mechanical Engineering: A Review. *4th International Conference on Materials Processing and Characterization. Materials Today: Proceedings, 2*, 2624 – 2630.

Bhuvenesh, R., Norizaman, M. H., & Manan, M. S. A. (2012). Surface roughness and MRR effect on manual plasma arc cutting machining. *International Journal of Mechanical Engineering*, *6*, 131–134.

Bini, R., Colosimo, B. M., Kutlu, A. E., & Monno, M. (2007). Experimental study of the features of the kerf generated by a 200 A high tolerance plasma cutting system. *Journal of Materials Processing Technology*, *196*(1-3), 345–355. doi:10.1016/j.jmatprotec.2007.05.061

Bit, A. K., Biswal, M. P., & Alam, S. S. (1993). Fuzzy programming approach to multiobjective solid transportation problem. *Fuzzy Sets and Systems*, *57*(2), 183–194. doi:10.1016/0165-0114(93)90158-E

Bjornstad, J. F. (1990). Predictive likelihood: A review. *Statistical Science*, *5*(2), 242–254. doi:10.1214/ss/1177012175

Bjornstad, J. F. (1996). On the generalization of the likelihood function and the likelihood principle. *Journal of the American Statistical Association*, *91*, 791–806.

Blaž, M., & Iztok, T. (2017). URANS prediction of flow fluctuations in rod bundle with split-type spacer grid. *International Journal of Heat and Fluid Flow*, *64*, 10–22. doi:10.1016/j.ijheatfluidflow.2017.01.008

Bloemhof, J. M., & Soysal, M. (2017). Sustainable food supply chain design. In *Sustainable Supply Chains* (pp. 395–412). Springer International Publishing. doi:10.1007/978-3-319-29791-0_18

Bobbili, R., Madhu, V., & Gogia, A. K. (2013). Effect of Wire-EDM Machining Parameters on Surface Roughness and Material Removal Rate of High Strength Armor Steel. *Materials and Manufacturing Processes*, *28*(4), 364–368. doi:10.1080/10426914.2012.736661

Boukherroub, T., Bouchery, Y., Corbett, C. J., Fransoo, J. C., & Tan, T. (2017). Carbon footprinting in supply chains. In *Sustainable Supply Chains* (pp. 43–64). Springer International Publishing. doi:10.1007/978-3-319-29791-0_3

Brandimarte, P. (1993). Routing and scheduling in a flexible job shop by tabu search. *Annals of Operations Research*, *41*(3), 157–183. doi:10.1007/BF02023073

Branke, J. (2000). *Efficient evolutionary algorithms for searching robust solutions*. ACDM. doi:10.1007/978-1-4471-0519-0_22

Brucker, P., & Schlie, R. (1990). Job-shop scheduling with multi-purpose machines. *Computing*, *45*(4), 369–375. doi:10.1007/BF02238804

Callister, W. D. J., & Rethwich, D. G. (2010). Material Science and Engineering: An Introduction (8th ed.). Wiley.

Campbell, J. B., & Wynne, R. H. (2011). *Introduction to remote sensing*. Guilford Press.

Canter, L. W. (1996). *Nitrates in groundwater*. CRC Press.

Cesàro, E. (1890). Sur la multiplication des séries. *Bulletin des Sciences Mathématiques, 14*(2), 114–120.

Chang, D., & Tavoularis, S. (2007). Numerical simulation of turbulent flow in a 37–rod bundle. *Nuclear Engineering and Design, 237*(6), 575–590. doi:10.1016/j.nucengdes.2006.08.001

Chankong, V., & Haimes, Y. Y. (1983). *Multiobjective Decision Making: Theory and Methodology*. New York: Elsevier Science Publishing.

Charnes, A., & Cooper, W. W. (1961). *Management models and industrial applications of linear programming*. New York: Wiley.

Charnes, A., Cooper, W. W., & Ferguson, R. (1955). Optimal estimation of executive compensation by linear programming. *Management Science, 1*(2), 138–151. doi:10.1287/mnsc.1.2.138

Cheng, S., Chen, H., & Zhang, X. (2017). CFD analysis of flow field in a 5×5 rod bundle with multi grid. *Annals of Nuclear Energy, 99*, 464–470. doi:10.1016/j.anucene.2016.09.053

Chen, S. M., & Wang, C. Y. (2013). Fuzzy decision making systems based on interval type-2 fuzzy sets. *Inf. Sci., 242*, 1–21. doi:10.1016/j.ins.2013.04.005

Chiang, K. T. (2008). Modeling and analysis of the effects of machining parameters on the performance characteristics in the EDM process of Al2O3+TiC mixed ceramic. *International Journal of Advanced Manufacturing Technology, 37*(5-6), 523–533. doi:10.1007/s00170-007-1002-3

Choi, B., & Rhee, F. (2009). Interval type-2 fuzzy membership function generation methods for pattern recognition. *Inform. Sci., 179*(13), 2102–2122. doi:10.1016/j.ins.2008.04.009

Choudhary, R., Kumar, H., & Grag, R. K. (2010). Analysis and evaluation of heat affected zones in electric discharge machining of EN-31 die steel. *Indian Journal of Engineering Materials Sciences, 17*, 91–98.

Choudhary, R., Kumar, P., & Singh, J. (2012). Analysis of Electro Discharged Machined Surfaces of AISI D3 Steel by Composite Tool Electrode. *International Journal of Surface Engineering & Materials Technology, 2*(2), 3–8.

Cinosi, N., Walker, S. P., Bluck, M. J., & Issa, R. (2014). CFD simulation of turbulent flow in a rod bundle with spacer grids (MATIS-H) using STAR-CCM+. *Nuclear Engineering and Design, 279*, 37–49. doi:10.1016/j.nucengdes.2014.06.019

Combest, D. P., Ramachandran, P. A., & Dudukovic, M. P. (2011). On the Gradient Diffusion Hypothesis and Passive Scalar Transport in Turbulent Flows. *Industrial & Engineering Chemistry Research, 50*(15), 8817–8823. doi:10.1021/ie200055s

Das, A., Bera, U. K., & Manoranjan, M. (2016). A breakable multi-item multi stage solid transportation problem under budget with Gaussian type-2 fuzzy parameters. *Applied Intelligence, 45*(3), 923–951. doi:10.1007/s10489-016-0794-y

Das, A., & Mishra, J. P. (2017). Experimental investigation on surface modification of aluminum by electric discharge coating process using TiC/Cu green compact tool-electrode. *Mining Science and Technology, 16*, 601–623.

Das, R., & Pradhan, M. K. (2014). General Regression Neural Network and Back Propagation Neural Network Modeling for Predicting Radial Overcut in EDM: A Comparative Study. *Mechatronic and Manufacturing Engineering, 8*(4), 799–805.

Datta, S., Biswal, B.B., & Mahapatra, S.S. (2017). A Novel Satisfaction Function and Distance-Based Approach for Machining Performance Optimization During Electro-Discharge Machining on Super Alloy Inconel 718. *Arabian Journal for Science and Engineering, 42*, 1999–2020. doi:10.1007/s13369-017-2422-5

Datta, S., & Mahapatra, S. S. (2010). Modeling, simulation and parametric optimization of wire EDM process using response surface methodology coupled with grey-Taguchi technique. *International Journal of Engineering Science and Technology*, *2*(5), 162–183. doi:10.4314/ijest.v2i5.60144

Defersha, F. M., & Chen, M. (2010). A parallel genetic algorithm for a flexible job-shop scheduling problem with sequence dependent setups. *International Journal of Advanced Manufacturing Technology*, *49*(1-4), 263–279. doi:10.1007/s00170-009-2388-x

Deng, J. (1989). Introduction to Grey System. *Journal of Grey System*, *1*(1), 1–24.

Dereli, T., Baykasoglu, A., Altun, K., Durmusoglu, A., & Türksen, I. B. (2011). Industrial applications of type-2 fuzzy sets and systems: A concise review. *Computers in Industry*, *62*(2), 125–137. doi:10.1016/j.compind.2010.10.006

Dominguez-Ontiveros, E., & Hassan, Y. A. (2014). Experimental Study of a Simplified 3x3 Rod Bundle Using DPTV. *Nuclear Engineering and Design*, *279*, 50–59. doi:10.1016/j.nucengdes.2014.04.037

Dubois, D., & Prade, H. (1979). Fuzzy real algebra: Some results. *Fuzzy Sets and Systems*, *2*(4), 327–348. doi:10.1016/0165-0114(79)90005-8

Dubois, D., & Prade, H. (1979). Operations in a Fuzzy-Valued Logic. *Information and Control*, *43*(2), 224–240. doi:10.1016/S0019-9958(79)90730-7

Durodola, J. F., Ramachandra, S., Li, N., & Thite, A. N. (2017). A pattern Recognition artificial neural network method for random fatigue loading life prediction. *International Journal of Fatigue*, *99*, 55–67. doi:10.1016/j.ijfatigue.2017.02.003

Dutta, A., & Jana, D. K. (2017). Expectations of the reductions for type-2 trapezoidal fuzzy variables and its application to a multi-objective solid transportation problem via goal programming technique. *Journal of Uncertainty Analysis and Applications*, *5*(3), 1–21.

Dutta, D., & Sen, M. (2017). A new method for solving multi-item multi-objective solid fixed charged shipment model with type-2 fuzzy variables. *Global and Stochastic Analysis*, *4*(1), 35–47.

Dutta, H., & Rhoades, B. E. (Eds.). (2016). *Current Topics in Summability Theory and Applications*. Singapore: Springer. doi:10.1007/978-981-10-0913-6

Egilmez, G., & Park, Y. S. (2014). Transportation related carbon, energy and water footprint analysis of US manufacturing: An eco-efficiency assessment. *Transportation Research Part D, Transport and Environment*, *32*, 143–159. doi:10.1016/j.trd.2014.07.001

Ehrgott, M., & Wiecek, M. (2005). Multiobjective programming. In J. Figueira, S. Greco, & M. Ehrgott (Eds.), *Multiple Criteria Decision Analysis. State of the Art Surveys* (pp. 667–722). Springer. doi:10.1007/0-387-23081-5_17

Elachi, C., & Van Zyl, J. J. (2006). *Introduction to the physics and techniques of remote sensing* (Vol. 28). John Wiley & Sons. doi:10.1002/0471783390

El-Sayed, S. M., EL-Wahed, W. F. A., & Ismail, N. A. (2008). A hybrid genetic scatter search algorithm for solving optimization problems. *Constraints*, *500*, 1.

El-Wakeel, A. S. (2014). Design optimization of PM couplings using hybrid Particle Swarm Optimization-Simplex Method (PSO-SM) Algorithm. *Electric Power Systems Research*, *116*, 29–35. doi:10.1016/j.epsr.2014.05.003

Engle, R. F., & Russell, J. R. (1998). Autoregressive conditional duration: A new model for irregularly spaced transaction data. *Econometrica*, *66*(5), 1127–1162. doi:10.2307/2999632

Ermoliev, Y., Makowski, M., & Marti, K. (2012). Robust management of heterogeneous systems under uncertainties. In *Managing Safety of Heterogeneous Systems* (pp. 1–16). Springer Berlin Heidelberg. doi:10.1007/978-3-642-22884-1_1

Ermoliev, Y., Makowski, M., & Marti, K. (Eds.). (2012). *Managing safety of heterogeneous systems: decisions under uncertainties and risks* (Vol. 658). Springer Science & Business Media. doi:10.1007/978-3-642-22884-1

Ermoliev, Y., & von Winterfeldt, D. (2012). Systemic risk and security management. In *Managing safety of heterogeneous systems* (pp. 19–49). Springer Berlin Heidelberg. doi:10.1007/978-3-642-22884-1_2

Eschenauer, H., Koski, J., & Osyczka, A. (1990). *Multicriteria Design Optimization.* Berlin: Springer-Verlag. doi:10.1007/978-3-642-48697-5

Escobar, L. A., & Meeker, W. Q. (1999). Statistical prediction based on censored life data. *Technometrics*, *41*(2), 113–124. doi:10.1080/00401706.1999.10485632

European Commission. (2005). *Common implementation strategy for the water framework directive* (2000-60-EC). Groundwater summary report. Technical report on groundwater body characterization, monitoring and risk assessment issues as discussed at the WG C workshop in 2003-2004.

European Environment Agency. (2003). *Europe's Water: An Indicator-based Assessment.* Topic Rep. 1. Copenhagen, Denmark: Author.

Experiments and CFD Code Applications to Nuclear Reactor Safety (XCFD4NRS).

Fisher, R. A. (1951). *Design of experiments.* Oliver and Boyd.

Fonseca, C., & Fleming, P. (1995). An overview of evolutionary algorithms in multiobjective optimization. *Evolutionary Computation*, *3*(1), 1–18. doi:10.1162/evco.1995.3.1.1

Gabus, A., & Fontela, E. (1972). *World problems, an invitation to further thought within the framework of DEMATEL.* Geneva, Switzerland: Battelle Geneva Research Center.

Gane, N., Rogozinski, M. W., Polivka, F., Doolette, A. G., & Ramakrishnan, S. (1994). *Quality of cut in air plasma cutting.* Paper presented at the Washington Technology Industry Association 42nd annual National Welding Conference, Melbourne, Australia.

Gangadhar, A., Shunmugam, M. S., & Philip, P. K. (1991). Surface modification in electrodischarge processing with a powder compact tool electrode. *Wear*, *143*(1), 45–55. doi:10.1016/0043-1648(91)90084-8

Gao, J., Sun, L., & Gen, M. (2008). A hybrid genetic and variable neighborhood descent algorithm for flexible job shop scheduling problems. *Computers & Operations Research*, *35*(9), 2892–2907. doi:10.1016/j.cor.2007.01.001

Garg, R., & Mittal, S. (2014). Optimization by Genetic Algorithm. *International Journal of Advanced Research in Computer Science and Software Engineering*, *4*(4), 587–589.

Gariboldi, E., & Previtali, B. (2004). High tolerance plasma arc cutting of commercially pure titanium. *Journal of Materials Processing Technology*, *160*(1), 77–89. doi:10.1016/j.jmatprotec.2004.04.366

Geng, Y., Zhang, J., Zhou, Q., Xu, C., & Zhao, Y. Q. (2011). *Fuzzy synthetic evaluation of Weihe water quality.* Academic Press.

Gen, M., Ida, K., Li, Y., & Kubota, E. (1995). Solving bicriteria solid transportation problem with fuzzy numbers by a genetic algorithm. *Computers & Industrial Engineering*, *29*(1-4), 537–541. doi:10.1016/0360-8352(95)00130-S

Genovese, A., Acquaye, A. A., Figueroa, A., & Koh, S. L. (2017). Sustainable supply chain management and the transition towards a circular economy: Evidence and some applications. *Omega*, *66*, 344–357. doi:10.1016/j.omega.2015.05.015

Geweke, J., & Amisano, G. (2012). Prediction and misspecified models. *The American Economic Review*, *102*(3), 482–486. doi:10.1257/aer.102.3.482

Gilan, S. S., Sebt, M. H., & Shahhosseini, V. (2012). Computing with words for hierarchical competency based selection of personal in construction companies. *Applied Soft Computing*, *12*(2), 860–871. doi:10.1016/j.asoc.2011.10.004

Girotto, F., Alibardi, L., & Cossu, R. (2015). Food waste generation and industrial uses: A review. *Waste Management (New York, N.Y.)*, *45*, 32–41. doi:10.1016/j.wasman.2015.06.008 PMID:26130171

Glover, F. (1977). Heuristics for integer programming using surrogate constraints. *Decision Sciences*, *8*(1), 156–166. doi:10.1111/j.1540-5915.1977.tb01074.x

Glover, F., Laguna, M., & Marti, R. (2000). Fundamentals of Scatter Search and Path Relinking. *Controls and Cybernetics*, *39*(3), 653–684.

Goldberg, D. E. (1998). The Design of Innovation: Lessons from Genetic Algorithms. Lessons for the Real World. University of Illinois at Urbana-Champaign.

Goldberg, D. E. (2006). Genetic algorithms in search, optimization & Machine learning. Pearson Education.

Goldberg, D. (1989). *Genetic Algorithms in Search and Machine Learning*. Reading Addison Wesley.

Gold, S., Hahn, R., & Seuring, S. (2013). Sustainable supply chain management in Base of the Pyramid food projects—A path to triple bottom line approaches for multinationals? *International Business Review*, *22*(5), 784–799. doi:10.1016/j.ibusrev.2012.12.006

González, M. A., Vela, C. R., & Varela, R. (2015). Scatter search with path relinking for the flexible job shop scheduling problem. *European Journal of Operational Research*, *245*(1), 35-45. doi:10.1016/j.ejor.2015.02.052

Gordon, H. R., Brown, J. W., & Evans, R. H. (1988). Exact Rayleigh scattering calculations for use with the Nimbus-7 coastal zone color scanner. *Applied Optics*, *27*(5), 862–871. doi:10.1364/AO.27.000862 PMID:20523701

Govindan, K., Kannan, D., & Shankar, K. M. (2014). Evaluating the drivers of corporate social responsibility in the mining industry with multi-criteria approach: A multi-stakeholder perspective. *Journal of Cleaner Production*, *84*, 214–232. doi:10.1016/j.jclepro.2013.12.065

Govindan, K., Kannan, D., & Shankar, M. (2015). Evaluation of green manufacturing practices using a hybrid MCDM model combining DANP with PROMETHEE. *International Journal of Production Research*, *53*(21), 6344–6371. doi:10.1080/00207543.2014.898865

Govindan, K., Khodaverdi, R., & Jafarian, A. (2013). A fuzzy multi criteria approach for measuring sustainability performance of a supplier based on triple bottom line approach. *Journal of Cleaner Production*, *47*, 345–354. doi:10.1016/j.jclepro.2012.04.014

Grueninger, T., & Wallace, D. (1996). Multi-modal optimization using genetic algorithms (Technical Report 96.02). CAD lab, Massachusetts Institute of Technology, Cambridge, MA.

Grunert, K. G. (2005). Food quality and safety: Consumer perception and demand. *European Review of Agriculture Economics*, *32*(3), 369–391. doi:10.1093/eurrag/jbi011

Gullu, A., & Atici, U. (2006). Investigation of the effects of plasma arc parameters on the structure variation of AISI 304 and St 52 steels. *Materials & Design*, *27*(10), 1157–1162. doi:10.1016/j.matdes.2005.02.014

Gustavsson, J., Cederberg, C., Sonesson, U., Van Otterdijk, R., & Meybeck, A. (2011). Global food losses and food waste. *Food and Agriculture Organization of the United Nations, Rom.*

Guu, Y. H. (2005). AFM surface imaging of AISI D2 tool steel machined by the EDM process. *Applied Surface Science*, *242*(3-4), 245–250. doi:10.1016/j.apsusc.2004.08.028

Guu, Y. H., Hocheng, H., Chou, C. Y., & Deng, C. S. (2003). Effect of electrical discharge machining on surface characteristics and machining damage of AISI D2 tool steel. *Materials Science and Engineering*, *358*(1-2), 37–43. doi:10.1016/S0921-5093(03)00272-7

Habib, S. S. (2009). Study of the parameters in electrical discharge machining through response surface methodology approach. *Applied Mathematical Modelling*, *33*(12), 4397–4407. doi:10.1016/j.apm.2009.03.021

Hahn, G. J., & Meeker, W. Q. (1991). *Statistical Intervals: A Guide for Practitioners*. New York: Wiley. doi:10.1002/9780470316771

Hahn, G. J., & Nelson, W. (1973). A survey of prediction intervals and their applications. *Journal of Quality Technology*, *5*, 178–188.

Hanson, A. S., & Todreas, N. (1977). *Fluid mixing studies in a hexagonal 61-pin, wire wrapped rod bundle (tech. report)*. doi:10.2172/5159920

Harik, G. (1995). Finding multimodal solutions using restricted tournament selection. In *Proceeding of Sixth International Conference on Genetic Algorithms.*

Haron, C. H. C., Ghani, J. A., Burhanuddin, Y., Seong, Y. K., & Swee, C. Y. (2008). Copper and graphite electrodes performance in electrical-discharge machining of XW42 tool steel. *Journal of Materials Processing Technology*, *201*(1-3), 570–573. doi:10.1016/j.jmatprotec.2007.11.285

Havinga, T., & Verbruggen, P. (2017). Understanding complex governance relationships in food safety regulation: The RIT model as a theoretical lens. *The Annals of the American Academy of Political and Social Science*, *670*(1), 58–77. doi:10.1177/0002716216688872

Herlin, I., Nakonechnyi, O., & Zhuk, S. (2012, September). Minimax optical flow estimation from a sequence of 2D images. *XX International Conference on Problems of Decision Making under Uncertainties (PDMU).*

Hernandez, R. R., Easter, S. B., Murphy-Mariscal, M. L., Maestre, F. T., Tavassoli, M., Allen, E. B., & Allen, M. F. et al. (2014). Environmental impacts of utility-scale solar energy. *Renewable & Sustainable Energy Reviews*, *29*, 766–779. doi:10.1016/j.rser.2013.08.041

Herrero, Á., Sedano, J., Baruque, B., Quintián, H., & Corchado, E. (2015). *Proceedings of the 10th International Conference on Soft Computing Models in Industrial and Environmental Applications*. Springer.

Hitchcock, F. L. (1941). The distribution of product from several sources to numerous localities. *Journal of Mathematical Physics*, *20*(1-4), 224–230. doi:10.1002/sapm1941201224

Hofstede, G. J., Spaans, L., Schepers, H., Trienekens, J. H., & Beulens, A. J. M. (2004). *Hide or confide: the dilemma of transparency*. Academic Press.

Ho, K. H., & Newman, S. T. (2003). State of the art electrical discharge machinings (EDM). *International Journal of Machine Tools & Manufacture*, *43*(13), 1287–1300. doi:10.1016/S0890-6955(03)00162-7

Holland, J. (1975). Adaptation in Natural and Artificial Systems. University of Michigan Press.

Holland, J. H. (1975). *Adaptation in natural and artificial systems*. Ann Arbor: University of Michigan Press.

Horn, B. K., & Schunck, B. G. (1981). Determining optical flow. *Artificial Intelligence*, *17*(1-3), 185–203. doi:10.1016/0004-3702(81)90024-2

Ho, S. K., Aspinwall, D. K., Voice, W., Box, P. O., & De, D. (2007). Use of powder metallurgy (PM) compacted electrodes for electrical discharge surface alloying / modification of Ti – 6Al – 4V alloy. *Journal of Materials Processing Technology*, *191*(1-3), 123–126. doi:10.1016/j.jmatprotec.2007.03.003

Hsu, C. W., Kuo, T. C., Chen, S. H., & Hu, A. H. (2013). Using DEMATEL to develop a carbon management model of supplier selection in green supply chain management. *Journal of Cleaner Production*, *56*, 164–172. doi:10.1016/j.jclepro.2011.09.012

Hurink, J., Jurisch, B., & Thole, M. (1994). Tabu search for the job-shop scheduling problem with multi-purpose machines. *Operations-Research-Spektrum*, *15*(4), 205–215. doi:10.1007/BF01719451

Hwang, C., Paidy, S., Yoon, K., & Masud, A. S. M. (1980). Mathematical programming with multiple objectives: A tutorial. *Computers & Operations Research*, *7*(1-2), 5–31. doi:10.1016/0305-0548(80)90011-8

Ikeno, T., & Kajishima, T. (2010). Analysis of dynamical flow structure in a square arrayed rod bundle. *Nuclear Engineering and Design*, *240*(2), 305–312. doi:10.1016/j.nucengdes.2008.07.012

International Atomic Energy Agency (IAEA). (2003). IAEA-TECDOC-1332: Safety margins of operating reactors (tech. report).

Ismail, M. I. S., & Taha, Z. (2011). Experimental design and performance analysis in plasma arc surface hardening, *World Academy of Science. Engineering and Technology*, *56*, 1052–1058.

Jae-Ho, J., Min-Seop, S., & Kwi-Lim, L. (2017). RANS based CFD methodology for a real scale 217-pin wire-wrapped fuel assembly of KAERI PGSFR. *Nuclear Engineering and Design*, *313*, 470–485. doi:10.1016/j.nucengdes.2017.01.007

Jahan, M. P., Wong, Y. S., & Rahman, M. (2009). A study on the fine-finish die-sinking micro-EDM of tungsten carbide using different electrode materials. *Journal of Materials Processing Technology*, *209*(8), 3956–3967. doi:10.1016/j.jmatprotec.2008.09.015

Jaharah, A. G., Liang, C. G., Wahid, S. Z. M. N., Rahman, A., & Hassan, C. H. C. (2008). Performance of copper electrode in electrical discharge machining (EDM) of AISI H13 harden steel. *International Journal of Mechanical and Materials Engineering*, *3*, 25–29.

Jameson, E. C. (2001). *Electrical discharge machining*. Society of Manufacturing Engineers.

Jason, R. E., & Shepard, C. N. (2006). *Water Suppliers*. Academic Press.

Jena, A. R., Pradhan, M. K., Das, R., Acharjya, D. P., & Mishra, M. (2014). Prediction of Radial Overcut in Electro-Discharge Machining through Artificial Neural Network. In *Advancements and Current Trends in Industrial Mechanical and Production Engineering*.

Jia, P., Govindan, K., & Kannan, D. (2015). Identification and evaluation of influential criteria for the selection of an environmental shipping carrier using DEMATEL: A case from India. *International Journal of Shipping and Transport Logistics*, *7*(6), 719–741. doi:10.1504/IJSTL.2015.072684

Jimenez, F., & Verdegay, J. L. (1998). Uncertain solid transportation problems. *Fuzzy Sets and Systems*, *100*(1-3), 45–57. doi:10.1016/S0165-0114(97)00164-4

Ju-long, D. (1982). Control problems of grey systems. *Systems & Control Letters, 1*(5), 288–294. doi:10.1016/S0167-6911(82)80025-X

Julong, D. (1988). Introduction to Grey System Theory. *Journal of Grey System, 1*, 1–24.

Justel, A., Peña, D., & Zamar, R. (1997). A multivariate Kolmogorov–Smirnov test of goodness of fit. *Statistics & Probability Letters, 35*(3), 251–259. doi:10.1016/S0167-7152(97)00020-5

Kacem, I., Hammadi, S., & Borne, P. (2002). Pareto-optimality approach for flexible job-shop scheduling problems: Hybridization of evolutionary algorithms and fuzzy logic. *Mathematics and Computers in Simulation, 60*(3), 245–276. doi:10.1016/S0378-4754(02)00019-8

Kadar, P. (2013). Application of Optimization Techniques in the Power System Control. *Acta Polytechnica Hungarica., 10*(5), 221–236.

Kadirgama, K., Noor, M. M., Harun, W. S. W., & Aboue-El-Hossein, K. A. (2010). Optimization of heat affected zone by partial swarm optimization in air plasma cutting operation. *Journal of Scientific and Industrial Research, 69*, 439–443.

Karaboga, D. (2005). *An idea based on honey bee swarm for numerical optimization, Technical Report-TR06*. Erciyes University, Engineering Faculty, Computer Engineering Department.

Karnik, N. N., & Mendel, J. M. (1999). Applications of type-2 fuzzy logic systems to forecasting of time-series. *Inform. Sci, 120*(1-4), 89–111. doi:10.1016/S0020-0255(99)00067-5

Karthikeyan, S., Asokan, P., Nickolas, S., & Page, T. (2015). A hybrid discrete firefly algorithm for solving multi-objective flexible job shop scheduling problems. *International Journal of Bio-inspired Computation, 7*(6), 386–401. doi:10.1504/IJBIC.2015.073165

Katherasan, D., Elias, J. V., Sathiya, P., & Haq, A. N. (2012). Flux Cored Arc Welding Parameter Optimization Using Swarm Optimization Algorithm. *Procedia Engineering, 38*, 3913–3296. doi:10.1016/j.proeng.2012.06.449

Kaufman, Y. J., & Tanre, D. (1992). Atmospherically resistant vegetation index (ARVI) for EOS-MODIS. *IEEE Transactions on Geoscience and Remote Sensing, 30*(2), 261–270. doi:10.1109/36.134076

Kecman, V. (2001). *Learning and Soft Computing Support Vector Machines. In Neural Networks and Fuzzy Logic Models*. MIT Press.

Keskin, Y., Halka, H., & Kizil, S. M. (2006). An experimental study for determination of the effects of machining parameters on surface roughness in electrical discharge machining (EDM). *International Journal of Advanced Manufacturing Technology, 28*(11-12), 1118–1121. doi:10.1007/s00170-004-2478-8

Keufmann, A., & Gupta, M. M. (1985). *Introduction to Fuzzy Arithmetic: Theory and Applications*. New York: Van Nostrand Reinhold.

Khan, M. Z. R., & Bajpai, A. K. (2013). Genetic Algorithm and Its Application In Mechanical Engineering. *International Journal of Engineering Research & Technology, 2*(5), 677-683.

Khanra, A. K., Sarkar, B., Bhattachary, B., Pathak, L. C., & Godkhindi, M. M. (2007). Performance of ZrB2–Cu composite as an EDM electrode. *Journal of Materials Processing Technology, 183*(1), 122–126. doi:10.1016/j.jmatprotec.2006.09.034

Kirkpatrick, S., Gelatt, C. D., & Vecchi, M. P. (1983). Optimization by Simulated Annealing. *Science, 220*(4598), 671–680. doi:10.1126/science.220.4598.671 PMID:17813860

Kiyak, M., & Cakir, O. (2007). Examination of machining parameters on surface roughness in EDM of tool steel. *Journal of Materials Processing Technology, 191*(1-3), 141–144. doi:10.1016/j.jmatprotec.2007.03.008

Koch, H., & Grünewald, U. (2009). A comparison of modelling systems for the development and revision of water resources management plans. *Water Resources Management*, *23*(7), 1403–1422. doi:10.1007/s11269-008-9333-x

Kogan, F., Stark, R., Gitelson, A., Jargalsaikhan, L., Dugrajav, C., & Tsooj, S. (2004). Derivation of pasture biomass in Mongolia from AVHRR-based vegetation health indices. *International Journal of Remote Sensing*, *25*(14), 2889–2896. doi:10.1080/01431160410001697619

Kopachevsky, I., Kostyuchenko, Y. V., & Stoyka, O. (2016). Land Use Drivers of Population Dynamics in Tasks of Security Management and Risk Assessment. *International Journal of Mathematical, Engineering and Management Sciences*, *1*, 18–24.

Korkut, I., Kasap, M., & Seker, U. (2004). Determination of optimum cutting parameters during machining of AISI 304 austenitic stainless steel. *Materials & Design*, *25*(4), 303–305. doi:10.1016/j.matdes.2003.10.011

Koshy, P., Jain, V. K., & Lal, G. K. (1993). Experimental investigations into electrical discharge machining with a rotating disk electrode. *Precision Engineering*, *15*(1), 6–15. doi:10.1016/0141-6359(93)90273-D

Kostyuchenko, Y. V., Kopachevsky, I., Yuschenko, M., Solovyov, D., Marton, L., & Levynsky, S. (2012). Spectral reflectance indices as indirect indicators of ecological threats. In Sustainable Civil Infrastructures–Hazards, Risk, Uncertainty. Research Publishing.

Kostyuchenko, Y. V., Movchan, D., Artemenko, I., & Kopachevsky, I. (2016). Stochastic Approach to Uncertainty Control in Multiphysics Systems: Modeling of Carbon Balance and Analysis of GHG Emissions Using Satellite Tools. In Mathematical Concepts and Applications in Mechanical Engineering and Mechatronics. IGI Global. Doi:10.4018/978-1-5225-1639-2.ch017

Kostyuchenko, Y. V., Yuschenko, M., Movchan, D., Zagorodny, A. G., & Yermoliev, Y. M. (2013). Regional risk analysis based on multisource data statistics of natural disasters. In Integrated modeling of food, energy and water security management for sustainable social, economic and environmental developments, 229-238.

Kostyuchenko, Y. V. (2015). Geostatistics and remote sensing for extremes forecasting and disaster risk multiscale analysis. In *Numerical Methods for Reliability and Safety Assessment* (pp. 439–458). Springer International Publishing. doi:10.1007/978-3-319-07167-1_16

Kostyuchenko, Y. V. (2018). On the Methodology of Satellite Data Utilization in Multi-Modeling Approach for Socio-Ecological Risks Assessment Tasks: A Problem Formulation. *International Journal of Mathematical, Engineering and Management Sciences*, *3*(1), 1–8.

Kostyuchenko, Y. V., Bilous, Y., Kopachevsky, I., & Solovyov, D. (2013). Coherent risk measures assessment based on the coupled analysis of multivariate distributions of multisource observation data. *Proceedings of 11-th International Probabilistic Workshop*, 183-192.

Kostyuchenko, Y. V., Movchan, D., Kopachevsky, I., & Bilous, Y. (2015, November). Robust algorithm of multi-source data analysis for evaluation of social vulnerability in risk assessment tasks. In *SAI Intelligent Systems Conference (IntelliSys)* (pp. 944-949). IEEE. doi:10.1109/IntelliSys.2015.7361256

Krauss, T., & Meyer, L. (1996). Characteristics of turbulent velocity and temperature in a wall channel of a heated rod bundle. *Experimental Thermal and Fluid Science*, *12*(1), 75–86. doi:10.1016/0894-1777(95)00076-3

Kuhn, H. W., & Tucker, A. W. (1995). Nonlinear Programming. In *Proceedings of the 2nd Berkley Symposium, Mathematics and Statistical Probability* (pp. 481-492).

Kumar, A., Pant, S., & Singh, S. B. (2016). Reliability Optimization of Complex System by Using Cuckoos Search algorithm. In Mathematical Concepts and Applications in Mechanical Engineering and Mechatronics (pp. 95–112). Hershey, PA: IGI Global.

Kumar, A., Pant, S., Ram, M., & Singh, S. B. (2017) On Solving Complex Reliability Optimization Problem Using Multi-Objective Particle Swarm Optimization. In *Mathematics Applied to Engineering* (pp. 115-131).

Kumar, A., & Joshi, L. K. (2011). MODWT Based Time Scale Decomposition Analysis of BSE and NSE Indexes Financial Time Series. *Int. Journal of Math. Analysis*, *5*(27), 1343–1352.

Kumar, A., Joshi, L. K., Pal, A. K., & Shukla, A. K. (2010). A New Approach for Generating Parametric Orthogonal Wavelets. *Journal of Wavelet Theory and Applications*, *4*(1), 1–8.

Kumar, A., Pant, S., & Joshi, L. K. (2016). Wavelet Variance, Covariance and Correlation Analysis of BSE and NSE Indexes Financial Time Series. *International Journal of Mathematical. Engineering and Management Sciences*, *1*(1), 26–33.

Kumar, A., Pant, S., & Ram, M. (2016). *System Reliability Optimization Using Gray Wolf Optimizer Algorithm. Quality and Reliability Engineering International*. John Wiley & Sons Ltd.

Kumar, A., Pant, S., & Ram, M. (2016). System reliability optimization using grey wolf optimizer algorithm. *Quality and Reliability Engineering International*. doi:10.1002/qre.2107

Kumar, A., Pant, S., Ram, M., & Singh, S. B. (2017). *On Solving Complex Reliability Optimization Problem Using Multi-Objective Particle Swarm Optimization. In Mathematics Applied to Engineering* (pp. 115–131). Elsevier.

Kumar, A., Pant, S., & Singh, S. B. (2017). *Reliability Optimization of Complex System by Using Cuckoos Search algorithm*. Mathematical Concepts and Applications in Mechanical Engineering and Mechatronics; doi:10.4018/978-1-5225-1639-2.ch005

Kumar, S., Singh, B., Qadri, M. A., Kumar, Y. S., & Haleem, A. (2013). A framework for comparative evaluation of lean performance of firms using fuzzy TOPSIS. *International Journal of Productivity and Quality Management*, *11*(4), 371–392. doi:10.1504/IJPQM.2013.054267

Kundu, P., Kar, S., & Maiti, M. (2012). Multi-objective multi-item solid transportation problem in fuzzy environment. *Applied Mathematical Modelling*, *37*(4), 2028–2038. doi:10.1016/j.apm.2012.04.026

Kundu, P., Kar, S., & Maiti, M. (2014). Fixed charge transportation problem with type-2 fuzzy variables. *Inf. Sci.*, *255*, 170–186. doi:10.1016/j.ins.2013.08.005

Kundu, P., Kar, S., & Maiti, M. (2014). Multi-objective solid transportation problems with budget constraint in uncertain environment. *International Journal of Systems Science*, *45*(8), 1668–1682. doi:10.1080/00207721.2012.748944

Kundu, P., Kar, S., & Maiti, M. (2015). Multi-item solid transportation problem with type-2 fuzzy parameters. *Applied Soft Computing*, *31*, 61–80. doi:10.1016/j.asoc.2015.02.007

Kuppan, P., Rajadurai, A., & Narayanan, S. (2008). Influence of EDM process parameters in deep hole drilling of Inconel 718. *International Journal of Advanced Manufacturing Technology*, *38*(1-2), 74–84. doi:10.1007/s00170-007-1084-y

Kuriachen, B., Somashekhar, K. P., & Mathew, J. (2015). Multi response optimization of micro- wire electrical discharge machining process. *International Journal of Advanced Manufacturing Technology*, *76*(1-4), 91–104. doi:10.1007/s00170-014-6005-2

Kurikose, S., & Shanmugham, M. S. (2005). Multi objective optimization of wire EDM process by non-dominated sorting genetic algorithms. *Journal of Materials Processing Technology*, *170*, 133–141. doi:10.1016/j.jmatprotec.2005.04.105

Landis, W. G., & Thomas, J. F. (2009). Regional risk assessment as a part of the long-term receiving water study. *Integrated Environmental Assessment and Management*, *5*(2), 234–247. doi:10.1897/IEAM_2008-062.1 PMID:19127980

Lawless, J. F. (1973). On estimation of safe life when the underlying distribution is Weibull. *Technometrics*, *15*(4), 857–865. doi:10.1080/00401706.1973.10489118

Lee, H. L., Padmanabhan, V., & Whang, S. (2004). Information distortion in a supply chain: the bullwhip effect. *Management Science*, *50*(12), 1875-1886.

Lee, S. H., & Li, X. P. (2001). Study of the effect of machining parameters on the machining characteristics in electrical discharge machining of tungsten carbide. *Journal of Materials Processing Technology*, *115*(3), 344–358. doi:10.1016/S0924-0136(01)00992-X

Lejeune, M., & Faulkenberry, G. D. (1982). A simple predictive density function. *Journal of the American Statistical Association*, *77*(379), 654–657. doi:10.1080/01621459.1982.10477866

Liao, Y. S., Chuang, T. J., & Yu, Y. P. (2014). Study of machining parameters optimization for different materials in WEDM. *International Journal of Advanced Manufacturing Technology*, *70*(9-12), 2051–2058. doi:10.1007/s00170-013-5458-z

Liao, Z., & Rittscher, J. (2007). A multi-objective supplier selection model under stochastic demand conditions. *International Journal of Production Economics*, *105*(1), 150–159. doi:10.1016/j.ijpe.2006.03.001

Li, J., Huang, G. H., Zeng, G., Maqsood, I., & Huang, Y. (2007). An integrated fuzzy-stochastic modeling approach for risk assessment of groundwater contamination. *Journal of Environmental Management*, *82*(2), 173–188. doi:10.1016/j.jenvman.2005.12.018 PMID:16574309

Li, L., Niu, Z. W., & Zheng, G. M. (2016). Ultrasonic Electrodeposition of Cu – SiC Electrodes for EDM. *Materials and Manufacturing Processes*, *31*(1), 37–41. doi:10.1080/10426914.2015.1025968

Li, L., Wong, Y. S. U., Fuh, J. Y. H., & Lu, L. (2001b). EDM performance of TiC/copper-based sintered electrodes. *Materials & Design*, *22*(8), 669–678. doi:10.1016/S0261-3069(01)00010-3

Li, L., Wong, Y. S., Fuh, J. Y. H., & Lu, L. (2001a). Effect of TiC in copper-tungsten electrodes on EDM performance. *Journal of Materials Processing Technology*, *113*(1-3), 563–567. doi:10.1016/S0924-0136(01)00622-7

Lillesand, T., Kiefer, R. W., & Chipman, J. (2014). *Remote sensing and image interpretation*. John Wiley & Sons.

Liu, Y. G., Wang, N. L., Wang, L. G., Zhao, Y. Q., & BoWu, X. (2013). Application of GIS in regional ecological risk assessment of water resources. *Environmental Engineering & Management Journal*, *12*(7).

Liu, B. (2009). *Theory and Practice of Uncertain Programming* (3rd ed.). UTLAB; http://orsc.edu.cn/liu/up.pdf doi:10.1007/978-3-540-89484-1

Liu, B., & Iwamura, K. (1998). Chance constrained programming with fuzzy parameters. *Fuzzy Sets and Systems*, *94*(2), 227–237. doi:10.1016/S0165-0114(96)00236-9

Liu, B., & Liu, Y. K. (2002). Expected value of fuzzy variable and fuzzy expected value models. *IEEE Transactions on Fuzzy Systems*, *10*(4), 445–450. doi:10.1109/TFUZZ.2002.800692

Liu, G., Liu, X., & Cheng, S. (2013). Food security: Curb Chinas rising food wastage. *Nature*, *498*(7453), 170–170. doi:10.1038/498170c PMID:23765482

Liu, S. (2006). Fuzzy total transportation cost measures for fuzzy solid transportation problem. *Applied Mathematics and Computation*, *174*(2), 927–941. doi:10.1016/j.amc.2005.05.018

Liu, Z. Q., & Liu, Y. K. (2010). Type-2 fuzzy variables and their arithmetic. *Soft Computing*, *14*(7), 729–747. doi:10.1007/s00500-009-0461-x

Li, X., Ruan, D., & van der Wal, A. J. (1998). Discussion on soft computing at FLINS96. *International Journal of Intelligent Systems*, *13*(2-3), 287–300. doi:10.1002/(SICI)1098-111X(199802/03)13:2/3<287::AID-INT10>3.0.CO;2-4

Li, Y., Ida, K., & Gen, M. (1997). Improved genetic algorithm for solving multiobjective solid transportation problem with fuzzy numbers. *Computers & Industrial Engineering*, *33*(3-4), 589–592. doi:10.1016/S0360-8352(97)00199-X

Li, Y., Ida, K., Gen, M., & Kobuchi, R. (1997). Neural network approach for multicriteria solid transportation problem. *Computers & Industrial Engineering*, *33*(3-4), 465–468. doi:10.1016/S0360-8352(97)00169-1

Luo, H., Huang, G. Q., Zhang, Y., Dai, Q., & Chen, X. (2009). Two-stage hybrid batching flowshop scheduling with blocking and machine availability constraints using genetic algorithm. *Robotics and Computer-integrated Manufacturing*, *25*(6), 962–971. doi:10.1016/j.rcim.2009.06.001

Lyalko, V. I., Kostyuchenko, Yu. V., Elistratova, L., & Artemenko, I. (2015). *The greenhouse effect and climate change in Ukraine: assessment and effect*. Kiev: Naukova Dumka. (in Ukrainian)

Mahapatra, S. S., Patnaik, A., & Patnaik, P. (2006). Parametric analysis and optimization of cutting parameters for turning operations. *Proceedings of the Int. Conference on Global Manufacturing and Innovation*, 1–6.

Mahdavi, M., Fesanghary, M., & Damangir, E. (2007). An improved harmony search algorithm for solving optimization problems. *Applied Mathematics and Computation*, *188*(2), 1567–1579. doi:10.1016/j.amc.2006.11.033

Maity, K. P., & Bagal, D. K. (2015). Effect of process parameters on cut quality of stainless steel of plasma arc cutting using hybrid approach. *International Journal of Advanced Manufacturing Technology*, *78*(1-4), 161–175. doi:10.1007/s00170-014-6552-6

Mangla, S. K., Kumar, P., & Barua, M. K. (2015). Flexible decision modeling for evaluating the risks in green supply chain using fuzzy AHP and IRP methodologies. *Global Journal of Flexible Systems Management*, *16*(1), 19–35. doi:10.1007/s40171-014-0081-x

Mangla, S. K., Kumar, P., & Barua, M. K. (2015). Prioritizing the responses to manage risks in green supply chain: An Indian plastic manufacturer perspective. *Sustainable Production and Consumption*, *1*, 67–86. doi:10.1016/j.spc.2015.05.002

Mangla, S., Kumar, P., & Barua, M. K. (2014). An evaluation of attribute for improving the green supply chain performance via DEMATEL method. *International Journal of Mechanical Engineering & Robotics Research*, *1*(1), 30–35.

Manning, L., Baines, R. N., & Chadd, S. A. (2006). Quality assurance models in the food supply chain. *British Food Journal*, *108*(2), 91–104. doi:10.1108/00070700610644915

Marichelvam, M. K., & Geetha, M. (2014a). Solving tri-objective multistage hybrid flow shop scheduling problems using a discrete firefly algorithm. *International Journal of Intelligent Engineering Informatics*, *2*(4), 284–303. doi:10.1504/IJIEI.2014.067190

Marichelvam, M. K., & Prabaharan, T. (2014). Performance evaluation of an improved hybrid genetic scatter search (IHGSS) algorithm for multistage hybrid flow shop scheduling problems with missing operations. *International Journal of Industrial and Systems Engineering*, *16*(1), 120–141. doi:10.1504/IJISE.2014.057946

Marsaglia, G., Tsang, W. W., & Wang, J. (2003). Evaluating Kolmogorovs Distribution. *Journal of Statistical Software*, *8*(18), 1–4. doi:10.18637/jss.v008.i18

Martínez, Y., & Albiac, J. (2006). Nitrate pollution control under soil heterogeneity. *Land Use Policy*, *23*(4), 521–532. doi:10.1016/j.landusepol.2005.05.002

Marucheck, A., Greis, N., Mena, C., & Cai, L. (2011). Product safety and security in the global supply chain: Issues, challenges and research opportunities. *Journal of Operations Management*, *29*(7), 707–720. doi:10.1016/j.jom.2011.06.007

Mastrolilli, M., & Gambardella, L. M. (2000). Effective neighborhood functions for the flexible job shop problem. *Journal of Scheduling*, *3*(1), 3–20. doi:10.1002/(SICI)1099-1425(200001/02)3:1<3::AID-JOS32>3.0.CO;2-Y

McFeeters, S. K. (1996). The use of the Normalized Difference Water Index (NDWI) in the delineation of open water features. *International Journal of Remote Sensing*, *17*(7), 1425–1432. doi:10.1080/01431169608948714

Mee, R., & Kushary, D. (1994). Prediction limits for the Weibull distribution utilizing simulation. *Computational Statistics & Data Analysis*, *17*(3), 327–336. doi:10.1016/0167-9473(92)00072-Y

Mendel, J. M. (2007). Computing with words and its relationship with fuzzistics. *Inform. Sci.*, *177*(4), 988–1006. doi:10.1016/j.ins.2006.06.008

Meyer, L. 2010. From discovery to recognition of periodic large scale vortices in rod bundles as source of natural mixing between sub channels, - A Review. *Nuclear Engineering and Design*, *240*(6), 1575 – 1588. doi:10.1016/j.nucengdes.03.014

Miettinen, K., & Mäkelä, M. M. (1994). A Nondifferentiable Multiple Criteria Optimization Method Applied to Continuous Casting Process. In A. Fasano & M. Primicerio (Ed.), *Proceedings of the Seventh European Conference on Mathematics in Industry (ECMI 9)*, Stuttgart (pp. 255-262).

Miguel, L. F. F., & Miguel, L. F. F. (2012). Novel Metaheuristic Algorithms Applied to Optimization of Structures. *WSEAS Transaction on Applied and Theoretical Mechanics*, *7*(3), 210–220.

Mikuž, B., & Tiselj, I. (2016). Wall-resolved Large Eddy Simulation in grid-free 5 x 5 rod bundle of MATiS-H experiment. *Nuclear Engineering and Design*, *298*, 64–77.

Ming, W., Hou, J., Zhang, Z., Huang, H., Xu, Z., Zhang, G., & Huang, Y. (2015). Integrated ANN-LWPA for cutting parameter optimization in WEDM. *International Journal of Advanced Manufacturing Technology*, 1–18.

Minitab User Manual Release 16. (2010). State College, PA: MINITAB Inc.

Mishra, P. K. (2012). *Nonconventional Machining*. New Delhi: Narosa Publishing House.

Mohanty, C. P., Mahapara, S. S., & Singh, M. R. (2017). An intelligent approach to optimize the EDM process parameters using utility concept and QPSO algorithm. *Engineering Science and Technology, an International Journal*, *20*, 552–562.

Mohd Abbas, N., Solomon, D. G., & Fuad Bahari, M. (2007, June). A review on current research trends in electrical discharge machining (EDM). *International Journal of Machine Tools & Manufacture*, *47*(7-8), 1214–1228. doi:10.1016/j.ijmachtools.2006.08.026

Montgomery, D. C. (2001). *Design and analysis of experiments*. New York: John Wiley.

Morse, J. N. (1980). Reducing the size of the nondominated set: Pruning by clustering. *Computers & Operations Research*, *7*(1-2), 55–66. doi:10.1016/0305-0548(80)90014-3

Moussiopoulos, N., Berge, E., Bohler, T., Leeuw, F., Gronskei, K. E., Mylona, S., & Tombrou, M. (1997). Ambient Air Quality, Pollutant Dispersion and Transport Models, European Topic Centre on Air Quality, Topic Report 19. European Environment Agency, EU Publications.

Mueller, J., Mueller, J. L., Pietras, C., Hooker, S. B., Clark, D. K., Frouin, A. M. R., & Fargion, G. S. (2002). Ocean Optics Protocols For Satellite Ocean Color Sensor Validation, Revision 3, volumes 1 and 2. *NASA tech. memo, 210004.*

Mukherjee, R., Chakraborty, S., & Samanta, S. (2012). Selection of wire electrical discharge machining process parameters using non-traditional optimization algorithms. *Applied Soft Computing*, *12*(8), 2506–2516. doi:10.1016/j.asoc.2012.03.053

Nabyvanets, B. I., Osadchy, V. I., Osadcha, N. M., & Nabyvanets, Yu. B. (2007). *Analytical chemistry of surface water.* Kiev: Naukova Dumka. (in Ukrainian)

Nagarjan, A., & Jeyaraman, K. (2010). Solution of chance constrained programming problem for multi-objective interval solid transportation problem under stochastic environment using fuzzy approach. *International Journal of Computers and Applications*, *10*(9), 19–29. doi:10.5120/1510-1690

Nahmias, S. (1978). Fuzzy variable. *Fuzzy Sets and Systems*, *1*(2), 97–101. doi:10.1016/0165-0114(78)90011-8

National Research Council. (1992). *Nuclear Power: Technical and institutional options for the future.* National Academies Press.

Nayak, B. B., Mahapatra, S. S., Chatterjee, S., & Abhishek, K. (2015). Parametric Appraisal of WEDM using Harmony Search Algorithm. *Materials Today: Proceedings*, *2*(4-5), 2562–2568. doi:10.1016/j.matpr.2015.07.207

Nechval, N. A., Purgailis, M., Berzins, G., Cikste, K., Krasts, J., & Nechval, K. N. (2010). Invariant embedding technique and its applications for improvement or optimization of statistical decisions. In K. Al-Begain, D. Fiems & W. Knottenbelt (Eds.), Analytical and Stochastic Modeling Techniques and Applications. LNCS (Vol. 6148, pp. 306–320). Berlin, Heidelberg: Springer-Verlag. doi:10.1007/978-3-642-13568-2_22

Nechval, N. A., Purgailis, M., Nechval, K. N., & Bruna, I. (2012). Optimal prediction intervals for future order statistics from extreme value distributions. In *Proceedings of the World Congress on Engineering WCE '12, LNECS* (Vol. 3, pp. 1340–1345).

Nechval, N.A., Nechval, K.N., & Vasermanis, E.K. (2003). Effective state estimation of stochastic systems. *Kybernetes (An International Journal of Systems & Cybernetics)*, *32*, 666–678.

Nechval, N. A., Berzins, G., & Danovics, V. (2016). Optimization of statistical decisions for age replacement problems via a new pivotal quantity averaging approach. *American Journal of Theoretical and Applied Statistics*, *5*, 21–28. doi:10.11648/j.ajtas.s.2016050201.14

Nelson, W. (1982). *Applied Life Data Analysis.* New York: Wiley. doi:10.1002/0471725234

Nelson, W. (2000). Weibull prediction of a future number of failures. *Quality and Reliability Engineering International*, *16*(1), 23–26. doi:10.1002/(SICI)1099-1638(200001/02)16:1<23::AID-QRE283>3.0.CO;2-Q

Noorul Haq, A., Saravanan, M., Vivekraj, A. R., & Prasad, T. (2006). A scatter search approach for general flowshop scheduling problem. *International Journal of Advanced Manufacturing Technology*, *31*(7-8), 731–736. doi:10.1007/s00170-005-0244-1

Nordman, D. J., & Meeker, W. Q. (2002). Weibull prediction for a future number of failures. *Technometrics*, *44*(1), 15–23. doi:10.1198/004017002753398191

Ojha, A., Das, B., Mondal, S., & Maiti, M. (2009). An entropy based solid transportation problem for general fuzzy costs and time with fuzzy equality. *Mathematical and Computer Modelling*, *50*(1-2), 166–178. doi:10.1016/j.mcm.2009.04.010

Osadchy, V. I., Nabyvanets, B. I., Osadcha, N. M., & Nabyvanets, Yu. B. (2008). *Hydrochemical guide. Surface water of Ukraine. Hydrochemical calculations. Methods of analysis.* Kiev: Naukova Dumka. (in Ukrainian)

Osyczka, A. (1984). *Multicriterion Optimization in Engineering - with FORTRAN Programs*. Chichester: Ellis Horwood.

Ozek, C., Caydas, U., & Unal, E. (2012). A fuzzy model for predicting surface roughness in plasma arc cutting of AISI 4140 steel. *Materials and Manufacturing Processes*, *27*(1), 95–102. doi:10.1080/10426914.2011.551952

Packham, I. S. J., & Parmee, I. C. (2000). Data analysis and Visualization of Cluster-Orient Genetic Algorithm Output. *Proceedings of the International Conference on Information Visualization*, 173-178. doi:10.1109/IV.2000.859752

Pandian, P., & Anuradha, D. (2010). A new approach for solving solid transportation problems. *Appl. Math. Sci.*, *4*, 3603–3610.

Pant, S., & Singh, S. B. (2011). Particle Swarm Optimization to Reliability Optimization in Complex System. In *Proceedings of the IEEE Int. Conf. on Quality and Reliability*, Bangkok, Thailand, Sept 14-17 (pp. 211-215). doi:10.1109/ICQR.2011.6031711

Pant, S., Anand, D., Kishor, A., & Singh, S.B. (2015). A Particle Swarm Algorithm for Optimization of Complex System Reliability. *International Journal of Performability Engineering*, *11*(1), 33-42.

Pant, S., Kumar, A., & Ram, M. (2017). Reliability Optimization: A Particle Swarm Approach. In Advances in Reliability and System Engineering (pp. 163-187). Springer International Publishing.

Pant, S., Kumar, A., Kishor, A., Anand, D., & Singh, S. B. (2015). Application of a Multi-Objective Particle Swarm optimization Technique to Solve Reliability Optimization Problem. In *Proceeding of IEEE Int. Conf. on Next generation Computing Technologies*, September 4-5 (pp. 1004-1007). doi:10.1109/NGCT.2015.7375272

Pant, S., Anand, D., Kishor, A., & Singh, S. B. (2015). A Particle Swarm Algorithm for Optimization of Complex System Reliability. *International Journal of Performability Engineering.*, *11*(1), 33–42.

Pant, S., Kumar, A., & Ram, M. (2017a). Reliability Optimization: A Particle Swarm Approach. Advances in Reliability and System Engineering. *Springer International Publishing*, *2017*, 163–187.

Pant, S., Kumar, A., & Ram, M. (2017b). *Flower pollination algorithm development: a state of art review. International Journal of System Assurance Engineering and Management*. doi:10.1007/s13198-017-0623-7

Pant, S., Kumar, A., & Singh, S. B. (2017). *Reliability Optimization of Complex Systems Using Cuckoo Search Algorithm. In Mathematical Concepts and Applications in Mechanical Engineering and Mechatronics*. Hershey, PA: IGI Global.

Pant, S., Kumar, A., Singh, S. B., & Ram, M. (2017c). A Modified Particle Swarm Optimization Algorithm for Nonlinear Optimization. *Nonlinear Studies.*, *24*(1), 127–138.

Papargyropoulou, E., Lozano, R., Steinberger, J. K., & Wright, N., & bin Ujang, Z. (. (2014). The food waste hierarchy as a framework for the management of food surplus and food waste. *Journal of Cleaner Production*, *76*, 106–115. doi:10.1016/j.jclepro.2014.04.020

Pappu, J. S. M., & Gummadi, N. (2017). Artificial Neural Network and regression coupled genetic algorithm to optimize parameters for enhanced xylitol production by Debaryomycesnepalensis in bioreactor. *Biochemical Engineering Journal*, *120*, 136–145. doi:10.1016/j.bej.2017.01.010

Parfitt, J., Barthel, M., & Macnaughton, S. (2010). Food waste within food supply chains: Quantification and potential for change to 2050. *Philosophical Transactions of the Royal Society of London. Series B, Biological Sciences*, *365*(1554), 3065–3081. doi:10.1098/rstb.2010.0126 PMID:20713403

Patel, J. K. (1989). Prediction intervals–a review. *Communications in Statistics. Theory and Methods*, *18*(7), 2393–2465. doi:10.1080/03610928908830043

Patil, N. G., & Brahmankar, P. K. (2010). Determination of material removal rate in wire electro-discharge machining of metal matrix composites using dimensional analysis. *International Journal of Advanced Manufacturing Technology, 51*(5), 599–610. doi:10.1007/s00170-010-2633-3

Patil, S. K., & Kant, R. (2014). A hybrid approach based on fuzzy DEMATEL and FMCDM to predict success of knowledge management adoption in supply chain. *Applied Soft Computing, 18*, 126–135. doi:10.1016/j.asoc.2014.01.027

Patowari, P. K., Saha, P., & Mishra, P. K. (2010). Artificial neural network model in surface modification by EDM using tungsten – copper powder metallurgy sintered electrodes. *International Journal of Advanced Manufacturing Technology, 51*(5-8), 627–638. doi:10.1007/s00170-010-2653-z

Patowari, P. K., Saha, P., & Mishra, P. K. (2011). Taguchi analysis of surface modification technique using W-Cu powder metallurgy sintered tools in EDM and characterization of the deposited layer. *International Journal of Advanced Manufacturing Technology, 54*(5-8), 593–604. doi:10.1007/s00170-010-2966-y

Patowari, P. K., Saha, P., & Mishra, P. K. (2015). An experimental investigation of surface modification of C-40 steel using W – Cu powder metallurgy sintered compact tools in EDM. *International Journal of Advanced Manufacturing Technology, 80*(1-4), 343–360. doi:10.1007/s00170-015-7004-7

Patra, K., Jha, A. K., Szalay, T., Ranjan, J., & Monostori, L. (2017). Artificial Neural Network Based tool condition monitoring in micro mechanical peck drilling used thrust force signals. *Precision Engineering, 48*, 279–291. doi:10.1016/j.precisioneng.2016.12.011

Pedrycz, W., & Gomide, F. (2007). *Fuzzy systems engineering: toward human-centric computing*. John Wiley & Sons. doi:10.1002/9780470168967

Petropoulos, G., Vaxevanidis, N. M., & Pandazaras, C. (2004). Modeling of surface finish in electro-discharge machining based upon statistical multi-parameter analysis. *Journal of Materials Processing Technology, 155-156*, 1247–1251. doi:10.1016/j.jmatprotec.2004.04.189

Petrovic, D. (2001). Simulation of supply chain behaviour and performance in an uncertain environment. *International Journal of Production Economics, 71*(1), 429–438. doi:10.1016/S0925-5273(00)00140-7

Petrovic, D., Roy, R., & Petrovic, R. (1998). Modelling and simulation of a supply chain in an uncertain environment. *European Journal of Operational Research, 109*(2), 299–309. doi:10.1016/S0377-2217(98)00058-7

Pezzella, F., Morganti, G., & Ciaschetti, G. (2008). A genetic algorithm for the flexible job-shop scheduling problem. *Computers & Operations Research, 35*(10), 3202–3212. doi:10.1016/j.cor.2007.02.014

Phadke, M. S. (2012). *Quality Engineering using Robust Design*. New Delhi: Pearson.

Pinedo, M. (1995). *Scheduling: theory, algorithms, and systems*. Englewood Cliffs: Prentice-Hall.

Pisinger, D., & Ropke, S. (2007). A General Heuristic for Vehicle Routing Problems. *Computers & Operations Research, 34*(8), 2403–2435. doi:10.1016/j.cor.2005.09.012

Podila, K., & Rao, Y. (2016). CFD modelling of turbulent flows through 5×5 fuel rod bundles with spacer-grids. *Annals of Nuclear Energy, 97*, 86–95. doi:10.1016/j.anucene.2016.07.003

Podila, K., Rao, Y. F., Krause, M., & Bailey, J. (2014). A CFD simulation of 5 X 5 rod bundles with split-type spacers. *Progress in Nuclear Energy, 70*, 167–175. doi:10.1016/j.pnucene.2013.08.012

Polasky, S., Carpenter, S. R., Folke, C., & Keeler, B. (2011). Decision-making under great uncertainty: Environmental management in an era of global change. *Trends in Ecology & Evolution*, *26*(8), 398–404. doi:10.1016/j.tree.2011.04.007 PMID:21616553

Pradhan, M. K. (2010). *Experimental investigation and modelling of surface integrity, accuracy and productivity aspect in EDM of AISI D2 steel* [Ph.D. thesis]. National Institute of Technology, Rourkela, India.

Pradhan, D., & Kumar, M. (2012). Multi-objective optimization of MRR, TWR and Radial Overcut of EDMed AISI D2 tool steel using response surface methodology, grey relational analysis and entropy measurement. *Journal for Manufacturing Science & Production*, *12*(1), 51–63. doi:10.1515/jmsp-2012-0004

Pradhan, M. K. (2012). Multi-objective optimization of MRR, TWR and radial overcut of EDMed AISI D2 tool steel using response surface methodology, grey relational analysis and entropy measurement. *J. Manuf. Science and Production*, *12*(1), 51–63.

Pradhan, M. K., & Biswas, C. K. (2008). Neuro-fuzzy model on material removal rate in electrical discharge machining in AISI D2 steel. *Proceedings of the 2nd International and 23rd All India Manufacturing Technology, Design and Research Conference*, *1*, 469-474

Pradhan, M. K., & Biswas, C. K. (2010). Investigating the effect of machining parameters on EDMed components a RSM approach. *Journal of Mechanical Engineering*, *7*, 47–64.

Pradhan, M. K., & Das, R. (2015). Application of a general regression neural network for predicting radial overcut in electrical discharge machining of AISI D2 tool steel. *International Journal of Machining and Machinability of Materials*, *17*(3-4), 355–369. doi:10.1504/IJMMM.2015.071998

Pradhan, M. K., & Das, R. (2015). Application of ann modeling of radial overcut in electrical discharge machining. *Journal of Manufacturing Technology Research*, *7*(1/2), 39.

Puertas, I., & Luis, C. J. (2003). A study on the machining parameters optimization of electrical discharge machining. *Journal of Materials Processing Technology*, *143-144*, 521–526. doi:10.1016/S0924-0136(03)00392-3

Puertas, I., Luis, C. J., & Villa, G. (2005). Spacing roughness parameters study on the EDM of silicon carbide. *Journal of Materials Processing Technology*, *164-165*, 1590–1596. doi:10.1016/j.jmatprotec.2005.01.004

Qin, R., Liu, Y. K., & Liu, Z. Q. (2011). Methods of critical value reduction for type-2 fuzzy variables and their applications. *Journal of Computational and Applied Mathematics*, *235*(5), 1454–1481. doi:10.1016/j.cam.2010.08.031

Qiu, H. L., Sanchez-Azofeifa, A., & Gamon, J. A. (2007). Ecological applications of remote sensing at multiple scales. In *Functional Plant Ecology* (2nd ed.). CRC Press. doi:10.1201/9781420007626.ch22

Radovanovic, M., & Madic, M. (2011). Modeling the plasma arc cutting using ANN. *Nonconventional Technologies Review*, *4*, 43–48.

Rahimi-Vahed, A. R., Javadi, B., Rabbani, M., & Tavakkoli-Moghaddam, R. (2008). A multi-objective scatter search for a bi-criteria no-wait flow shop scheduling problem. *Engineering Optimization*, *40*(4), 331–346. doi:10.1080/03052150701732509

Rahman, M.M., Khan, K., A.R., K., Maleque, M.A., & Bakar, R.A. (2010). Parametric optimization in EDM of Ti-6Al-4V using copper tungsten electrode and positive polarity: A Statistical Approach. *Mathematical Methods and Techniques in Engineering and Environmental Science*, *6*, 23–29.

Rajkumar, M., Asokan, P., Anilkumar, N., & Page, T. (2011). A GRASP algorithm for flexible job-shop scheduling problem with limited resource constraints. *International Journal of Production Research*, *49*(8), 2409–2423. doi:10.1080/00207541003709544

Ramakrishnan, S., Shrinet, V., Polivka, F. B., Kearney, T. N., & Koltun, P. (2000). Influence of gas composition on plasma arc cutting of steel. *Journal of Physics. D, Applied Physics*, *33*(18), 2288–2299. doi:10.1088/0022-3727/33/18/313

Ramaswamy, H., & Blunt, L. (2002). 3D surface characterisation of elctropolished EDMed surface and quantitative assessment of process variables using Taguchi Methodology. *International Journal of Machine Tools & Manufacture*, *42*(10), 1129–1133. doi:10.1016/S0890-6955(02)00057-3

Ramaswamy, H., & Blunt, L. (2004). Effect of EDM process parameters on 3D surface topography. *Journal of Materials Processing Technology*, *148*(2), 155–164. doi:10.1016/S0924-0136(03)00652-6

Ramirez, C. (2011). Simulation of the bird age structured population growth based on an interval type-2 fuzzy cellular structure. *Inform. Sci*, *181*(3), 519–535. doi:10.1016/j.ins.2010.10.011

Ranjbar, M., De Reyck, B., & Kianfar, F. (2009). A hybrid scatter search for the discrete time/resource trade-off problem in project scheduling. *European Journal of Operational Research*, *193*(1), 35–48. doi:10.1016/j.ejor.2007.10.042

Rao, B. T., & Krishna, A. G. (2014). Selection of optimal process parameters in WEDM while machining Al7075/SiCp metal matrix composites. *International Journal of Advanced Manufacturing Technology*, *73*(1-4), 299–314. doi:10.1007/s00170-014-5780-0

Rehme, K. (1973). Pressure drop correlations for fuel element spacers. *Nuclear Technology*, *17*, 15.

Rehme, K. (1987). The structure of turbulent flow through rod bundles. *Nuclear Engineering and Design*, *99*, 141–154. doi:10.1016/0029-5493(87)90116-6

Reynolds, O. (1895). On the dynamical theory of incompressible viscous fluids and the determination of the criterion. *Philosophical Transactions of the Royal Society of London*, *186*(0), 123–164. doi:10.1098/rsta.1895.0004

Romanenko, O. V., Arsan, O. M., Kipnis, L. S., & Sytnyk, Yu. M. (2015). *Ecological problems of the Kiev reservoirs and adjoining territories*. Kiev: Naukova Dumka. (in Ukrainian)

Roosenburg, N., & Eekeks, J. (1995). *Product Design: Fundamentals and Methods*. John Wiley & Sons Inc.

Rosenman, M. A., & Gero, J. S. (1985). Reducing the pareto optimal set in multicriteria optimization (With application to Pareto optimal dynamic programing). *Engineering Optimization*, *8*(3), 189–206. doi:10.1080/03052158508902489

Ross, P. J. (1996). *Taguchi Techniques for Quality Engineering* (2nd ed.). New York: McGraw Hill.

Routara, B. C., Bandyopadhyay, A., & Sahoo, P. (2007). Use of desirability function approach for optimization of multiple performance characteristics of the surface roughness parameters in CNC turning. *Proceedings of the International Conference on Mechanical Engineering*.

Roy, R. K. (1990). *A primer on the Taguchi method Dearborn*. Michigan Society of Manufacturing Engineers.

Sahoo, P., Routara, B. C., & Bandyopadhyay, A. (2009). Roughness modelling and optimization in EDM using response surface method for different work piece materials. *International Journal of Machining and Machinability of Materials*, *5*(2/3), 321–346. doi:10.1504/IJMMM.2009.023398

Sahu, A. K., Mohanty, P. P., & Sahoo, S. K. (2017). Electro discharge machining of Ti-alloy (Ti6Al4V) and 316L Stainless Steel and Optimization of process parameters by Grey relational analysis(GRA) method. In D.I. Wimpenny et al. (Eds.), Advances in 3D Printing & Additive Manufacturing Technologies (pp. 65–78). Singapore: Springer Science+Business Media.

Saidi-Mehrabad, M., & Fattahi, P. (2007). Flexible job shop scheduling with tabu search algorithms. *International Journal of Advanced Manufacturing Technology*, *32*(5-6), 563–570. doi:10.1007/s00170-005-0375-4

Salama, M., Garba, J., Demsetz, L., & Udwadia, F. (1988). Simultaneous Optimization of Controlled Structures. *Computational Mechanics*, *3*(4), 275–282. doi:10.1007/BF00368961

Salonitis, K., & Vatousianos, S. (2012). Experimental investigation of the plasma arc cutting process. *Procedia CIRP*, *3*, 287–292. doi:10.1016/j.procir.2012.07.050

Samuel, M. P., & Philip, P. K. (2007). Power metallurgy tool electrodes for electrical discharge machining. *International Journal of Machine Tools & Manufacture*, *37*(11), 1625–1633. doi:10.1016/S0890-6955(97)00006-0

Saravanan, M., & Noorul Haq, A. (2010). A scatter search algorithm for scheduling optimization of job shop problems. *International Journal of Product Development*, *10*(1-3), 259–272. doi:10.1504/IJPD.2010.029996

Saxena, A. (2014). Thermal-hydraulic numerical simulation of fuel sub-assembly for Sodium-cooled Fast Reactor [Ph.D. Thesis]. CEA Cadarache, Aix-Marseille University.

Schaffer, J. (1985). Multiple objective optimization with vector evaluated genetic algorithms. In *Proceeding of International Conference on Genetic Algorithms*, Pittsburgh.

Schell, E. D. (1955). Distribution of a product by several properties. *Proceedings 2nd Symposium in Linear Programming*. DCS/Comptroller, HQ US Air Force.

Schlaifer, R., & Raiffa, H. (1961). *Applied statistical decision theory*. Academic Press.

Seale W. J. (1979). Turbulent Diffusion of Heat between Connected Flow Passages, Part 1: Outline of Problem and Experimental Investigation. *Nuclear Engineering and Design*, *54*(2), 183-195. (79)90166-310.1016/0029-5493

Seuring, S., & Müller, M. (2008). From a literature review to a conceptual framework for sustainable supply chain management. *Journal of Cleaner Production*, *16*(15), 1699–1710. doi:10.1016/j.jclepro.2008.04.020

Simao, J., Aspinwall, D., El-Menshawy, F., & Meadows, K. (2002). Surface alloying using PM composite electrode materials when electrical discharge texturing hardened AISI D2. *Journal of Materials Processing Technology*, *127*(2), 211–216. doi:10.1016/S0924-0136(02)00144-9

Sitton, D. J., Zeinali, Y., & Story, A. (2017). Rapid soil classification using artificial neural network for use in constructing compressed earth blocks. *Construction & Building Materials*, *138*, 214–221. doi:10.1016/j.conbuildmat.2017.02.006

Somashekhar, K. P., Ramachandran, N., & Mathew, J. (2010). Optimization of Material Removal Rate in Micro-EDM Using Artificial Neural Network and Genetic Algorithms. *Materials and Manufacturing Processes*, *25*(6), 467–475. doi:10.1080/10426910903365760

Sonker, S., & Munjal, A. (2017a). A note on boundness conditions of absolute summability ϕ-|A|k factors. In *Proceedings of the International Conference on Advances in Science and Technology* (pp. 208-210).

Sonker, S., & Munjal, A. (2016a). Sufficient conditions for triple matrices to be bounded. *Nonlinear Studies*, *23*(4), 531–540.

Sonker, S., & Munjal, A. (2016b). Absolute Summability Factor $\varphi - |C, 1; \delta|_k$ of Infinite Series. *International Journal of Mathematical Analysis*, *10*(23), 1129–1136. doi:10.12988/ijma.2016.6690

Sonker, S., & Munjal, A. (2017b). Approximation of the function $f \in Lip\ (\alpha, p)$ using infinite matrices of Cesàro submethod. *Nonlinear Studies*, *24*(1), 113–125.

Souquet, J., Liu, H., Liu, J., Fan, X., Liddell, J., Levison, P., ... Thomas, O. R. T. (n.d.). Tracking the movement of individual adsorbent particles in expanded beds. In *Book of Abstracts* (p. 76). Academic Press.

Soysal, M., Bloemhof, J., & van der Vorst, J. G. (2012). A Review of Quantitative Models for Sustainable Food Logistics Management: Challenges and Issues. *Proceedings in Food System Dynamics*, 448-462.

Spears, W. M., & De Jong, K. A. (1991). On the Virtues of Parameterized Uniform Crossover. *Proceedings of the 4th International Conference on Genetic Algorithms*, 230-236.

STAR-CCM+. (2015). Retrieved from http://www.cd-adapco.com/products/STAR-CCMplus/

Stark, J. (2015). Product lifecycle management. In *Product Lifecycle Management* (pp. 1–29). Springer International Publishing. doi:10.1007/978-3-319-17440-2_1

Steuer, R. (1986). *Multiple criteria optimization: theory, computation and application*. New York: John Wiley & Sons, Inc.

Straub, D. (2009). Stochastic modeling of deterioration processes through dynamic Bayesian networks. *Journal of Engineering Mechanics*, *135*(10), 1089–1099. doi:10.1061/(ASCE)EM.1943-7889.0000024

Sun, H., & Lee, S. (2005). Response surface approach to aerodynamic optimization design of helicopter rotor blade. *International Journal for Numerical Methods in Engineering*, *64*(1), 125–142. doi:10.1002/nme.1391

Suter, G. W. II. (2016). *Ecological risk assessment*. CRC Press.

Swinburn, B., Vandevijvere, S., Kraak, V., Sacks, G., Snowdon, W., Hawkes, C., ... & L'abbé, M. (2013). Monitoring and benchmarking government policies and actions to improve the healthiness of food environments: a proposed Government Healthy Food Environment Policy Index. *Obesity Reviews*, *14*(S1), 24-37.

Swinburn, B., Kraak, V., Rutter, H., Vandevijvere, S., Lobstein, T., Sacks, G., & Magnusson, R. et al. (2015). Strengthening of accountability systems to create healthy food environments and reduce global obesity. *Lancet*, *385*(9986), 2534–2545. doi:10.1016/S0140-6736(14)61747-5 PMID:25703108

Syswerda, G. (1989). Uniform Crossover in Genetic Algorithms. *Proceedings of the 3rd International Conference on Genetic Algorithms*, 2-9.

Taguchi, G. (1990). *Introduction to Quality Engineering*. Tokyo: Asian Productivity Organization.

Tavakkoli-Moghaddam, R., Javadian, N., Khorrami, A., & Gholipour-Kanani, Y. (2010). Design of a scatter search method for a novel multi-criteria group scheduling problem in a cellular manufacturing system. *Expert Systems with Applications*, *37*(3), 2661–2669. doi:10.1016/j.eswa.2009.08.012

Teulet, P., Girard, L., Razafinimanana, M., Gleizes, A., Bertrand, P., Camy-Peyret, F., & Richard, F. et al. (2006). Experimental study of an oxygen plasma cutting torch: II. Arc material interaction, energy transfer and anode attachment. *Journal of Physics. D, Applied Physics*, *39*(8), 1557–1573. doi:10.1088/0022-3727/39/8/015

Thomas, A. M., White, G. R., Plant, E., & Zhou, P. (2017). Challenges and practices in Halal meat preparation: A case study investigation of a UK slaughterhouse. *Total Quality Management & Business Excellence*, *28*(1-2), 12–31. doi:10.1080/14783363.2015.1044892

Tom, M. V. (2017). *Transportation System Engineering, Genetic Algorithm*. Indian Institute of Technology Bombay.

Tosun, Ö., & Marichelvam, M. K. (2016). Hybrid bat algorithm for flow shop scheduling problems. *International Journal of Mathematics in Operational Research*, *9*(1), 125–138. doi:10.1504/IJMOR.2016.077560

Trienekens, J. H., Wognum, P. M., Beulens, A. J., & van der Vorst, J. G. (2012). Transparency in complex dynamic food supply chains. *Advanced Engineering Informatics*, *26*(1), 55–65. doi:10.1016/j.aei.2011.07.007

Trippe, G., & Weinberg, D. (1979b). Non-isotropic eddy viscosities in turbulent flow through rod bundles. In S. Kakac & D. B. Spalding (Eds.), *Turbulent Forced Convection in Channels and Bundles* (Vol. 1, p. 505). Hemisphere, Washington.

Trupp, A. C., & Azad, R. S. (1975). The structure of turbulent flow in triangular array rod bundles. *Nuclear Engineering and Design*, *32*(1), 47–84. doi:10.1016/0029-5493(75)90090-4

Tsai, H. C., Yan, B. H., & Huang, F. Y. (2003). EDM performance of Cr / Cu-based composite electrodes. *International Journal of Machine Tools & Manufacture*, *43*(3), 245–252. doi:10.1016/S0890-6955(02)00238-9

Tsai, W. H., & Chou, W. C. (2009). Selecting management systems for sustainable development in SMEs: A novel hybrid model based on DEMATEL, ANP, and ZOGP. *Expert Systems with Applications*, *36*(2), 1444–1458. doi:10.1016/j.eswa.2007.11.058

Ueno, M., Fujita, N., Kimura, Y., & Nakata, N. (2016). Journal of Materials Processing Technology Evaluation of coating and wear characteristics of roll surface coated with TiC by electrical discharge coating. *Journal of Materials Processing Technology*, *236*, 9–15. doi:10.1016/j.jmatprotec.2016.04.025

Unitek Miyachi Group. (1999). Welding Material Control. *Technical Application Brief*, *2*, 1–5.

van Weele, A., & van Tubergen, K. (2017). Responsible purchasing: moving from compliance to value creation in supplier relationships. In *Sustainable Supply Chains* (pp. 257–278). Springer International Publishing. doi:10.1007/978-3-319-29791-0_11

Varin, C., & Vidoni, P. (2006). Pairwise likelihood inference for ordinal categorical time series. *Computational Statistics & Data Analysis*, *51*(4), 2365–2373. doi:10.1016/j.csda.2006.09.009

Varin, C., & Vidoni, P. (2009). Pairwise likelihood inference for general state space models. *Econometric Reviews*, *28*(1-3), 170–185. doi:10.1080/07474930802388009

Vasant, P. M., Weber, G., & Diew, V. N. (Eds.). (2016). Handbook of Research on Modern Optimization Algorithms and Applications in Engineering and Economics. IGI Global.

Vasant, P. M., Weber., G., & Diew, V. N. (Eds.). (2016). Handbook of Research on Modern Optimization Algorithms and Applications in Engineering and Economics. IGI Global.

Vasant, P. M. (Ed.). (2014). *Handbook of Research on Artificial Intelligence Techniques and Algorithms*. IGI Global.

Vasant, P. M., Barsoum, N., & Webb, J. (Eds.). (2011). *Innovation in Power, Control, and Optimization: Emerging Energy Technologies*. IGI Global.

Vasilev, K. V. (2003). Plasma-arc cutting – A promising method of thermal cutting. *Welding International*, *17*(2), 147–151. doi:10.1533/wint.2003.3096

Vo, D. N., & Schegner, P. (2012). An Improved Particle Swarm Optimization for Optimal Power Flow. In P. M. Vasant (Ed.), *Meta-Heuristics Optimization Algorithms in Engineering, Business, Economics, and Finance* (pp. 1–40). IGI Global.

Walker, C., Manera, A., Niceno, B., Simiano, M., & Prasser, H.-M. (2010). Steady-state RANS-simulations of the mixing in a T-junction. *Nuclear Engineering and Design*, *240*(9), 2107–2115. doi:10.1016/j.nucengdes.2010.05.056

Walley, P. (1991). *Statistical Reasoning with Imprecise Probabilities*. London: Chapman and Hall. doi:10.1007/978-1-4899-3472-7

Wang, J., Zhu, Z., He, C., & Yang, F. (2011). Effect of dual swirling plasma arc cutting parameters on kerf characteristics. *International Journal of Material*, *4*, 39–43.

Wang, T. R., Sun, G. N., & Liu, S. Y. (2011). Relationship between Spatiotemporal Variation of Water Pollution and Runoff Volume of Mainstream Section of the Weihe River in Shaanxi Province. *Arid Zone Research*, *28*(4), 599–615.

Waters, E. D. (1961). *Fluid mixing experiments with a wire wrapped 7-rod bundle fuel assembly*. Annular Space.

Whipple, J. M., Voss, M. D., & Closs, D. J. (2009). Supply chain security practices in the food industry: Do firms operating globally and domestically differ? *International Journal of Physical Distribution & Logistics Management*, *39*(7), 574–594. doi:10.1108/09600030910996260

Wognum, P. N., Bremmers, H., Trienekens, J. H., van der Vorst, J. G., & Bloemhof, J. M. (2011). Systems for sustainability and transparency of food supply chains–Current status and challenges. *Advanced Engineering Informatics*, *25*(1), 65–76. doi:10.1016/j.aei.2010.06.001

Wu, D., & Tan, W. (2006). A simplified type-2 fuzzy logic controller for real-time control. *ISA Transactions*, *45*(4), 503–516. doi:10.1016/S0019-0578(07)60228-6 PMID:17063934

Wu, W. W. (2012). Segmenting critical factors for successful knowledge management implementation using the fuzzy DEMATEL method. *Applied Soft Computing*, *12*(1), 527–535. doi:10.1016/j.asoc.2011.08.008

Wu, W. W., Lan, L. W., & Lee, Y. T. (2011). Exploring decisive factors affecting an organizations SaaS adoption: A case study. *International Journal of Information Management*, *31*(6), 556–563. doi:10.1016/j.ijinfomgt.2011.02.007

Xiaopeng, F. (2007). *Engineering design using genetic algorithms*. Retrospective Theses and Dissertations. Paper 15943.

Xia, X., Govindan, K., & Zhu, Q. (2015). Analyzing internal barriers for automotive parts remanufacturers in China using grey-DEMATEL approach. *Journal of Cleaner Production*, *87*, 811–825. doi:10.1016/j.jclepro.2014.09.044

Xing, L. N., Chen, Y. W., Wang, P., Zhao, Q. S., & Xiong, J. (2010). A knowledge-based ant colony optimization for flexible job shop scheduling problems. *Applied Soft Computing*, *10*(3), 888–896. doi:10.1016/j.asoc.2009.10.006

Xu, W. J., Fang, J. C., & Lu, Y. S. (2002). Study on ceramic cutting by plasma arc. *Journal of Materials Processing Technology*, *129*(1-3), 152–156. doi:10.1016/S0924-0136(02)00600-3

Xu, Y., Wang, L., Wang, S. Y., & Liu, M. (2015). An effective teaching–learning-based optimization algorithm for the flexible job-shop scheduling problem with fuzzy processing time. *Neurocomputing*, *148*, 260–268. doi:10.1016/j.neucom.2013.10.042

Yakovleva, N., Sarkis, J., & Sloan, T. (2012). Sustainable benchmarking of supply chains: The case of the food industry. *International Journal of Production Research*, *50*(5), 1297–1317. doi:10.1080/00207543.2011.571926

Yan, B. H., Tsai, H. C., & Huang, F. Y. (2005). The effect in EDM of a dielectric of a urea solution in water on modifying the surface of titanium. *International Journal of Machine Tools & Manufacture*, *45*(2), 194–200. doi:10.1016/j.ijmachtools.2004.07.006

Yang, X. (2009). Harmony Search as a Metaheuristic Algorithm. In Z.W. Geem (Eds.), Music-Inspired Harmony Search Algorithm: Theory and Applications, Studies in Computational Intelligence. Springer Berlin. doi:10.1007/978-3-642-00185-7_1

Yang, L. J. (2001). Plasma surface hardening of ASSAB 760 steel specimens with Taguchi optimization of the processing parameters. *Journal of Materials Processing Technology*, *113*(1-3), 521–526. doi:10.1016/S0924-0136(01)00583-0

Yang, L. J. (2007). The effect of specimen thickness on the hardness of plasma surface hardened ASSAB 760 steel specimens. *Journal of Materials Processing Technology*, *185*(1-3), 113–119. doi:10.1016/j.jmatprotec.2006.03.117

Yang, L., & Feng, Y. (2007). A bicriteria solid transportation problem with fixed charge under stochastic environment. *Applied Mathematical Modelling*, *31*(12), 2668–2683. doi:10.1016/j.apm.2006.10.011

Yang, L., & Liu, L. (2007). Fuzzy fixed charge solid transportation problem and algorithm. *Applied Soft Computing*, *7*(3), 879–889. doi:10.1016/j.asoc.2005.11.011

Yang, X. S., & He, X. (2013). Firefly Algorithm: Recent Advancement and Applications. *International Journal Swarm Intelligence*, *1*(1), 36–50. doi:10.1504/IJSI.2013.055801

Yazdani, M., Amiri, M., & Zandieh, M. (2010). Flexible job-shop scheduling with parallel variable neighborhood search algorithm. *Expert Systems with Applications*, *37*(1), 678–687. doi:10.1016/j.eswa.2009.06.007

Yeh, L. J., Chang, Y. C., & Chiu, M. C. (2004). Article. *Journal of Marine Science and Technology*, *12*(3), 189–199.

Ylonen, A. T. (2013). High-resolution flow structure measurements in a rod bundle [Ph.D. Thesis]. Lappeenranta University of Technology.

Yu-Zhen, Y. U., Xin-yi, R., Feng-Shan, D., & Jun-jie, S. (2012). Application of Improved PSO Algorithm in Hydraulic Pressing System Identification. *Journal of Iron and Steel Research International*, *19*(9), 29–35. doi:10.1016/S1006-706X(13)60005-9

Zadeh, L. A. (1965). Fuzzy Sets. *Information and Control*, *8*(3), 338–353. doi:10.1016/S0019-9958(65)90241-X

Zadeh, L. A. (1975). The Concept of a Linguistic Variable and its Application to Approximate Reasoning-I. *Information Sciences*, *8*(3), 199–249. doi:10.1016/0020-0255(75)90036-5

Zailani, S., Arrifin, Z., Abd Wahid, N., Othman, R., & Fernando, Y. (2010). Halal traceability and halal tracking systems in strengthening halal food supply chain for food industry in Malaysia (a review). *Journal of Food Technology*, *8*(3), 74–81. doi:10.3923/jftech.2010.74.81

Zajac, A., & Pfeifer, T. (2006). Restricting the heat-affected zone during the plasma cutting of high-alloy steels. *Welding International*, *20*(1), 5–9. doi:10.1533/wint.2006.3534

Zannetti, P. (1970). Numerical simulation modeling of air pollution: an overview. *WIT Transactions on Ecology and the Environment, 1.*

Zavardehi, S. (2013). Solving a fuzzy fixed charge solid transportation problem by metaheuristics. *Mathematical and Computer Modelling*, *57*(5-6), 1543–1558. doi:10.1016/j.mcm.2012.12.031

Zaw, H. M., Fuh, J. Y. H., Nee, A. Y. C., & Lu, L. (1999). Formation of a new EDM electrode material using sintering techniques. *Journal of Materials Processing Technology*, *89*(90), 182–186. doi:10.1016/S0924-0136(99)00054-0

Zhang, G., Gao, L., & Shi, Y. (2010, June). A genetic algorithm and tabu search for multi objective flexible job shop scheduling problems. In *Proceedings of the 2010 International Conference on Computing, Control and Industrial Engineering (CCIE)* (Vol. 1, pp. 251-254). IEEE. doi:10.1109/CCIE.2010.71

Zhang, G., Shao, X., Li, P., & Gao, L. (2009). An effective hybrid particle swarm optimization algorithm for multi-objective flexible job-shop scheduling problem. *Computers & Industrial Engineering*, *56*(4), 1309–1318. doi:10.1016/j.cie.2008.07.021

Zhang, G., Zhang, Z., Ming, W., Guo, J., Huang, Y., & Shao, X. (2014). The multi-objective optimization of medium-speed WEDM process parameters for machining SKD11 steel by the hybrid method of RSM and NSGA-II. *International Journal of Advanced Manufacturing Technology*, *70*(9-12), 2097–2109. doi:10.1007/s00170-013-5427-6

Zhang, H. P., & Gen, M. (2005). Multistage-based genetic algorithm for flexible job-shop scheduling problem. *Journal of Complexity, 11*, 223–232.

Zhang, J. H., Lee, T. C., & Lau, W. S. (1997). Study on the electro-discharge machining of a hot pressed aluminium oxide based ceramic. *Journal of Materials Processing Technology*, *63*(1-3), 908–912. doi:10.1016/S0924-0136(96)00012-X

Zhang, Q., Fan, Z., & Pan, D. (1999). A ranking approach for interval numbers in uncertain multiple attribute decision making problems. *Syst. Eng. Theory Pract*, *5*, 129–133.

Zheng, C., & Bennett, G. D. (2002). *Applied contaminant transport modeling* (Vol. 2). New York: Wiley-Interscience.

Zilioli, E., & Brivio, P. A. (1997). The satellite derived optical information for the comparative assessment of lacustrine water quality. *The Science of the Total Environment*, *196*(3), 229–245. doi:10.1016/S0048-9697(96)05411-3 PMID:9260309

Zimmermann, M. (2015). Development and Application of a Model for the Cross-Flow Induced by Mixing Vane Spacers in Fuel Assemblies [Ph.D. Thesis]. Karlsruhe Institute of Technology.

Zimmermann, H.-J. (1978). Fuzzy programming and linear programming with several objective functions. *Fuzzy Sets and Systems*, *1*(1), 45–55. doi:10.1016/0165-0114(78)90031-3

Zitzler, E., Deb, K., & Thiele, L. (2000). Comparison of multi-objective evolutionary algorithms: Empirical results. *Evolutionary Computation*, *8*(2), 173–195. doi:10.1162/106365600568202 PMID:10843520

Zygmund, A. (1988). *Trigonometric series*. Cambridge University Press.

About the Contributors

Goutam Kumar Bose is currently working as Head of the Department and Professor in Mechanical Engineering Department, Haldia Institute of Technology, Haldia, India. He has obtained Ph.D in Production Engineering from the Jadavpur University, Kolkata, India. He has obtained a Masters in Engineering in Mechanical Engineering from the Bengal Engineering & Science University, Shibpur, India. He has worked as an Assistant Professor in the Department of Mechanical Engineering of the College of Engineering & Management, Kolaghat for ten years. He was an Engineer in R & D Centre of M/s Hindustan Motors Ltd. West Bengal, India. His active areas of interests are Metal Cutting, Non-conventional machining, Industrial and Production Management and optimization techniques. He has published research papers in the Journals of International repute. He has presented papers at several international conferences in India and abroad.

D. K. Chandraker joined Reactor Design and Development Group (RD&DG) of Bhabha Atomic Research Centre (BARC) Trombay, Mumbai in 1993 and has over 23 years of research experiences in the field of thermal hydraulic design and safety of nuclear reactors. He received his Bachelor of Engineering Degree in Mechanical Engineering from Awadesh Pratap Singh University (APSU), Rewa, M.P, India in 1990, Master of Technology Degree from BHU, Varanasi, UP, India (Presently IIT BHU) in 1992 and Ph.D. in Nuclear Engineering from Tokyo Institute of Technology, Japan in 2012. He was received awards and JSPS RONPOKU Fellowship, (Tokyo Institute of Technology, Japan), during 2009-2012 for conducting research programme on the thermal hydraulics of nuclear reactors. Dr. Chandraker is involved in analytical and experimental investigations on the thermal hydraulics and safety of nuclear reactor and has over 20 journal publications and 100 papers on international/national conferences. Dr. Chandraker is also a faculty at Homi Bhabha National Institute (HBNI) Mumbai and is a member of various professional bodies. He is a recipient of DAE group achievement award for excellence in Science and Technology during years 2014 and 2015.

Raja Das is Assistant Professor (Senior) in the School of Advanced Sciences, V.I.T. University, Vellore, Tamil Nadu, India. He received his M.Sc in Applied Mathematics from National Institute of Technology, Rourkela, India and Ph.D. from Sambalpur University, Odisha, His area of research interest includes modelling, optimization and Soft Computing. He has published more than 25 research papers in the international journal and presented more than 15 articles at different international/national conferences.

Nehal Dash (B.Tech, Mechanical Engineering, SRM University, Chennai), presently pursuing M.E. (Design of Mechanical Equipment, BIT Mesra). His areas of interests are Product and Process Design, Material Engineering, Rapid Prototyping and Automation. He has 1 Book, 1 Book Chapter, 4 international Journal and 2 International Conference publication to his credit. He has been awarded for Best Research Project and Best Final Year Project during his undergraduate study at SRM University.

Sanghamitra Debta (B.Tech, Mechanical Engineering, ITER SOA University, Bhubaneswar), presently pursuing M.E. (Design of Mechanical Equipment, BIT Mesra). Her areas of interests are Product and Process Design, Strength of materials, Material Engineering, and Automation. She has 1 Book, 1 Book Chapter, 4 international Journal and 2 International Conference to her credit.

Dhiman Dutta completed his M.Sc. degree in mathematics from Tezpur University, Assam, India during the month of May 2014. Later, he has joined as a research scholar in the Department of Mathematics, National Institute of Technology Silchar, Assam, India on July 2014 and pursuing Ph.D. till now. His research area comprises of type-2 fuzzy set, type-2 fuzzy relation and the application of type-2 fuzzy variables in the solid transportation problem.

M. Geetha received her B.Sc. degree in Mathematics from Madurai Kamaraj University, Madurai, India, in 2005 and her M.Sc. degree in Mathematics from Madurai Kamaraj University, Madurai, India, in 2007. She has obtained her M.Phil. degree in Mathematics from Madurai Kamaraj University, Madurai, India, in 2008. She has been working as Assistant Professor in the department of Mathematics, Kamaraj College of Engineering and Technology, Virudhunagar, India since 2008. Her main research interests are Graph theory and optimization techniques. She has published many research papers in the Journals of International repute.

Ivan Kopachevsky is Research Scientist of Scientific Centre for Aerospace Research of the Earth, National Academy of Sciences of Ukraine. He received an MS in Aerospace Engineering and Geodesy in 2008. His studies directed to application of satellite observations to risk analysis, uncertainty control, and socio-ecological security assessment. Ivan Kopachevsky is author and co-author of more than 20 scientific papers and book chapters in fields of remote sensing, Earth sciences (ecology and climate change), engineering and social sciences.

Yuriy V. Kostyuchenko is Leading Research Scientist of Scientific Centre for Aerospace Research of the Earth, National Academy of Sciences of Ukraine. He received an MS in Astrophysics and Stellar Astronomy from Moscow State University in 1993. After that he joined National Academy of Sciences of Ukraine as a Scientist. He received his PhD in Geophysics and Remote Sensing in 1997 from the Scientific Centre for Aerospace Research of the Earth of National Academy of Sciences of Ukraine. In 2011-2014 he served on position of Assistant Professor of the Geomorphology and Soil Science Department of the Faculty of Geography, Kiev University. Since 2000 he serves as Executive secretary of Ukrainian National Member Organization of International Institute for Applied Systems Analysis (IIASA). His research interests include remote sensing application for environmental and socio-ecological security analysis. Yuriy V. Kostyuchenko is a member of American Society of Civil Engineers (ASCE), American Meteorological Society (AMS), American Statistical Society (ASA), and International Association for Promotion Geoethics (IAPG). He was honored by special award of Space Research Council of National Academy of Sciences of Ukraine in 2003 for fundamental research in remote sensing, and "Makarov" Medal of National Space Agency of Ukraine in 2012 for development of advanced technologies of remote sensing. Yuriy V. Kostyuchenko is author and co-author of more than 70 scientific papers and book chapters in fields of remote sensing, Earth sciences (ecology, climate change), engineering, computer and social sciences.

Kaushik Kumar, B.Tech (Mechanical Engineering, REC (Now NIT), Warangal), MBA (Marketing, IGNOU) and Ph.D (Engineering, Jadavpur University), is presently an Associate Professor in the Department of Mechanical Engineering, Birla Institute of Technology, Mesra, Ranchi, India. He has 14 years of Teaching & Research and over 11 years of industrial experience in a manufacturing unit of Global repute. His areas of teaching and research interest are Biomechanics, Optimization, Non-conventional machining, CAD / CAM, Rapid Prototyping and Composites. He has 9 Patents, 12 Books, 5 Edited Book Volumes, 22 Book Chapters, 114 international Journal publications, 19 International and 1 National Conference publications to his credit. He is on the editorial board and review panel of 14 International and 1 National Journals of repute. He has been felicitated with many awards and honors.

S. S. Mahapatra is a Professor in the Department of Mechanical Engineering, National Institute of Technology Rourkela, India. He has more than thirty years of experience in teaching and research. His current area of research includes Multi-criteria Decision-Making, Quality Engineering, Assembly Line Balancing, Group Technology, Neural Networks, and Non-traditional Optimization and Simulation. He has published more than three hundred papers in peer-reviewed international journals. He has written few books related to his research work. He is currently dealing with few sponsored projects.

Sachin Kumar Mangla is working in the field of Green Supply Chain/Smart Manufacturing/Machine Learning/Risk Management/Simulation/Optimization/Reverse Logistics/Renewable Energy Systems/ MCDM. His teaching interests are: Total Quality Management & Six Sigma; Smart Manufacturing; Supply Chain Network Design and Management/Green/Sustainable Supply Chain Management; Decision Support Models; Process Planning and Sequencing; Operations Management. He has published/ presented more than 75 papers in repute international/national journals (RSER, TRE-D, JCP, PPC, IJPR, IJPE, PPC, RCR, IJOR, IJLSM, and JFSM) and conferences (POMS, SOMS, IIIE, GLOGIFT). He is also Editorial Board member of several international journals of repute. He has an h-index 13, i10-index 16, Google Scholar Citations of approximately 450.

M. K. Marichelvam received his B.E. degree in Mechanical Engineering from Madurai Kamaraj University, Madurai, India, in 2000. He has obtained a Masters in Engineering in Industrial Engineering from Madurai Kamaraj University, Madurai, India, in 2002. He has obtained Ph.D. in Mechanical Engineering from Anna University, Chennai, India. He is currently working as Assistant Professor in Mechanical Engineering Department, Mepco Schlenk Engineering College, Sivakasi, India. His active areas of interests are Manufacturing Scheduling, Multi-objective optimization, Industrial and Production Management and optimization techniques. He has published many research papers in the Journals of International repute. He has presented papers at several international conferences.

Alka Munjal is a Senior Research Fellow under the DST-INSPIRE scheme in the Department of Mathematics, NIT Kurukshetra, Haryana. She received her B.Sc. degree in science from Kumaun University, Nainital, India, in 2011, the M.Sc. degree major in mathematics and minor in computer science from G. B. Pant University of Agriculture and Technology, Pantnagar, India, in 2013. She did her M.Sc. with the research area reliability theory and published 2 research papers in repudiated journals. She choose the topic Summability Techniques to pursue her Ph.D. and published 7 research articles in which 1 in SCI, 4 are in SCOPUS indexed journals. Also, she presented her works at national and international conferences.

Lata Nautiyal holds a PhD in Computer Science by the Gurukul Kangri University, Haridwar and is associate professor in Department of Computer Science and Engineering, Graphic Era University, Dehradun. Her main area of interest is the study of cloud computing, component-based software from a development, testing and reliability point of view. Her doctoral thesis explored the same fields. She has published 18 international journal publications and presented number of papers in national and international conferences.

Konstantin N. Nechval (Ph.D. in Mechanical Engineering) is an Associate Professor in the Department of Applied Mathematics at the Transport and Telecommunication Institute, Riga, Latvia. He has authored and coauthored more than 79 papers. His research interests include stochastic processes, pattern recognition, operations research, statistical decision theory, adaptive control, and mechanical engineering. Dr. K. Nechval was awarded the "CASYS'07 Best Paper Award" for his paper: "Dual Control of Education Process" presented at the Eight International Conference on Computing Anticipatory Systems (Liege, Belgium, August 6-11, 2007) and the "MM2009 Best Paper Award" for his paper: "Optimal Statistical Decisions in a New Product Lifetime Testing" presented at the Fourth International Conference on Maintenance and Facility Management (Rome, Italy, April 22-24, 2009).

Nicholas A. Nechval (Ph.D. in Automatic Control and Systems Engineering, and Dr. Habil. Sc. in Radio Engineering) is a Professor in the Department of Mathematics at the Baltic International Academy, Riga, Latvia, and Principal Investigator in the BVEF Research Institute at the University of Latvia, Riga, Latvia. Before his current position, from 1993 to 1999, Dr. Nechval was a Professor of Mathematics and Computer Science and the Head of the Research Laboratory at the Riga Aviation University. From 1999 to 2014, Dr. Nechval was a Professor and Head (Statistics Department and EVF Research Institute) at the University of Latvia, Riga, Latvia. He was also a Principal Investigator in the Institute of Mathematics and Computer Science at the University of Latvia. Dr. Nechval, in 1991, was awarded a Silver Medal of the Exhibition Committee (Moscow, Russia) for his research on the problem of Prevention of Collisions between Aircraft and Birds. His research interests include mathematics, econometrics, stochastic processes, pattern recognition, digital radar signal processing, operations research, statistical decision theory, adaptive control, and mechanical engineering. Professor Nechval has authored and coauthored more than 455 papers and 9 books, including the book: *Aircraft Protection from Birds* (Moscow: Russian Academy of Science, 2007) coauthored with V. D. Illyichev (Academician of the Russian Academy of Science), and the book: *Improved Decisions in Statistics* (Riga: Izglitibas Soli, 2004) coauthored with Prof. E. K. Vasermanis (University of Latvia). This book was awarded the "2004 Best Publication Award" by the Baltic Operations Research Society. Dr. Nechval is also an Editorial Board Member of several international journals.

Iurii Negoda is Research Scientist of Institute of Geological Sciences of National Academy of Sciences of Ukraine. He received an MS in Hydrogeology and Engineering Geology in 1992. His studies directed to application of mathematical modeling to environmental protection, assessment of ground water contamination, remediation methods of petroleum contaminated geological environment. Iurii Negoda is author and co-author of about 30 scientific papers in fields of hydrogeology (ground water pollution analysis), engineering and computer sciences.

Pritam Pain is currently working as Professor in Mechanical Engineering Department, JLD Engineering and Management College. He has obtained his Masters in Technology in the Mechanical Engineering from Haldia Institute of Technology, Haldia, India. He ranked 1st in Mechanical Engineering from West Bengal University of Technology during his Master Degree (2016). He completed his B.Tech degree in Mechanical Engineering from West Bengal University of Technology (Dr. Sudhir Chandra Sur Degree Engineering College), Kolkata, India. His areas of interest are non-traditional and modern machining technology and optimization by employing nature inspired advanced techniques.

Pravin P. Patil is working in the field of Finite Element Analysis, Optimization and Soft Computing Techniques. His teaching interests are: Finite element method, Mechanical vibrations, Machine design, Dynamics of machines and Strength of materials. His research interests are: Mechanical Vibration, Fluid structure interaction, Finite element modeling, Mass flow sensors, Soft computing modelling and Tribology in automotive engines. He has 16 years of teaching and research experience. He has published/ presented more than 60 papers in repute international/national journals. He is also reviewer for IEEE Transactions/ Elsevier Journals and Member of various international professional societies like ASME, IEEE etc. He guided 03 PhD thesis and 11 M.Tech thesis.

Mohan Kumar Pradhan is Assistant Professor in the Department of Mechanical Engineering, and Head of Production Engineering Lab. and CAM lab. of the Maulana Azad National Institute of Technology, Bhopal, India. He received his M. Tech and PhD from National Institute of Technology, Rourkela, India. He has over 16 years of teaching and research experience in manufacturing and 6 years of postdoctoral research experience. He has advised over 50 graduates, 20 Post graduate and four PhD students. Dr. Pradhan's research interests include manufacturing, non-traditional machining, metrology, micro-machining, hybrid machining, and process modelling and optimization. Dr. Pradhan has more than 55 refereed publications and nearly 60 technically edited papers, which were published in conference proceedings, edited two books, Five Conference Proceedings and five journals as Guest editor authored ten book chapters. Dr. Pradhan is Charted Engineer, a life fellow of IIPE and life member of ISTE, IACSIT, IAENG and MIE (I). Dr Pradhan's biographical information has appeared in the Marquis Who's Who in The World 2011 (28th Edition) as distinguished as one of the leading achievers from around the country, in 2014 (31st Edition)! also included in this exclusive reference directory distinguished as one of the world's foremost achievers in your field and the official 2017 "Who's Who in Asia.

Apurba Kumar Roy, B.E. (Mechanical Engineering, REC (Now NIT), Jaipur), M.E. (Mechanical Engineering, Jadavpur University, Kolkata) and PhD (Engineering IIT Kharagpur), is an Associate Professor at the Department of Mechanical Engineering, Birla Institute of Technology, Mesra, Ranchi, India. He has over 27 years of teaching, research and industrial experience. His areas of interests are Fluid Dynamics, Turbo machines, Multiphase Flow, CFD, Optimization and Non-Conventional Energy, Direct Energy Conversion. He has 1 Book, 2 Book Chapters, 34 international Journal, 5 International and 15 National Conference publications to his credit. He is on the editorial board and review panel of 1 International Journal of repute.

Anshuman Kumar Sahu is currently pursuing his PhD in the Department of Mechanical Engineering, National institute of Technology, Rourkela, Odisha, India. He is currently working on non-conventional machining, rapid tooling and rapid manufacturing.

Mausumi Sen received her M.Sc. degree in mathematics in 1996 and Ph.D. in Mathematics in 2004, from Gauhati University. She is working in the Department of Mathematics, National Institute of Technology Silchar for more than eleven years. She has taught several core courses in Mathematics at undergraduate, postgraduate, and doctorate levels. She has published many papers in national and international journals of repute and she is a reviewer of many international journals. Till now six doctoral students received degree under her supervision. She also received project grant from DST (GOI).

Yogesh Kumar Sharma is a research scholar of Mechanical Engineering at Graphic Era University of Dehradun; Uttarakhand. He received his B.Tech and M.Tech degree in Mechanical Engineering from ICFAI and Graphic Era University, Dehradun, Uttarakhand. His research interests are focused on Sustainable Food Supply Chain Management, Optimization Techniques and Dairy industry. He is a senior member of Indian Institution of Industrial Engineering (IIIE), Navi Mumbai. Yogesh Kumar Sharma is author and co-author of more than 08 research papers and book chapters in the field of Sustainable Food Supply Chain Management. He has presented papers at several national and international conferences in India. He has 04 years of teaching and research experience in Manufacturing, Materials and Mechanical Engineering with special emphasis in supply chain management.

Preeti Shivach received her PhD in Computer Science from Gurukul Kangri University, Haridwar. Currently, she is holding the position of Assistant Professor in Department of Computer Applications, Graphic Era University, Dehradun. She published 8 research papers in various international journals and presented number of papers in national and international conferences. Her area of research interest includes mobile agent and machine learning.

S. L. Sinha is Professor in Mechanical Engineering at National Institute of Technology Raipur (India). She has received her Ph.D. from Indian Institute of Technology, Kharagpur, W. B. She has about 32 years of teaching, 26 years of research and 11 years of administrative experience in Govt. Polytechnic, Durg /Govt. Engineering College, Raipur /NIT, Raipur, CG. She has participated in many conferences / seminars to present research paper, to deliver keynote speech or to chair the academic sessions of national and international level. She has published 3 book chapters, 36 papers in International/National Journals and about 58 papers in International/National Conferences. She has also been Nodal Officer of Chhattisgarh-Indian Institute of Technology, Kanpur, Knowledge Sharing Program since 2005. She has been conferred with "Young Scientist Award" in Young Scientist Congress organized at Pt. Ravishankar University, Raipur by Madhya Pradesh Council of Science & Technology, Bhopal, M.P. in 1991.

Smita Sonker is currently working as Assistant Professor in Department of Mathematics, NIT Kurukshetra, Haryana. She has obtained Ph.D in Summability Theory from the Department of Mathematics, IIT Roorkee, India. She has obtained a Masters in Mathematics from the P.P.N. College, Kanpur University, Kanpur, India. Her research areas of interests are Approximation Theory, Summability Theory, Absolute Summability, Operator Theory, Fourier analysis and Functional Analysis. She has published many research papers in the reputed National/International Journals. She has presented many papers at several international conferences in India and abroad.

Yulia Stoyka is Research Scientist of Institute of Hydrobiology of National Academy of Sciences of Ukraine. She received an MS in General and Vertebrate Zoology in 1997. Her studies directed to application of biotesting methods for estimation of water contamination, and development of approaches in field of ecophysiology and cytogenetics of aquatic animals for environmental protection. Yulia Stoyka is author and co-author of about 20 scientific papers in fields of hydrobiology and ecology.

Surbhi Uniyal is a research scholar of Mechanical Engineering at Graphic Era University of Dehradun; Uttarakhand. She received his B.Tech and M.Tech degree in Mechanical Engineering both from Graphic Era University, Dehradun, Uttarakhand. Her research interests are focused on Sustainable Consumption and Production, Operational Research. She has 08 years of teaching and research experience in Manufacturing, Materials and Mechanical Engineering with special emphasis in fluid mechanics and operational research. She has presented papers at several national and international conferences in India.

Shashi Kant Verma is Research Scholar in Mechanical Engineering Department at National Institute of Technology Raipur (India) since August 2014. He has also work as project trainee in BARC, Trombay, Mumbai India. He has received his Bachelor of Engineering Degree in Mechanical Engineering from Govt. Engineering College Jagdalpur, Master in Technology Degree from Visvesvaraya National Institute of Technology (VNIT), Nagpur (India). He has received his Diploma in Mechanical Engineering from Govt. Polytechnic Durg, Chhattisgarh. He has published various papers in International Journals (SCI and Scopus index) and International Conferences. He has published various book chapters in International level in the field of Nuclear Engineering. He has been conferred with "Young Scientist Award" in Young Scientist Congress organized at Chhattisgarh Council of Science & Technology Raipur (C.G.) and Chhattisgarh Swami Vivekananda Technical University Bhilai (C.G.) held from February 28th to March 1st, 2017.

Index

I

L

M

N

O

P

R

S

T

W

gramcontent.com/pod-product-compliance
Source LLC
TN
62138300926
1LV00010B/2569
78152253035 0 *